生态文明视域下的
生物学研究

《天水师范学院60周年校庆文库》编委会 | 编

光明日报出版社

图书在版编目（CIP）数据

生态文明视域下的生物学研究 /《天水师范学院 60 周年校庆文库》编委会编. -- 北京：光明日报出版社，2019.9

ISBN 978-7-5194-5510-1

Ⅰ.①生… Ⅱ.①天… Ⅲ.①生物学—文集 Ⅳ.①Q-53

中国版本图书馆 CIP 数据核字（2019）第 189388 号

生态文明视域下的生物学研究

SHENGTAI WENMING SHIYU XIA DE SHENGWUXUE YANJIU

编　　者：《天水师范学院 60 周年校庆文库》编委会

责任编辑：郭玫君　　　　责任校对：赵鸣鸣

封面设计：中联学林　　　　责任印制：曹　诤

出版发行：光明日报出版社

地　　址：北京市西城区永安路 106 号，100050

电　　话：010-67017249（咨询）　63131930（邮购）

传　　真：010-67078227，67078255

网　　址：http：//book. gmw. cn

E - mail：guomeijun@ gmw. cn

法律顾问：北京德恒律师事务所龚柳方律师

印　　刷：三河市华东印刷有限公司

装　　订：三河市华东印刷有限公司

本书如有破损、缺页、装订错误，请与本社联系调换，电话：010-67019571

开　　本：170mm×240mm

字　　数：350 千字　　　　印　　张：19.5

版　　次：2019 年 9 月第 1 版　　　　印　　次：2019 年 9 月第 1 次印刷

书　　号：ISBN 978-7-5194-5510-1

定　　价：89.00 元

总　序

春秋代序，岁月倥偬，弦歌不断，薪火相传。不知不觉，天水师范学院就走过了它60年风雨发展的道路，迎来了它的甲子华诞。为了庆贺这一重要历史时刻的到来，学校以“守正·奋进”为主题，筹办了缤纷多样的庆祝活动，其中“学术华章”主题活动，就是希冀通过系列科研活动和学术成就的介绍，建构学校作为一个地方高校的公共学术形象，从一个特殊的渠道，对学校进行深层次也更具力度的宣传。

《天水师范学院60周年校庆文库》（以下简称《文库》）是“学术华章”主题活动的一个重要构成。《文库》共分9卷，分别为《现代性视域下的中国语言文学研究》《“一带一路”视域下的西北史地研究》《“一带一路”视域下的政治经济研究》《“一带一路”视域下的教师教育研究》《“一带一路”视域下的体育艺术研究》《生态文明视域下的生物学研究》《分子科学视域下的化学前沿问题研究》《现代科学思维视域下的数理问题研究》《新工科视域下的工程基础与应用研究》。每卷收录各自学科领域代表性科研骨干的代表性论文若干，集中体现了师院学术的传承和创新。编撰之目的，不仅在于生动展示每一学科60年来学术发展的历史和教学改革的面向，而且也在于具体梳理每一学科与时俱进的学脉传统和特色优势，从而体现传承学术传统，发扬学术精神，展示学科建设和科学研究的成就，砥砺后学奋进的良苦用心。

《文库》所选文章，自然不足以代表学校科研成绩的全部，近千名教职员工，60年孜孜以求，几代师院学人的学术心血，区区九卷书稿300多篇文章，个中内容，岂能一一尽显？但仅就目前所成文稿观视，师院数十年科研的旧貌新颜、变化特色,也大体有了一个较为清晰的眉目。

首先,《文库》真实凸显了几十年天水师范学院学术发展的历史痕迹,为人们全面了解学校的发展提供了一种直观的印象。师院的发展,根基于一些基础老学科的实力,如中文、历史、数学、物理、生物等,所以翻阅《文库》文稿,可以看到这些学科及其专业辉煌的历史成绩。张鸿勋、雒江生、杨儒成、张德华……,一个一个闪光的名字,他们的努力,成就了天水师范学院科研的初始高峰。但是随着时代的发展和社会需求的变化,新的学科和专业不断增生,新的学术成果也便不断涌现,教育、政法、资环等新学院的创建自是不用特别说明,单是工程学科方面出现的信息工程、光电子工程、机械工程、土木工程等新学科日新月异的发展,就足以说明学校从一个单一的传统师范教育为特色的学校向一个兼及师范教育但逐日向高水平应用型大学过渡的生动历史。

其次,《文库》具体显示了不同历史阶段不同师院学人不同的学术追求。张鸿勋、雒江生一代人对于敦煌俗文学、对于《诗经》《尚书》等大学术对象的文献考订和文化阐释,显见了他们扎实的文献、文字和学术史基本功以及贯通古今、熔冶正反的大视野、大胸襟,而雍际春、郭昭第、呼丽萍、刘雁翔、王弋博等中青年学者,则紧扣地方经济社会发展做文章,彰显地域性学术的应用价值,于他人用力薄弱或不及处,或成就了一家之言,或把论文写在陇原大地,结出了累累果实,发挥了地方高校科学研究服务区域经济社会发展的功能。

再次,《文库》直观说明了不同学科特别是不同学人治学的不同特点。张鸿勋、雒江生等前辈学者,其所做的更多是个人学术,其长处是几十年如一日,埋首苦干,皓首穷经,将治学和修身融贯于一体,在学术的拓展之中同时也提升了自己的做人境界。但其不足之处则在于厕身僻地小校之内,单兵作战,若非有超人之志,持之以恒,广为求索,自是难以取得理想之成果。即以张、雒诸师为例,以其用心用力,原本当有远愈于今日之成绩和声名,但其诸多未竟之研究,因一人之逝或衰,往往成为绝学,思之令人不能不扼腕以叹。所幸他们之遗憾,后为国家科研大势和学校科研政策所改变，经雍际春、呼丽萍等人之中介，至如今各学科纷纷之新锐，变单兵作战为团队攻坚，借助于梯队建设之良好机制运行，使一人之学成一众之学，前有所行，后有所随，断不因以人之故废以方向之学。

还有，《文库》形象展示了学校几十年科研变化和发展的趋势。从汉语到外语，变单兵作战为团队攻坚，在不断于学校内部挖掘潜力、建立梯队的同时，学校的一些科研骨干如邢永忠、王弋博、令维军、李艳红、陈于柱等，也融入了更大和更高一级的学科团队，从而不仅使个人的研究因之而不断升级，而且也带动学校的科研和国内甚至国际尖端研究初步接轨，让学校的声誉因之得以不断走向更远也更高更强的区域。

当然，前后贯通，整体比较，缺点和不足也是非常明显的，譬如科研实力的不均衡，个别学科长期的缺乏领军人物和突出的成绩；譬如和老一代学人相比，新一代学人人文情怀的式微等。本《文库》的编撰因此还有另外的一重意旨，那就是立此存照，在纵向和横向的多面比较之中，知古鉴今，知不足而后进，让更多的老师因之获得清晰的方向和内在的力量，通过自己积极而坚实的努力，为学校科研奉献更多的成果，在区域经济和周边社会的发展中提供更多的智慧，赢得更多的话语权和尊重。

六十年风云今复始，千万里长征又一步。谨祈《文库》的编撰和发行，能引起更多人对天水师范学院的关注和推助，让天水师范学院的发展能够不断取得新的辉煌。

是为序。

李正元 安涛

2019 年 8 月 26 日

目 录
CONTENTS

基因概念的发展

张德华*

基因是遗传学研究的中心问题。从这个意义上说，遗传学是研究基因的科学。它研究基因的结构与功能，基因的遗传与突变，基因的表达与调控等等。然而，人们对基因本质的认识却是在遗传学的发展中逐步深化的。

遗传因子

在遗传学的奠基时期，孟德尔在他的《植物杂交试验》的著名论文中，首先提出遗传因子的概念。他把假设的控制生物性状的遗传单位叫做遗传因子。当时的遗传因子是一个十分抽象的概念，人们既不了解它在细胞中的确切位置，更不知道它的化学本质是什么，而只能从它的遗传效应感知它的存在。按照孟德尔的观点，遗传因子是呈颗粒式遗传的，它具有高度的稳定性。两个相对的遗传因子在一起时，不会相互混杂或沾染。孟德尔的遗传因子学说，为以后的大量科学实验所证明，于是逐渐取代了曾流行一时的“融合遗传”的理论。

本世纪初（1909 年），丹麦学者约翰森（Johannsen）用“基因”（gene）一词代替了孟德尔的遗传因子。从此，“基因”一词为遗传学界沿用至今。同时他还提出了“基因型”与“表现型”这两个含义不同的术语。初步阐明了基因与性状的关系。

功能、交换、突变三位一体的基因概念

1903 年，萨顿（Sutton）和博维里（Boveri）首先发现了遗传过程中染色体与遗传因子行为的相似性，例如，在性细胞中，染色体和遗传因子都成单存在；在体细胞中，染色体成双存在，遗传因子也是如此。于是，他们提出遗传因子就在染色体上的著名假设。随后不久，摩尔根等人以果蝇为材料进行了大量试验研究，不但

* 作者简介：张德华（1940— ），男，山东维坊人，曾为天水师范学院生物工程与技术学院教授，学士，主要从事植物分类与进化研究。

进一步证实了染色体是基因载体的假设,而且还发现基因在染色体上呈直线排列。他们认为,基因是遗传的功能单位,它能产生特定的表型效应;基因又是一个独立的结构单位,在同源染色体之间可以发生基因的互换,但交换只能发生在基因之间而不能发生于基因之内,基因可以发生突变,由一个等位形式变为另一等位形式,因而基因又是突变单位。这就是本世纪四十年代以前所流行的所谓“功能、交换、突变”三位一体的基因概念。这种认识,把基因与染色体联系起来,说明了基因的物质性,基因存在的场所及排列方式,基因就不再是一个抽象概念了。但这时 仍不了解基因的化学本质是什么,以及它是如何控制生物性状的。

基因是 DNA 分子的-个区段

1944 年,艾弗里(Avery)等人通过肺炎双球菌的转化试验,首次证明了生物的遗传物质是 DNA,于是基因的化学本质得到了阐明。1953 年,沃森(Watson)和克里克(Crick)又确定了 DNA 分子的双螺旋结构。进一步的研究证明,基因就是 DNA 分子的一个区段。每个基因平均由 1000 个左右的碱基对组成。一个 DNA 分子可以包含几个乃至几千个基因。基因的化学本质和分子结构的确定,具有划时代的意义,它为基因的复制、转录、表达和调控等方面的研究奠定了基础,开创了分子遗传学的新纪元。

基因的顺反子概念

1955 年,美国分子生物学家本泽(Benzer)通过对大肠杆菌的噬菌体 T_4 的 rⅡ区基因的深入研究,揭示了基因内部的精细结构。提出了基因的顺反子(Cistron)概念。他发现,在一个基因内部,可以发生若干不同位点的突变,倘若在一个基因内部发生两个以上位点的突变,其顺式和反式结构的表型效应是不同的。如图 1 所示,顺式是野生型,反式却是突变型,所以,基因就是一个顺反子。基因内部这些不同位点之间还可以发生交换和重组。所以,一个基因不是一个突变单位,也不是一个重组单位。一个基因中可包含许多突变单位和重组单位。本泽分别把它们称为突变子(muton)和重组子(recon)。显然,一个突变子或重组子可小到一个核苷酸对。

基因的顺反子概念冲破了传统的“功能、交换、突变”三位一体的基因概念,纠正了长期以来认为基因是不能再分的最小单位的错误看法,使人们对基因的认识有了显著的提高。

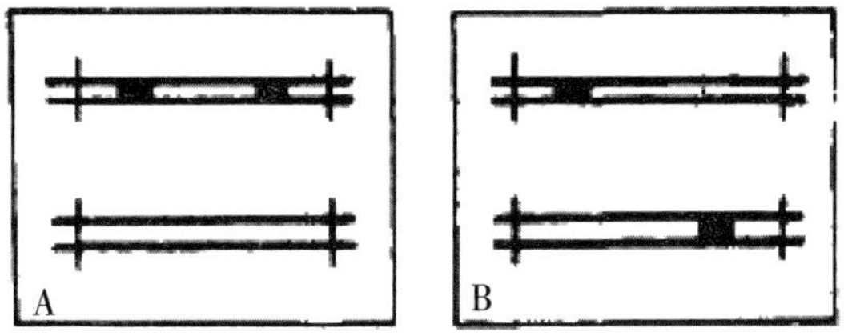

图 1 A 顺式 野生型；B 反式 突变型（黑方块表示突变位点）

结构基因与调控基因的划分

在遗传学研究的早期，人们曾注意到一个基因可以控制一种多肽链的合成，提出了"一个基因一个多肽"的学说。但是，进入 60 年代以后，人们在原核生物中首先发现，并不是所有基因都能为蛋白质多肽链编码。于是便把能为多肽链编码的基因称为结构基因（Structural gene），包括编码结构蛋白和酶蛋白的基因，也包括编码阻遏蛋白或激活蛋白的调节基因。

除结构基因而外，有些基因是只转录而不转译的，例如 tRNA 基因和 rRNA 基因。还有些 DNA 区段，其本身并不进行转录，但对其邻近的结构基因的转录起控制作用，被称做启动基因（启动子）和操纵基因（操作子）。操纵基因与其控制下的一系列结构基因组成一个功能单位叫做操纵子（operon）。

就其功能而言，调节基因、启动基因和操纵基因都属于调控基因。这些基因的发现，大大拓宽了人们对基因功能及相互关系的认识。

重迭基因和断裂基因

70 年代中期，随着 DNA 序列分析技术的发展，相继发现了基因在 DNA 分子上相互重迭的现象，以及基因内部存在间隔序列的现象。

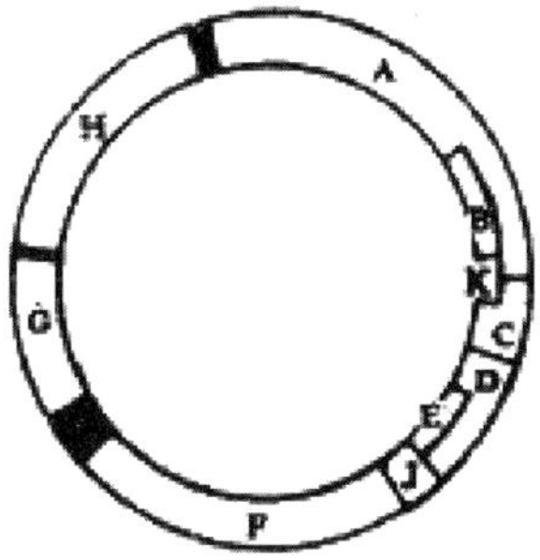

图 2 ΦX174 的基因重迭示意图

重迭基因 1977 年英国学者桑格(Sanger)在对单链 DNA 病毒 φX174 进行序列分析时发现,这个由 5375 个核苷酸组成的小型环状的 DNA 分子竟能为九种蛋白质编码。经进一步研究后知道,原来有好几个基因在 DNA 分子上是相互重迭的,即它们共同用一些核苷酸。如图 2 所示,基因 B 位于基因 A 中;基因 K 的一部分与 A 重迭,另一部分又与基因 C 重迭;基因 E 位于基因 D 中。这些基因虽然各有一部分核昔酸相重迭,但它们的读码结构不同,故不会转译出一段氨基酸顺序相同的多肽链。例如,基因 E 的起始密码是从基因 D 中间一个密码 TAT 中的 A 开始起读;基因 D 的最后一个核苷酸 A 同时又是基因 J 的第一个核苷酸(图 3)。

图 3　ΦX174 部分基因重迭及读码结构

基因之间发生相互重迭的现象至少对那些 DNA 含量较少的原始生物具有积极意义,它可以使有限的 DNA 序列携带更多的遗传信息,是生物对它的遗传物质经济而合理的利用。但其缺点也是不言而喻的,在这些共用序列中倘若一旦发生某个核苷酸的代换,受株连的将不只是一个结构基因。所以,对这些生物来说,突变将会产生更为严重的影响。

断裂基因 所谓断裂基因,指的是在这些基因内部存在一个或几个间隔序列,把一个完整的基因分隔成几个互不邻接的区段。断裂基因最初由查姆邦(Chambon)和波盖特(berget)于 70 年代中期在猿猴病毒(SV40)及腺病毒中首先发现。近年来的研究表明,断裂基因在真核生物中,尤其在高等真核生物中相当普遍地存在。

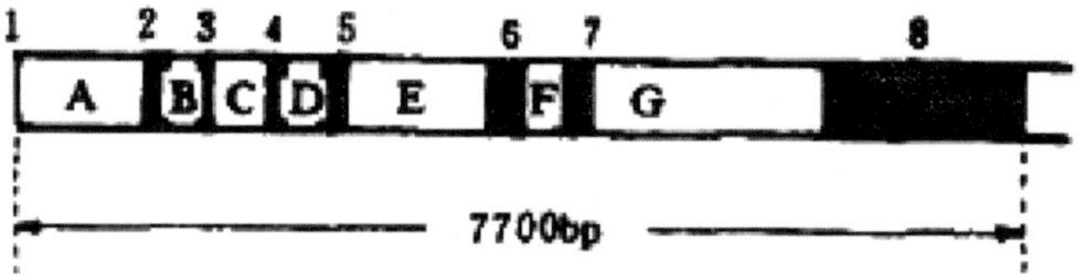

图 4　鸡卵清蛋白基因中的内含子

(白色区段)和外星子(黑色区役)

那么,一个被间隔序列分隔为数段的结构基因是如何编码组成一条完整的多肽链的? 原来这些基因在转录时,首先转录成为一个包括间隔序列在内的较大型的 RNA 前体,然后,其中所含的间隔序列被一一切除,最后才拼接形成一个成熟的 RNA 分子。人们通常把基因内部这些间隔序列称为内含子(intron)而把基因中出现于成熟 RNA 中的有效区段叫做外显子(extron)。例如在鸡的卵清蛋白基因中就含有七个内含子,把一个完整的基因分成八段(图 4)。其成熟 mRNA 只有该基因总长度的四分之一。

基因中存在内含子的现象,被认为是基因进化的产物,又是继续进化的基础。它将为基因功能的发展赋于更大的潜力。例如,有实验表明,在同一个基因内,通过外显子区段的不同的排列组合,可以形成若干种不同的基因产物。近年来,有些研究者还发现,某些前体 RNA 中的内含子部分有催化自我剪接的功能,甚至还能催化其他生化反应的进行,是一种重要的生物催化剂,所以,基因中内含子的存在对生物是有重要意义的。

过去人们曾经认为,基因像念珠一样,一个接一个地排列在染色体上,是互不重迭的;基因是一个连续的核苷酸序列,是不能间断的,重迭基因和断裂基因的发现,刷新了这些陈旧的看法,使人们对基因的结构与功能的认识又提高到一个新的高度。

本文曾发表在 1992 年第 2 期生物学通报上。

小麦体细胞胚胎发生早期淀粉含量变化的研究

张德华　焦成瑾　王亚馥　崔凯荣　王仑山*

小麦体细胞胚胎发生早期淀粉消长的动态变化,似可大致分为 3 个阶段。1. 单个胚性细胞中含有密集而大型的淀粉粒,是淀粉积累的第一个高峰。2. 2—细胞,3—细胞以至多细胞原胚阶段细胞中淀粉含量逐渐降低,这是由于细胞分裂迅速,淀粉消耗大于积累的结果。3. 球形胚时期,淀粉含量再次升高,表明此阶段淀粉积累多于消耗。

淀粉是植物胚胎发育期间重要的代谢物质之一。植物组织培养中体细胞胚发生过程中的淀粉代谢的研究工作已有些报道[1-4]。由于小麦愈伤组织中体细胞胚发生频率不高,加之,随着继代进程,体细胞胚发生频率逐渐降低,这就为定时定点地追踪它的淀粉消长变化带来了一定困难,有关的研究还不多。

本实验以小麦幼胚为外植体,在附加 2,4 — D 的 MS 培养基上诱导出胚性愈伤组织, 经取样,固定,包埋,用半薄切片结合“PAS"反应染色等细胞组织学方法,系统观察了愈伤组织内,发育早期的体细胞胚中淀粉含量的消长变化,并对这些变化与体细胞胚发育的关 系进行了初步的探讨。

1　材料和方法

1.1　胚性愈伤组织的诱导

以春小麦品种陇春一号授粉 12 ~15 d 的幼胚为外植体,接种在附加 2,4 - D 2mg/L + KT 0.5mg/L + LH 300 mg/L + 3 % 蔗糖的 MS 培养基上,30d 后产生大量愈伤组织,每 25d 左右继代一次,经继代培养 2 ~3 次后,转入附加 2,4 - D 0.5 mg/L + KT 0.5 mg/L + LH 300 mg/L + 3 % 蔗糖的 MS 培养基上,25d 后分化出

---

* 作者简介:张德华(1940—　),男,山东维坊人,曾为天水师范学院生物工程与技术学院教授,学士,主要从事植物分类与进化研究。

淡黄色颗粒状和白色而无结构两种形态不同的愈伤组织,切片观察表明,它们分别属于胚性愈伤组织和非胚性愈伤组织。

1.2 半薄切片制样

挑取胚性愈伤组织,切成 1 mm^3 的小块,经戊二醛、锇酸双固定,乙醇脱水,Epon812 环氧树脂包埋,在 AO 切片机上作半薄切片,厚度为 2 ~ 3μm,先用亚甲蓝——天青 I 染色,再用高腆酸——席夫反应(PAS)对染,自然晾干,中性树胶封片,在 Olympus 光学显微下观察照相。

2 实验结果

2.1 胚性细胞与非胚性细胞中淀粉含量的差异

经切片观察,在愈伤组织内存在有集中分布的胚性细胞团,这类细胞体积小,近球形,排列紧密,核大,细胞质浓,着色深,不含液泡或仅含小液光;其周围的非胚性细胞,体积大,壁薄,形状不规则,细胞质较稀薄,染色较浅,常含大液泡。在这两类不同的细胞中,淀粉粒的分布呈现显著的差异:胚性细胞中含有大量的淀粉粒,而非胚性细胞中淀粉粒稀少,甚至观察不到淀粉粒的存在〔图版 1 ~2〕。

2.2 单个的胚性细胞与周围薄壁细胞中淀粉含量的差异

胚性细胞团中,并非所有细胞都能进一步发育,而是仅有个别的胚性细胞得到充分发

育的机会,体积增大,呈现圆球形,具有厚壁,与周围细胞缺少紧密联系,处于相对的“孤立”状态,成为单个的胚性细胞〔图版,3〕. 据观察,单个的胚性细胞还可能有另外一条发生途径,即由愈伤组织表层或近表层的薄壁细胞转化而来,它们也具有单个胚性细胞的典型特点如壁加厚,细胞质浓,核大,核仁明明显等,与周围的薄壁细胞有明显区别〔图版,4〕。单个的胚性细胞中的淀粉含量显著的多于邻近的薄壁细胞,也多于胚性细胞团中的那些小型的原始胚性细胞〔图版,3、4〕。它们含有大型而密集的淀粉粒,而且通常分布于核的周围〔图版,5〕。单个的胚性细胞进一步发育,逐渐发生极性化,表现为核向一端偏移,大液泡形成,淀粉粒的分布也表现出相应的极性化状态〔图版,6〕。

2.3 不同发育阶段的体细胞胚中淀粉量的比较

观察比较发现,单个的胚性细胞是体细胞胚发育的起始阶段,也是淀粉含量的第一个高峰期。单个的胚性细胞的第一次分裂,我们观察到两种情况:一是细胞不经伸长,便直接横裂为大小相似的 2 个细胞〔图版,8〕。另一种情况是细胞先伸长,然后横裂为一大一小两个细胞〔图版,7 〕。大的为基细胞,小的是顶细胞,胚体主要由它分裂而成。两细胞间大都以斜向壁相隔,细胞中仍有较多的淀粉积

累。随后,伴随着细胞的迅速分裂,经3一细胞期形成多细胞原胚。淀粉被大量消耗,以致在这些不同发育阶段的原胚中,淀粉含量逐渐减少,细胞中几乎看不到淀粉粒或者仅有稀疏的小颗粒淀粉存在〔图版,9~10〕。当体细胞胚发育到球形胚阶段,淀粉含量又有所回升,成为淀粉积累的第二个高峰。同时,球形胚周围的细胞大都失去正常结构,有的趋于解体,使胚体处于"隔离"状态〔图版,11~12〕。在成熟的球形胚中,淀粉亦呈现极性化分布,标志着球形胚细胞已开始分化〔图版,12〕。

球形胚进一步发育,可依次形成梨形胚,盾片胚,成熟胚。关于这些晚期阶段的体细胞胚中淀粉含量的消长变化情况,将在另文中进行讨论。

3 讨论

3.1 关于淀粉积累与体细胞胚发生及发育的关系

一些作者指出,体细胞胚的形成及器官发生与淀粉的消长紧密相关[1-4]。淀粉在胚胎发生和发育中的作用主要体现在两个方面:一是作为主要的能源物质,为细胞的分裂和分化提供能量;二是作为结构物质,它的降解产物可为蛋白质、核酸、纤维素等重要细胞结构物质的合成提供必要的碳架。在这方面体细胞胚与合子胚有相似之处。高锦华[5]等在水稻合子胚发育的研究中发现,淀粉的积累高峰,总是出现在蛋白质、核酸的合成高峰之前,表明淀粉的积累可为蛋白质、核酸等物质的合成奠定基础.我们还观察到,在愈伤组织的胚性细胞团中以及单个的胚性细胞中。都有大量的淀粉积累,表明淀粉的积累与胚性的发生及实现有一定关系.

3.2 小麦体细胞胚发育早期淀粉含量的消长变化

在小麦体细胞胚发育早期,有两个淀粉积累高峰:一个是单个的胚性细胞时期,另一

个出现在球形胚阶段,这个结果与有的报道不尽一致。从这两个淀粉积累高峰推测,这些阶段的淀粉合成大于分解的速度,其意义在于为下一阶段细胞的迅速分裂积累能量,正如有作者指出,淀粉的大量积累常常预示着新的发育状态即将来临[6]。本实验的观察结果,与这一看法相一致。

在不同阶段的多细胞原胚中,淀粉含量急剧减少,一个可能的原因是由于这时细胞分裂异常迅速,以致使淀粉的合成远远赶不上消耗的速度,从而使淀粉含量大幅度下降。

3.3 关于体细胞胚中淀粉的来源问题

单个的胚性细胞中有大量的淀粉积累,显然不大可能仅由细胞本身的可溶性

糖转化而来,它可能有赖于从周围细胞中吸收某些营养物质。我们观察到,伴随着单个的胚性细胞中大量的淀粉积累,一个相应的变化就是其周围细胞淀粉粒减少;球形胚阶段更是如此,它所处的“孤立”或“隔离”状态看来不是偶然的,实际是它从周围细胞中大量吸收营养物质导致某些细胞解体的必然结果,合子胚的发育中也有类似的现象。这是生物为保证种族繁衍而形成的一种适应性特征。

参考文献

[1]SabharwalPS. In vitro culture of nucelli embryos of Citrus aurantifolia swingle. Plant embryology. 1962,(2),239 -243

[2]Thorpe'T A. Starch accumulation in shoot formihg hobacco ca1lus. Science,1968,160,421 -422.

[3]Komar KN. Origin and structure of embryoiods a rising from the epidermal cell of *Ranunculus Scel-eratus L.* Cell Sci,1972,(11):77 -93.

[4]Halperin W. Embryos from somatic plant cell. Cell Biol,1972,(9):161 -191.

[5]商锦华,唐锡华. 高等植物胚胎的发育生物学研究V,水稻种胚发育过程中胚和胚乳内几种碳水化合物含量及淀粉酶活力的动态[J]. 植物生理学报,1982,8(3):258 -294.

[6]刘春明,姚敦义. 陆地棉体细胞胚胎发生及细胞组织学研究[J]. 植物学报,1991,33(5):278 -284.

图版说明

本文曾发表在1993年第2期兰州大学学报(自然科学版)上。

银杏组织培养研究
Ⅰ.不同培养条件对银杏愈伤组织形成的影响

张德华　张占甲　陈坤明　李仲芳　刘红岩　焦成瑾*

取银杏树干上萌生的幼嫩枝条的基尖、茎段和叶片分别接种到高生长素含量的 MS 培养基上,愈伤组织诱导频率可高达 90%。在低生长素浓度下,诱导频率较低,但能促进茎外植体顶芽和侧芽的生长。以银杏幼茎、幼叶作外植体,形成的愈伤组织形态结构不同:幼基愈伤组织较疏松,呈粒状,色浅黄,生长迅速,幼叶愈伤组织较致密,平滑,色黄绿,生长稍慢。叶片的放置方式很重要:下表皮接触培养基时,愈伤组织只从上表皮发生,而反向放置,上下表皮都不形成愈伤组织。

银杏(*Ginkgo biloba L.*)是我国珍贵的特产树种之一,其树姿优美,材质优良,同时又是重要的药用经济林木。银杏为雌雄异株,在生产实践中,雌株具更大的经济价值。但用种子繁殖,速度较慢且无法控制种苗性别,因此用组培方法加速繁育银杏优系优株,尤其是有选择的繁殖优良雌株,是一个值得探讨的问题。关于裸子植物的组织培养和快速繁殖,已有不少研究[1]。阈国宁等以杉木幼苗的茎尖、成龄优树的茎芽、茎段为外植体,用组培方法繁育优良试管苗近万株[2]。梁玉堂等曾用返幼的黑松针叶束经离体培养,获得再生植株[3]。朱尉华等在云南红豆杉、东北红豆杉、红豆杉及南方红豆杉的组织培养中,愈伤组织诱导频率可达 70%[4]。Attree S. M. 等在对黑云杉与白云杉的组培中,通过体细胞胚发生途径获得再生植株[8]。Von Arnold S. 在挪威云杉上也取得类似的结果[9]。罗紫娟等曾对银杏胚和茎段进行过组织培养工作,获得再生植株[5]。但采用银杏幼叶为外植体,以及对不同外植体和不同接种方式的培养效果进行系统的比较研究报道尚

* 作者简介:张德华(1940—　),男,山东维坊人,曾为天水师范学院生物工程与技术学院教授,学士,主要从事植物分类与进化研究。

少。本试验以银杏茎、叶为试材,研究影响银杏茎、叶愈伤组织诱导和分化的因素,以便为银杏试管快速繁殖提供资料。

1 材料和方法

6月初,从8－10年生银杏树干萌生的幼嫩枝条上,剪取幼叶、茎尖和茎段。先用清水冲洗干净,置于0.1%的升汞中浸泡10min,70%的乙醇中浸泡1min,用无菌水反复冲洗数次。将叶片剪成0.5cm×0.5cm的小方块,茎尖、茎段的长度约1cm,分别接种到附加不同浓度的NAA和6－FA(6一糖基氨基嘌呤)的MS培养基上。置于25±1℃,相对湿度78%,光照强度为2000 lx的组培室中培养。3周后获得大量愈伤组织,经继代培养一次后,转入MS＋6－FA 0.1mg/L＋NAA 0.02mg/L＋ YE 200mg/L及MS＋6－FA 2mg/L＋ 2,4－D 0.05 mg/L＋YE200 mg/L的分化培养基中,诱导愈伤组织分化。定时观察记载。

2 实验结果

2.1 不同外植体的培养效果

观察表明,采用银杏的茎尖和茎段作外植体效果相似。但它们与幼叶外植体比较,在相同的培养基上,产生愈伤组织的时间、愈伤组织的生长速度及愈伤组织的形态结构均表现出较大的差异,而愈伤组织诱导频率相似(表1)。茎尖、茎段经7d培养后,在接触培养基的切面茎皮部分开始膨大变厚,逐渐产生浅黄包愈伤组织.继续培养1周后,愈伤组织生长速度明显加快,迅速形成结构疏松,大颗粒状,表面微呈褐色的大块愈伤组织。叶片外植体经10d培养后,在边缘切口处产生许多瘤状突起,继而在上表皮形成稍致密的,颗粒较小的黄绿色愈伤组织。生长速度略低于前者,所形成的愈伤组织块亦较小,但不易褐化。

外植体的接种方式对银杏愈伤组织形成起重要作用,观察发现,当叶上表皮朝上以下表皮接触培养基时,愈伤组织可以90%以上的频率由上表皮发生;但反向放置,以上表皮接触培养基,上下表皮都不能形成愈伤组织,外植体逐渐干枯。当茎尖、茎段在培养基上立放时,愈伤化首先从接触培养基的下端断面茎皮发生,形成一圈浅黄色初始愈伤组织,随后愈伤组织逐渐长大,以至把整个外植体包埋起来。当横放于培养基上时,愈伤组织首先从贴近培养基的断面茎皮发生,然后扩展到其他部分,表明以银杏幼茎作外植体时,愈伤组织最初都由断面的茎皮发生,而且培养基中的诱导因素在脱分化中起关键作用。

表1 不同外植体在MS+6-FA0.5mg/L+NAA10mg/L培养基上的培养效果

外植体种类	茎尖	茎段	叶片*
接种外植体块数	20	40	37
形成愈伤组织数	19	37	35
成愈频率%	95	92.5	94.6
外植体脱分化所需天数(d)	7	7	10
愈伤组织生长速度	+ + +	+ + +	+ +
愈伤组织形态	较疏松,大颗粒状,浅黄色	较疏松,大颗粒状,浅黄色	较致密,颗粒小,黄绿色

只统计上表皮朝上的叶块,反向放置不能形成愈伤组织。

2.2 不同激素浓度及配比对愈伤组织的诱导作用

在MS培养基中附加不同浓度的生长素和细胞分裂素对成愈频率和愈伤组织的生长速度影响显著(表2)。实验表明,当MS中单独附加6-FA时,不能诱导愈伤组织产生,而单独添加NAA却能诱导出愈伤组织,但频率较低。对银杏茎叶外植体愈伤组织诱导和生长最适宜的培养基是MS+NAA 10mg/L + 6-FA 0.5mg/L,诱导频率可高达90%以上,而且愈伤组织生长迅速。茎叶外植体对激素的敏感程度并无显著差异。

表2 MS培养基中附加不同配比的NAA.6—FA对银杏茎叶外植体愈伤组织形成的影响

MS中附加激素(mg/L)			成愈频率(%)			愈伤组织生长速度	
NAA	6—FA	NAA/6—FA	茎尖	茎段	叶片*	茎愈伤组织	叶愈伤组织
0	0.5	0/0.5	0	0	0		
0.5	0	0.5/0	8.5	6.4	6.2	+	+
0.5	0.5	1/1	17.4	21.0	16.5	+ +	+
5	0.5	10/1	80.5	78.6	75	+ + +	+ +
10	0.5	20/1	95.0	92.5	94.6	+ + +	+ +
25	0.5	50/1	76.2	78.0	75.5	+ + +	+ +

同表1 * Similar to table 1.

3 讨论

3.1 外植体的影响

采用同一种植物的不同器官作外植体培养效果不同,已有不少报道[6],本实验证明,以银杏幼茎的茎尖、茎段为外植体,两者差异不大。茎、叶外植体比较,成愈频率基本相同,但它们所形成的愈伤组织的形态结构迥然不同:幼叶形成的愈伤组织结构紧密,表面较平滑,浅绿色,生长较慢;幼茎形成的愈伤组织生长迅速,结构疏松,表面呈大颗粒型,颜色浅黄,表层微褐色。这种差异可能由于不同器官的细胞形态结构和生理状况不同所致。特别值得注意的是,以银杏幼叶作外植体时,叶片的接种方式非常重要。当以下表皮接触培养基时,愈伤组织只从上表皮发生,而反向放置不能形成愈伤组织。这可能与银杏叶上下表皮的结构差异直接相关。据报道,银杏叶片上下表皮细胞形态结构有很大不同,上表皮细胞排列整齐,表面平滑,气孔极少;下表皮细胞排列不规则,细胞壁上有很多突起,气孔密布,甚至有 2 - 3 个气孔生在一起的现象。我们分析,银杏叶下表皮细胞的特征似乎有利于外植体从培养基中接受诱导因子,进入叶肉,传至上表皮,故愈伤组织可以从上表皮发生。当以上表皮接触培养基时,上表皮细胞的紧密排列及加厚角质层外壁可能会限制诱导因子的进入,因而外植体难脱分化,愈伤组织不能形成。

3.2 激素的影响

许多研究表明,在培养基中附加一定浓度的激素对诱导愈伤组织产生是必要的。但通常采用的激素浓度较低,一般不高于 2 - 3mg/L。我们的实验证明,银杏愈伤组织的诱导需要超常的高生长素浓度,而且适宜的生长素与细胞分裂素的配比也是至关重要的。当生长素浓度偏低时(NAA 0.5 mg/L),外植体脱分化困难,成愈频率很低,但却有利于茎外植体顶芽和侧芽的生长。当培养基中单独附加 6 - FA 时,愈伤组织不能产生,而单独附加 NAA 时,愈伤组织能够产生。表明 NAA 是银杏愈伤组织形成所必需的因素,而 6 - FA 的存在能增强 NAA 的诱导效果。

曾有作者指出,以成年木本植物作材料,用组培法再生植株是困难的[1]。在常规培养下情况确是如此。对银杏这种“活化石”植物而言,适当提高生长素浓度并辅之以不同激素的适当配合,愈伤组织诱导频率可达 90% 以上,且愈伤组织生长迅速。但能否顺利分化及再生植株,有待进一步研究。

参考文献

[1]王怀智. 植物组织培养与植树造林. 经济植物组织培养[M]. 北京:科学出版社,

1988:105 – 115.

[2]阙国宁.杉木的组织培养育苗经济植物组织培养[M]. 北京:科学出版社,1988:210 – 215.

[3]梁玉堂,邢世岩.几种影响黑松针叶束离体培养再生植株的因子[J]. 植物生理学通讯,1990(3):31 – 33.

[4]朱蔚华,陆俭.几种红豆杉植物愈伤组织诱导培养的观察[J]. 中药材,1991,14(9):5 – 7.

[5]罗紫娟.银杏茎段的组织培养[J]. 植物生理学通讯,1985(1):35 36.

[6]王喆之,张大力等.灯笼果组织培养的研究[J]. 西北植物学报,1991,11(1):44 – 49.

[7]李正理,贺晓.银杏叶表皮结构[J]. 植物学报,1989,31(6):427 – 431.

[8]Attree S M. Somatic embryo mature germination and plantle formation of *Picea glauca and P. mariana. Bot.* ,1990;68(12),2583 – 2589.

[9]Von Arnold S. Vegetative propagation of Picea abies L. V. international congress of plant tissue and cell culture. Abstr,. 1982:246.

本文曾发表在1994年第6期西北植物学报上。

汞对蚕豆根尖细胞微核的诱变效应

毛学文*

利用蚕豆根尖细胞微核方法，检测汞的诱变效应。结果表明，用 25 PPM 至 150 PPM 的不同浓度的 $HgCl_2$溶液处理蚕豆根尖，微核总数及微核率明显高于阴性对照组($P < 0.05$)，从实验数据方差分析得知，各组实验数据与对照组数据之间均存在着差异($t > 3.4$)。用 0 - 1.0 *PPM* 低浓度溶液处理，蚕豆根尖伤害不明显，说明 Hg^{2+}不是强诱变剂。

汞是常见的一种金属污染源。对水质污染和土壤污染有很大的影响。在日常生活中，汞的化合物、汞农药和人们的接触机会很多。关于汞对植物根尖细胞核的影响已有报道[1-2]。本文主要是在前人研究的基础上，进一步验证汞虽不是诱变剂，但对蚕豆根尖细胞微核的形成有一定的诱变效应。

材料和方法

1 蚕豆种子购买于华中师大生物系培植的松滋青皮豆。氯化汞购于西安化学玻璃试剂商店。用蒸馏水配 $Hgcl_2$溶液，浓度分别为 25、50、75、100、150PPM，用蒸馏水作阴性对照。

2 选择大小均匀的蚕豆，用自来水萌发，待根尖长 0.5 - 1cm 时，选出生长良好的蚕豆，每组 3 粒分别放入不同浓度的处理液中，处理 48 h，用蒸馏水冲洗两次，再放入蒸馏水内修复培养 24 h 后剪下根尖，用卡诺固定液固定 24 h，INHcl 160℃下水解 14 min，并经 Schiff 试剂染色，常规压片，高倍镜下每组观察 3 个根尖，每个根尖统计 1000 个细胞，统计微核千分率。求出各实验组微核率均值及标准差，对全部数据进行方差分析和 t 检验。

* 作者简介：毛学文(1945—)，男，山西沁水人，曾为天水师范学院生物工程与技术学院教授，学士，主要从事植物资源开发与应用研究。

结果

1 汞对蚕豆根尖和微核的诱变效应

本实验中各实验组与阴性对照组的微核率差异显著,表明本试验体系可行。在各组处理中,微核总数及微核率均明显高于阴性对照组,各实验组微核及标准差列于表1,在此基础上进行方差分析,结果列于表2,方差分析的 *F* 检验结果表明,对照组微核率与实验组微核率之间存在显著的差异。

表1 各实验组细胞微核率统计

组别编号	浓度(PPM)	根尖微核率			$\bar{x} \pm s\bar{x}$
CK	0	2.0	2.2	2.4	2.2 ±0.20
1	25	4.8	5.1	6.6	5.5 ±0.96
2	50	8.1	7.8	7.5	7.5 ±0.47
3	75	8.7	8.2	10.4	9.1 ±1.15
4	100	9.5	11.8	11.1	10.8 ±1.17
5	150	5.7	5.3	5.9	5.6 ±0.76

表2 实验数据方差分析

变异来源	自由度	平方和	均方	*F* 值
处理间	2	859	429.5	5.27*
误差	15	1 223	81.5	
总和	17	2 072		

* 差异显著

2 汞诱变蚕豆根尖微核的剂量效应关系

本试验汞处理的五个浓度均能诱导蚕豆根尖的微核的形成。在 25 PPM 至 100PPM 四个浓度组蚕豆根尖的微核率随着汞浓度的升高而增加,在浓度 100PPM 时,达到最高峰,细胞微核率达 10.8 ‰,汞溶液浓度与微核率之间呈明显的剂量效应关系。但继续增大氯化汞的浓度,蚕豆根尖微核率有所下降,这是因为高浓度的氯化汞作用强度的增加抑制了细胞的活动,使细胞增殖延迟或终止进行[3]。

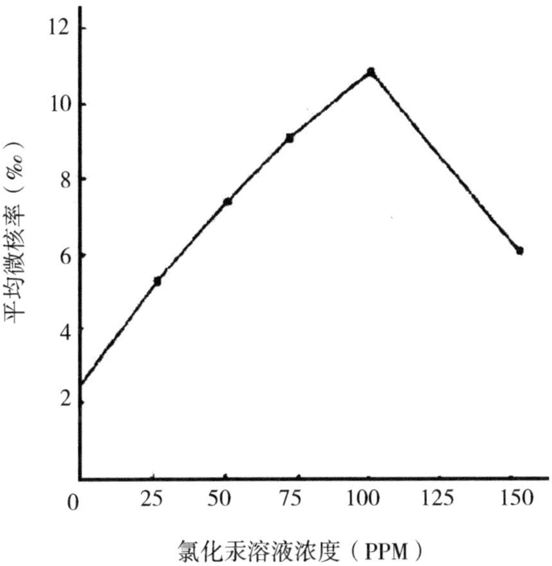

图氯化汞浓度与微核率的关系

讨论

汞的化合物是常用于教学、科研。汞农药广泛应用于农田。我们的实验结果表明汞对蚕豆根尖的诱变效应比明矾、煤焦油沥青等强诱变物质弱[4-5]。但是对蚕豆根尖的形态、细胞核和染色体等均产生较大的影响，汞对蚕豆根尖的微核效应与其浓度和处理时间密切相关，在低浓度汞溶液处理下，蚕豆根尖伤害不明显。随着工农业生产的发展，会有多种污染物质的工业废水、生活污水排入水系，造成水质污染、土壤污染。其中汞金属、汞农药最常见。据测定汞在水中的本底值一般不超过 0.1PPM、在水稻的培养液中含汞 7.4PPM 以上严重受害[6]。植物在生长的过程中不间断的受到水质污染和土壤污染的影响，所以汞对植物根尖尤其是蚕豆根尖细胞核的影响是很大的，同时对蚕豆根尖的生长、水分代谢都具有明显的抑制作用。

另外，由于受汞的诱导，蚕豆根尖内的 DNA、RNA 和蛋白质等大分子化合物均产生化学变化[7]。使控制染色体状态变化和细胞分裂的蛋白质的功能受到破坏[1]，影响了蚕豆根尖细胞 DNA 的复制。重金属汞在正常情况下，分布比较恒定，通常并不对人体构成威胁，但当食物受到“三废”污染，大量汞元素进入食品后，可使人体中毒[8]，至于汞是否对人体具有致突效应，尚需要进一步研究。

参考文献

[1]陈永喆,肖辅珍. Hg^{2+} 对蚕豆根尖细胞核的影响[J]. 北方植物学研究(第一集),1993:(2):13.

[2]王英彦,汤大友. 十五种化合物质的蚕豆根尖细胞核效应[J]. 中国环境科学,1986,6(2):19.

[3]沈光平. 微核与染色体畸变的相关性,1985,7(1):15.

[4]李建平. 蚕豆根尖细胞微核试验明矾的诱变效应[J]. 遗传,1989,11(1):8.

[5]祝庆蕃. 蚕豆根尖细胞微核试验煤焦油沥青水悬液的诱变效应[J]. 遗传,1989,11(2):7.

[6]李杰芬主编. 植物生物学[M]. 北京:北京师大出版社,1988 :219.

[7]CleaverJ E. 物理和化学致突物损伤后 DNA 切补修复的研究方法. 环境致变物、致癌生物学试验[M]. 北京:人民卫生出版社,1982 :255 - 256。

[8]刘志宗主编. 食品营养学[M]. 北京:中国轻工业出版社,1991 :208.

本文曾发表于 1998 年第 5 期癌变、畸变、突变。

不同光质对毛地黄愈伤组织诱导和增殖的效应

毛学文　陈　荃*

毛地黄(Digital is purpurea)又名洋地黄,是一种玄参科植物,它是一种极有药用价值的中草药,叶入药,主要作用在兴奋心肌,增加心肌收缩力,使收缩期的血液输出量大为增加,改善血液循环对心脏性水肿患者有利尿作用。在叶片中含有强心甙类,是很好的强心剂。由于栽培周期长,不能满足市场的需求,鉴于此,我们用组织培养方法对毛地黄愈伤组织进行诱导达到扩大药源的目的。本文主要报道了不同光质对毛地黄叶愈伤组织的诱导及增殖的效应。

表1　不同光质主要技术参数

技术参数	光质					
	黑暗	蓝光	绿光	黄光	红光	白光
波长峰(nm)		455	520	580	630	
光照度(lx)		280	1040	475	345	1410
功率(W)		80	80	80	80	80

材料取幼叶,将其在无菌条件下剪成0.5cm见方的小块,放置在不同浓度激素的MS培养基上培养,并用5种不同的光质进行光照,以黑暗培养为对照,主要技术参数见表1,各处理每天光照10小时,培养温度24±1℃,每隔2天取不同光质条件下培养的愈伤组织各2瓶,称重并计算每块愈伤组织的生长量,15天后统计并计算愈伤组织的诱导率。计算方法如下:

愈伤组织生长量=愈伤组织鲜重/愈伤组织块数-接种时每块外植体鲜重

愈伤组织诱导率=愈伤组织块数/接种外植体块数×100%

不同光质处理愈伤组织形态学变化为接种4天后,两端微微向上卷曲,6天后

* 作者简介:毛学文(1945—　),男,山西沁水人,曾为天水师范学院生物工程与技术学院教授,学士,主要从事植物资源开发与应用研究。

在切口处开始形成颗粒状愈伤组织，每种光下起始时间为黄光、白光 6 天，蓝光、绿光 7 天，红光黑暗 8 天、10 天后，愈伤组织急剧分裂，细胞体积增大，质地较密。只有黑暗条件下形成的愈伤组织质地疏松，15 天后各光质下愈伤组织的诱导率如表 2。

表 2　不同光质处理的愈伤组织的诱导率

光质	蓝光	绿光	黄光	红光	白光	黑暗
愈伤组织诱导率%	95.88	95.73	95.65	95.50	94.85	94.58

各种光质条件下的愈伤组织的诱导率均达到 94% 以上，差异不显著，其大小顺序为蓝光 > 绿光 > 黄光 > 红光 > 白光 > 黑暗。培养 20 天 ~ 22 天后，愈伤组织均开始变为褐色，25 天后逐渐老化死亡。

不同光质对愈伤组织的增殖曲线为 S 型。且有一定差异，前 10 天不太明显，10 天后，蓝光、黑暗和白光下愈伤组织增殖高于其它几种光，20 天后称重，各光质条件下的愈伤组织增殖量均达最大值，其中蓝光为 292mg，绿光为 261.8mg，黄光为 225.4mg，白光为 216.8mg，黑暗为 150mg，红光为 160.3mg，结果表明红光对毛地黄愈伤组织的诱导不如蓝、绿、黄、白光，但比黑暗有效。

不同光质对愈伤组织增殖的效应，由于研究者使用的材料不同，培养基的组分各异，研究的结果也不尽相同，何小弟（1993）研究红光对愈伤组织的生长促进作用明显，其效应值高于其它光处理的效应值，王维荣等（1991）认为黄瓜的愈伤组织在蓝光下生长量最高，绿光次之，其它光不明显，而红光对番茄愈伤组织诱导量最明显。元英进（1993）研究发现，蓝光对细胞的生长有促进作用而绿光有抑制作用。Sei bert 等在烟草叶组织培养研究表明蓝光有促进愈伤组织增重的作用。我们的实验结果与以上研究结果有相同的也有相异的。

参考文献

[1]张泓．植物学通报，1994（11）：12 – 13.
[2]何小弟．激光生物学，1993（2）：268 – 271.
[3]元英进．植物生理学通讯，1993（6）：47.

本文曾发表在 1997 年第 1 期植物学通报上

植物激素与营养液配施对平菇产量和品质的影响

张占甲　陈坤明　李仲芳　张德华　焦成瑾*

不同配比的 GA_3和 KT 与营养液配施，对平菇菌丝生长影响不同。GA_3和 KT 的适当配比，能促进菌丝分枝生长，也能显著提高平菇子实体的产量和品质，但维生素 C 的含量有所下降。GA_3和 KT 若与营养液配合施用，比单独施用更能发挥其促进作用。

Bandoni 曾经指出，许多真菌本身能大量产生 IAA、GA 等生长调节物质。这预示着人们利用植物生长物质，有目的地对食用菌进行调控是完全可能的[1]。近年来，植物生长物质在食用菌研究和生产上的应用越来越广泛。国内许多厂家以植物生长调节剂为主要原料配制而成的"蘑菇助长剂""川蘑菇增产灵""蘑菇健壮素"之类，应用于食用菌生产均有一定的增产效果[1,3,4,5]。

李人圭等发现低浓度的 IAA、NAA、KT 和 GA_3能够促进香菇菌丝体的生长。陈都珍的试验说明在 PDA 培养基上，使用 2 mg/L 的 NAA 和 IBA 可明显促进平菇菌丝生长，而且 IAA 和 GA_3有一定的抑制作用；在木屑培养基上，lmg/L 的各种生长激素，都不同程度地促进菌丝生长。吴帮平将 NAA，6 - BA 和 2，4 - D，于不同时期分别喷施在平菇上，都有明显增产。杨瑞长于初菇期、二潮菇期在香菇上分别喷施 6 - BA、6 - FA 和 IAA 均获得增产，而压块时处理无效。说明处理时期直接关系最终结果。肖崇明则证明高浓度(10 mg/L)的 KT 对金针菇菌丝生长有抑制作用，但却使其可溶性糖增加了 17%，众多研究者的工作证实，使用植物生长物质，能够促进食用菌产量和品质的改变。关键在于针对不同的食用菌，在适当的时期，选用适当种类，适当浓度的植物生长物质。

---

* 作者简介：张占甲(1941—　)，男，山东滨州人，曾为天水师范学院生物工程与技术学院副教授，学士，主要从事蔬菜与花卉无土栽培技术研究。

但众所周知,植物生长物质只能调节控制植物的生长发育,而不能代替植物生长所需要的营养物质[2]。在不增加营养的情况下,单纯依靠植物生长物质的调控作用去提高产量和品质,将很难取得好的效果。为了进一步揭示植物生长物质在食用菌生产上的应用价值和正确的使用方法,本文研究了不同配比的 GA_3 和 KT 与营养液配施,对平菇产量和品质的影响,并对平菇栽培中植物生长物质的调控作用与营养条件的关系进行某些探讨。

1 材料和方法

1.1 供试菌株

平菇(*Pleurotus ostreatus*)P-1,由西北农业大学引进。

1.2 试验方法

1.2.1 菌丝体生长培养基的配制 GA_3 和 KT 的浓度各设 3 个水平,自由配合成 9 个处理(表 1),将其分别加入到 PDA 平板培养基中。对照 A 加清水,对照 B 加营养液 M_1(M_1:每升含 $MgS0_4 \cdot 7H_20$:0.4g,$ZnS04 \cdot 7H_20$:0.2g,H_3B0_3:0.1 g,$Co(NH_2)_2$:3g,KH_2P0_4:2 g),共 11 个处理。每处理接种 5 个 PDA 平板,置 25℃恒温培养 6 天,每 24 小时测菌落直径,共测 5 次,求出不同处理中菌丝体的平均生长率

表 1 PDA 培养基上 GA_3 和 KT 的不同配比

处理	A	B	C	D	E	F	G	H	I	J	K
GA_3 (mg/L)	0	0	1	1	1	0.5	0.5	0.5	0.1	0.1	0.1
KT (mg/L)	0	0	5	2.5	1	5	2.5	1	5	2.5	1

注:1. 对照:A + H_20;2. 副对照:B 十 M_1。

1.2.2 产菇培养基的配制(1)GA_3、KT 拌料试验:采用木屑、敖皮,按常规配制基本培养基 N,除对照 A 外,在每 kg 培养基中加入一份 M_1,然后将表 2 中不同配比的 GA3、KT 组合,各配成 100ml 水溶液,分别拌入 C-K 处理的每 kg 干料中。N 加等量清水为对照 A,N 加 M_1 为对照 B。具体配比与施用方式见表 2。以上每个处理分装 8 个罐头瓶,每瓶装干料 125 g,高压灭菌,接种,于 25℃温度下培养观察。菌丝长满瓶后移至栽培室出菇。(2)出菇后喷洒 GA_3、KT 试验:当菇蕾长到 2cm 后,以表 2 中不同配比的 GA_3、KT 组合,先分别与 M_2(营养液 M_2:Vitamin B_1:

1mg,$MgSO_4·7H_2O$:0.5g,$ZnSO_4·7H_2O$:0.8g,KH_2PO_4:1g 作为 1000 ml 水中的加入量)配合在一起,各制成 1000 ml 水溶液,对 C′-K′各处理每隔一天喷洒菇蕾一次,每次喷湿为度。对照 A 喷等量清水,对照 B 喷等量 M_2 水溶液。C 至 K 各处理也同时喷 M_2 水溶液。待菇长成后采摘称其鲜重;再各取样 3 份,60 ℃恒温下真空干燥 56 小时,称其干重。

表 2　测定 GA_3、KT 不同配比、不同施用方法对平菇产量、品质的影响试验设计

处理	拌料			喷菇蕾		
	营养液	GA_3 (mg/L)	KT (mg/L)	营养液	GA_3 (mg/L)	KT (mg/L)
A	H_2O	0	0	H_2O	0	0
B	M_1	0	0	M_2	0	0
C	M_1	1	5	M_2	0	0
D	M_1	1	2.5	M_2	0	0
E	M_1	1	1	M_2	0	0
F	M_1	0.5	5	M_2	0	0
G	M_1	0.5	2.5	M_2	0	0
H	M_1	0.5	1	M_2	0	0
I	M_1	0.1	5	M_2	0	0
J	M_1	0.1	2.5	M_2	0	0
K	M_1	0.1	1	M_2	0	0
C′	M_1	0	0	M_2	1	5
D′	M_1	0	0	M_2	1	2.5
E′	M_1	0	0	M_2	1	1
F′	M_1	0	0	M_2	0.5	5
G′	M_1	0	0	M_2	0.5	2.5
H′	M_1	0	0	M_2	0.5	1
I′	M_1	0	0	M_2	0.1	5
J′	M_1	0	0	M_2	0.1	2.5
K′	M_1	0	0	M_2	0.1	1

1.3 营养成分的测定

对上述20个处理,采第一潮菇匀样后各称2g,用DNS法测定其总糖,以麦芽糖作标准曲线;用双缩脲试剂法,测定其蛋白质,以滴定卵清蛋白作标准曲线;维生素C参照孙永芳的2,4一二硝基苯肼比色法进行测定。

2 结果与分析

2.1 GA_3和KT不同配比对菌丝体生长的影响

试验结果证明,GA_3和KT对平菇菌丝伸长均有明显的抑制作用。凡施用GA_3、KT的各个处理,其菌丝日平均生长长度,均明显低于对照A;除E和H外,也都低于对照B(表3)。如将对照A除外,菌丝生长速度最快的是$KT/GA_3=l/1$的E处理;最慢的是$KT/GA_3=50/1$的I处理。这说明GA3、KT不同组合,对平菇菌丝体生长的影响不同。KT/GA3的比值越高,对菌丝伸长速度之抑制作用越强。

然而,KT含量较高的处理,菌丝分枝多,生长密度较大(如J和F处理)。不曾施用任何激素的对照A,虽然菌丝日平均伸长率最高,但是菌丝密度最小,相对而言,其菌丝生长总量还不如E处理多。

表3 GA_3,和KT不同配比对平菇菌丝体生长的影响

处理	A	B	C	D	E	F	G	H	I	J	K
日平均生长度(mm)	16.68	6.05	5.50	5.90	6.67	5.12	5.77	6.30	4.13	5.38	5.48
菌丝密度	+	++	++	++	+++	+++	++	++	+++	++	++

2.2 不同配比GA_3、KT和营养液对平菇产量的影响

从表4可以看出,施用GA3、KT不同组合,绝大多数处理的产量均显著提高,其中C′、F′、I′、C、G′、D、D′、F,J各个处理,分别比对照A增产52.88%,45.90%,40.31%,39.44%,37.00%,32.29%,30.54%,30.37%,26.53%。值得注意的是对照B比对照A仅高出17.80%,在20个处理中位居14。这说明营养液M1、M2单独施用,平菇产量的增加不显著。只有与GA3、KT配合施用,才能发挥更大作用。另一方面,根据吴帮平选用NAA,6-BA和2,4-D在不同时期喷施平菇,仅分别增产16.40%,18.50%和31.48%的试验结果来看[1],单独施用植物生长调节剂,其增产效果不如本试验中植物生长调节剂与营养液配合施用更为明显。

表 4　GA_3、KT 不同配比、不同施用方法对平菇产量的影响

处理	施用方法	总产量 (g)	位次
A(CK_1)		573	18
B(CK_2)		675	14
C	拌料 Add to medium	799	4
D		758	6
E		619	16
F		747	8
G		682	13
H		690	12
I		802	3
J		725	9
K		651	15
C′	喷菇蕾 Sprinkle to jmmature fruiting	876	1
D′		748	7
E′		691	11
F′		836	2
G′		785	5
H′		584	17
I′		701	10
J′		558	19
K′		514	20

产量数据经方差分析和显著性测定可以看出，不同施用方法之间，$F<1$，差异不显著；KT 不同水平间的差异也不显著（$F=2.91<F_{0.05}$）；而 GA_3不同水平间差异达极显著水平（$F=8.85>F_{0.01}$）。由此可见，影响平菇产量的主要因素是 GA_3，施用浓度以 1 mg/L 和 0.5 mg/L 为宜。但是必须看到，本试验中不是单独施用 GA_3，而是在施用营养液的基础上，将 GA_3与 KT 配合施用。所获得的增产效果是三者恰当配合的综合效应。与 GA_3配合的 KT 浓度在 2.5 mg/L 以下时，效果不佳。

2.3　不同处理对平菇主要营养成分和干物质积累的影响

从表 5 可以看出，凡是施用 GA_3和 KT 的处理，维生素 C 的含量均明显低于对照，而糖分、蛋白质和干湿重比的变化情况，则与 GA_3、KT 不同组合中 GA_3的含量密切相关。GA_3浓度较低(0.lmg/L)的三个处理中，糖分、蛋白质平均值和干湿重比的平均值，一般均高于对照（$\frac{I+J+K}{3}>B>A$）。而 GA_3浓度较高(1 mg/L)的三

个处理中,蛋白质、干湿重比的平均值,均低于对照 F 糖分平均值虽略高于对照,但明显低于 GA_3浓度较低的各处理。

在对照 B 中,糖分、蛋白质含量和干湿重比均高于对照 A,说明施用营养液 M1 和 Mz,对改善平菇品质均有一定促进作用,至于维生素 C 含量为什么对照 B 明显低于对照 A,尚有待进一步研究。

表 5 不同处理下每百克鲜菇的糖分、蛋白质、维生素 C 含量干物质的积累

处理	总糖(mg)	蛋白质(mg)	维生素 C(μg)	干湿重比(%)	干物质积累位次
A(CK_1)	7090	3075	1120	9.43	7
B(CK_2)	9080	3150	890	9.75	6
C	5460	2475	460	9.27	15
D	8600	2150	480	8.61	19
E	7700	2450	840	9.23	16
F	7800	2800	760	9.56	9
G	9140	3275	540	9.38	14
H	8700	2325	505	9.22	17
I	8190	4450	840	10.72	1
J	9180	4125	860	9.93	4
K	9140	4225	730	10.63	2
C′	8900	2950	560	9.21	18
D′	7600	3125	610	8.54	20
E′	7440	2850	820	9.41	12
F′	8630	3300	760	9.42	11
G′	9140	2275	740	9.40	13
H′	7150	2950	780	9.43	10
I′	9600	3375	880	10.02	3
J′	9000	3725	840	9.85	5
K′	9580	3850	670	9.63	8

3 结果与讨论

试验结果表明,施用 GA_3、KT 不同组合,对平菇菌丝体的伸长均有明显抑制作用(这与李人圭等在香菇上的试验结果[1]和本文作者在金针菇上的试验结果[5]大不一致)。这种抑制作用,在 KT 与 GA3 比值越大的组合中,效果越明显。

可见在抑制菌丝伸长的作用过程中,KT 及其浓度是主要因素。值得注意的是 KT 浓度较高(5 mg/L)的处理,菌丝分枝多,生长密度大,生长总量也大,由此不难看出,只要 GA3、KT 等植物生长物质配比恰当,就可以促进细胞内含物增多,加速细胞分裂的及菌丝体的分化,从而不断增殖出大量优质菌丝。这一点也是施用恰当配比的 GA3,KT 组合,之所以能使子实体生长阶段产量大幅度增加的原因所在。

在营养物质充足的前提下,施用恰当配比的 GA_3,KT,不仅能够大幅度提高平菇的产量, 而且可以提高糖、蛋白质和干物质的含量,但 Vc 含量下降。在 GA3,KT 不同组合中,较高浓度 的 GA_3(1mg/L)对促进平菇产量大幅度提高的效果尤为显著。但这种浓度的 GA_3却使蛋白质 含量明显下降,糖分含量和干湿重比降到副对照之下。而较低浓度的 GA_3(0. lmg/L)与 KT 恰当配合,虽然能使平菇糖分、蛋白质和干湿有明显的提高,但产量提高不明显,个别的处理甚至低于对照。显然,在筛选 GA_3的使用浓度时,存在着高产和优质的复杂矛盾。今后,需从研究植物生长物质对食用菌生长发育的调节机理入手,进一步探索能够促进平菇既高产又优质的最佳 GA_3使用浓度及其与 KT 的最佳配比。

在平菇上单独施用 GA_3、KT 等植物生长物质或者单独施用营养液,均不及将 GA_3、KT 与营养液配合施用效果显著。营养元素是植物生长发育必需的物质基础,无论高等植物或低等植物,其产量的增加和品质的改善,都必须以充足的营养为前提。GA3,KT 等植物生长物质,只能对食用菌的生长发育起调节控制作用,而不能代替生长发育所需要的营养元素。因些,我们认为选择适当种类、适当浓度的植物生长物质和营养物质配合施用,是今后研究促进食用菌高产优质的正确途径。

参考文献

[1]吴惧.植物生长调节剂在食用菌上的应用概况[J]. 中国食用菌,1992(5):5 – 6.

[2]潘瑞炽,董愚得.植物生理学:下册[M].2 版. 北京:高等教育出版社,1984;6:1 – 2.

[3]吴锦文,孙玉萍等.磨菇增产灵对磨菇产量影响的研究[J]. 中国食用菌,1989(6):18 – 19.

[4]沐晨,钟世彬等.花粉提取物 BR 对金针菇菌丝体生长的影响[J]. 中国食用菌,1989(6):12 – 13.

[5]张占甲,陈坤明,李仲芳.激素与营养液配施对金针菇生育的影响[J]. 食用菌,1994(2):3 – 5.

[6]肖崇明,杨申之.激动素对金针菇菌丝体生长的影响[J]. 中国食用菌,1988(1):16 – 17.

[7]F. B. 索尔兹伯里,C. 罗斯. 植物生理学[M]. 北京大学生物系,华北农业大学农学系,等,译. 北京科学出版社,1979:374 – 381.

注:本文曾发表在 1994 年第 6 期西北植物学报期刊上

天水古树的生物学复壮的技术方法

刘红岩　王炳岐*

近年来,由于生态环境破坏的日益严重,以及人们对古树的保护不力,再加上古树的"年事已高",所以出现了不同程度的毁坏,甚至死亡。为了使这些"历史老人"能恢复昔日容颜,本文提出古树的生物学复壮技术,对天水的古树进行彻底整修复壮,再现往日绿荫。

一、生物学复壮技术方法

1. 人工补充水分的复壮技术

由于古树对水分的吸收能力降低,一般都水分不足,所以对进行复壮的古树就必须采取人为的补给措施:首先,在树冠投影部位扩出一至两米处,采取机械的钻空0.8－1.5m深的洞,进行灌水,孔洞一般在1－2m之间梅花形排列钻孔,以便对古树进行水分补充。其次,定期给古树进行叶面喷水,尤其在夏季干旱的季节。第三,给树根浇水补充水分或安置管道型塑料软管进行季节性的滴灌,用渗透方法来补充一定数量的水分.有条件的单位应购置检查水分含量的仪器,进行经常性的检查,缺水后就能及时的予以补充。

2. 人工土壤通气复壮技术

对古树进行及时的空气补充,可以保证呼吸系统的平衡。提高古树的透气性要采取:①叶面光滑透气,清除不利叶枝透气的附着灰尘、烟尘及油气附着物。②土攘板结,则用开挖、回填的办法来增加透气性,对开挖后的土层要进行加工处理,用腐质土、树枝、叶等填充,增加土层透气性。③钻孔埋管来增加土壤的透气性,对不能埋管的部位应多钻10－50个/m的孔,深度在30－150cm,均能起到通气作用;对孔口进行有孔的陶器盖口,以便日常通气。④在酷热的夏天地面温

---

* 作者简介:刘红岩(1955—　),男,甘肃秦安人,曾为天水师范学院生物工程与技术学院教授,学士,主要从事古树生态与保护研究。

度过高,要适当的降低地面温度,喷洒一些水,或铺草. 甚至树叶、锯末等隔热物质,来降低地热对古树根条的侵害。在土层内适当增加树枝或陶质颗粒, 同样能达到透气效果。

3. 对古树进行营养补充复壮技术

古树正常生长需多种营养元素,常见的就有 19 种之多,主要的有氯、钙、镁、铁 、钾、钠、磷、锌、铜、钼、锰等,这些必须元素都得需要但不能过量,要有适当平衡比例,失比会影响古树的正常生长。在检查古树的土壤元素含量时要采取科学的测试手段,对土样进行适当的化验、分析, 再就是根据树势情况,包括叶枝、外观的观察来判断古树营养元素的缺少情况,再加以补充。在没有进行地下复壮的古树应采取叶面施肥,和地下施液肥方法。以目前有效的测量方法和化验方法来制定本区的基本标准, 在川平地、山台地、街巷道中生长旺盛的古树对土攘中的 19 种元素的吸收情况,进行化验分析。通过比较对即将复壮的古树工程提供有力的依据,在复壮过程中对缺少元素的多少给于适当的补充。从一般规律看,常青树裸子植物的输导系统比被子植物输导系统吸收微量元素的速度慢得多,古树和少年、青年树输导速度都有很大差异, 同类同时间古树吸收微量元素的速度都不相同,因为受土攘、环境、树强弱的影响. 吸收元素的多少同样与气候有一定关系,降水量、空气温度、燕发量较高时,吸收元素就快,反之吸收量就低,季节也是影响古树吸收元素的一个因素, 秋冬季节古树的输导系统吸收能力逐渐减弱和停止。这样就得根据古树的自然特点和规律,在复壮和养护工作中结合考虑, 在古树立地环境受到污染发生盐碱危害时,就该采取有放的补救措施,以减少盐碱的侵害。

二、古树在不同的立地环境下采取不同的复壮措施

川平地的古树生物学复壮技术

川平地古树大部分布在市内各公园、寺庙内,古树生长的地下环境比较复杂,地下的管线、砖石、建筑垃圾较多,土攘贫瘠,透气性差,营养面积不足,土壤污染较重,对这种类型的古树大都采取综合复壮措施,要有复壮沟连接通气系统,设置地下渗水井. 复壮沟内加复壮基质及其它要求的物质,形成一定数量的全肥,以满足古树对多种元素的需要。人流量多的古树周围, 以古树营养面积上铺设透气砖和种植草坪地被植物,加强平时养护,预防病虫害,加围栏支撑等。

低洼潮湿型环境的古树生物学复壮技术

低洼处古树所处环境多积水,造成树根通气不良而烂根。解决古树的复壮问题,主要以排水通气为主,加深复壮沟和渗水井。渗水井要比一般情况井深 40 -

50cm，以加大排水量，井的下部要设有2 -4个放射状渗水道，渗水道内填充陶粒或卵石，及砂粒，防 止被土堵塞。这种低洼潮温的地方经常会出现地下盐分沿毛细管上升造成反盐碱现象，所以可在复壮沟底铺一层卵石或粗砂阻止盐分上升。如果土壤中盐含量超过0.3 -0.5%，应赶快换土。

砂质土或假山上的古树生物学复壮技术

砂质土壤往往漏水缺肥，这些地域里的古树就必须在复壮的时候，填实土壤层下的孔隙，使水分和肥力不下漏，同时要加围护边，使水分存留，起到保水作用，多采用灌水措施。是否安置排水井，一般看情况而定，复壮沟要加入保水基质，增加土壤的持水保肥能力。复壮沟要多加入枝条，夏季在水管孔里灌水，灌水以满足复壮沟的下部空隙为止。如果灌水过多，可在另一端通气管中检查。

如果超出以上几种情况的特殊环境，要因地制宜，特殊情况特殊对待，具体分析，采取相应的复壮措施。

以上是对天水古树复壮技术和保护的一些措施，供参考。古树作为一种历史的遗留物，其价值是无法用经济衡量的，它有历史、入文、自然、生物、气候、土壤、植物、坏境造型艺术等众多领域的属性。古树寿命越长，其历史价值、人文价值、科研价值就越高，越珍贵。创造一个良好的古树生存环境，是延长古树寿命、复壮的 最基本要求和条件. 任何地方都想让自己本地的古树再现葱绿，以增加这个地方的自然、人文、科研、观赏、旅游等产业的发展，达到相互促进、相互发展的良性循环体系，为提高当地知明度有着不可替代的作用，天水也不例外，因此保护天水古树应成为所有天水人的一个共识。

本文曾发表在2003年的植物保护上。

天水植物资源调查及开发利用研究一食用与药用花卉

刘红岩　袁毅君*

一、花卉的观赏性与多用途发展

(1)花卉的药用。自古以来,人们就将花卉作为防治疾病、保健强身、延年益寿的常用药物。秦汉时的《神农本草经》中,即为菊花、百合、鸢尾等花作药用的记载,《本草纲目》记述了近千种草花及木本花卉的性昧、功能和主治病症。近百年来,花卉的药用 研究和应用得到进一步发展,人们发现许多花卉有着极其显著和广泛的药理作用,并在临床实践中逐步扩大其应用范围。《全国中草药汇编》一书中,列举了2200多种药物,其中花卉入药约占1/3,如金银花具有很好的清热解毒功放.对于热毒病症,无论是瘟病、痈肿、疮疡疔疖、毒痢脓血,疗效都较显著。鸢尾以根茎入药,可活血祛瘀、祛风利湿、解毒消积。美人蕉以根茎入药,活血利湿,安神降压。百合以鳞茎入药,可润肺止咳,清心安神,主治肺痨久咳,咳唾痰血,热后余热未清,虚烦惊悸,神志恍惚。

另外,花香也可治病。已发现300多种鲜花的香味中含有不同杀菌素,其中许多是对人体有益的,不同的花香对不同的疾病有辅助治疗功效。

(2)花卉的菜用:我国食用花卉的栽培历史悠久。早在唐代,人们就把桂花糕、菊花糕视为宴席珍品。清代《餐芳谱》就详细叙述了20多种鲜花食品的制作方法,有些方法至今不保留在我国 的"八大菜系"中。如山东的桂花丸子、莱莉汤、广东的菊花铲鱼、菊花风骨,还有晶莹清澈的江浙冰糖百合汤等。南京许多餐厅、酒家除选用牡丹、菊花、桂花等传统食用花卉外,又添夜丁香、紫荆、玫瑰等。

---

* 作者简介:刘红岩(1955—　),男,甘肃秦安人,曾为天水师范学院生物工程与技术学院教授,学士,主要从事古树生态与保护研究。

鲜花精心制作的小妙、焖炒、清炖、甜点等美味佳肴,一经推出,顾客盈门。这些独具特色的鲜花菜肴花香浓郁、滑嫩可口,又富含大量氨基酸和维生素,既有丰富的营养价值,又有外形美观新颖的特点。另外,花粉食品也正在兴起。花粉是“地球上最完美的食物”,目前已有“活性花粉冷冻口服液”投放市场。花粉制品不仅能增进食欲、增强体力、预防和治疗疾病。而且能延缓衰老,延长寿命,目前我国的花粉研究已经进入开发应用阶段,一些地方已有产品问世,如北京的花粉酥点心,杭州的保健蜜等。

(3)花卉的饮用:茉莉花、珠兰花、代代花、栀子花、桂花、玫瑰花、米兰花、荷花、梅花、兰花、蜡梅花等,都具有宜人的芳香,是窖制花茶的好材料。其中用茉莉花作原料熏制的花茶,既能保持浓纯爽口的茶味,又兼具馥郁宜人的花香,茶汤明净,鲜爽不浊,已被誉为有益于健康的饮料。山东荷泽百源鲜花酒业开发公司研制成功的荷泽牡丹鲜花酒.酒味清醇,花香温馨,酸甜适中,为色香味俱佳的营养型保健美酒。啤酒花则是啤酒酿造工业重要原料。近年来以花卉作原料制成的保健饮料已有几十种,常见的有玫瑰茄保健饮料、金莲花以及其他花卉清凉饮料等。

(4)花卉的香用:花卉在香料中占有重要地位,主要是从香花中提取出芳香油,用于香水、香精等日用化工产品制造。据统计,约有40%的植物鲜花中含有丰富的芳香物质。例如从白兰、桂花、茉莉、米兰等香花中提取的芳香油,可用于制作花香型化妆品。从水仙花中提取的芳香油,可用来制作高级香精。特别是从玫瑰花中提取的玫瑰泊,更为名贵,在国际市场上1公斤玫瑰香精,相当于2~3公斤黄金价值,而用香叶天竺葵的叶片提取的香精,其价值比玫瑰香精还高。

二、食用与药用花卉的开发

花是大自然美的使者,药给人类带来美的陶冶和享受。将观赏之花卉加菜肴之中,纪念品厚味去腻增鲜、使谈味提香提色,尤是花卉独特之处。故以花入菜,在中国古籍中多有零星记载。古人于秋季采集野花,既用于入药.也用以入馔,所谓“朝饮木兰之附露兮,夕餐秋菊之落英”的诗句,说明那时古人已用木兰花做饮料和用菊花做菜肴。北魏以后饮食著作中花卉入菜逐渐增多,同时也烹制方法。

古人以花入菜,取其色艳、香清、和味美,也考虑到花有祛病延年之功。中医认为,常饮菊花、莲花和椿芽之汁,有健身延年之效。实验证明,花卉含有多种营养成分,确有保健作用,当前盛行的花粉食品则足以说明这一点。有关专家认为仅仅花的香气,就有益于人的健康和智力。中国菜之美食,讲究色、香、味、形俱佳.从姹紫嫣红的花海中撷取色、香、味、形别有特色的花卉来丰富菜肴的绚丽风采,更令人垂涎,不愧为华夏子孙独有的美食瑰宝,目前食用花卉,正是市场热点,因此结合天水的环境优势,选择适宜的品种,发展食用花卉栽培大有作为。以下介绍几种较好的食用与药用花卉。

1. 蜡梅:属蜡梅科落叶灌木,蜡梅花是耐寒之花,不但是著名观赏植物,还可提取芳香油,具有解暑生津、顺气止咳的药效,其花蕾油可治烫伤,特别有寒冷的冬季,百花凋零,唯梅香飘逸是制作花馔的最佳鲜花。而且形态小巧,色泽鲜艳,质地柔软,香味清雅。在冬雪纷扬的季节里,踏雪闻香,采集梅花宋烹调佳肴,阵阵梅香萦绕优雅纤美的蜡悔,仿佛将凝冻的严寒融化,把无声的温馨和挚爱,送到充满情谊的餐桌上。

2. 木槿(Hisbiscus syeiacus),又名面花、朝开幕落花、嗽叭花、篱障花等,为锦葵科木槿属落叶灌木或小乔木,木槿原产中国,印度、叙利亚、朝鲜、日本也有分布。日本等国人民十分喜爱此花。朝鲜人民称其为"无穷花",作为美丽和幸福永存的象征,并将它定为国花。有米黄色、淡红色、纷披陆离,迎霞沐日,临风招展,光彩秀美受历代诗人赞扬。《诗经》中就将木槿花比作美女宋歌咏。唐李白《咏槿》有:"园花笑芳年,池草艳春色,犹不如槿花,蝉娟玉阶侧。

木槿枝叶繁茂,树姿优美,花朵绚丽,花期长。园林中,可单植、丛植点缀庭院。也可用作花篱,因枝条柔软,作围篱时可进行编织。枝条可塑性大,可扎成花蓝、狮子、老虎等模型美化园林。木槿对二氧化碳、氯气等有害气体有很强的抗性,又有很好的治尘功能,还可作为工矿和街道的绿化树种。

木槿花中富含大量人体所需的营养成分,据测定:每 100 克鲜花可食部分含水 92 克、蛋白质 2. 克、糖类 2.26 克、脂肪 0.3 克、热量 32 千卡、灰分 1.2 克、粗纤维 0.8 克;以及胡萝卜素、氨基酸和多种维生素等,具有防治病毒性疾病和降低胆固醇的作用。对高血压病患者常吃素木槿花菜汤,有良好的食疗效果。用嫩茎叶做菜,洗净后水焯,清水漂去涩味,可炒食或做汤,味道鲜美。名菜木槿豆腐汤即是。也可凉拌,具有清凉解毒的功效。

木槿花可食用,花朵调入面粉和葱花,入油锅煎,称为"面花" 食之松脆可口。木槿花煮豆腐,是为味道鲜美的木槿豆腐汤。木槿嫩叶可食做汤味美,也可代茶饮。

3. 牛蒡(Arctium lappa)又名大力子、黑萝卜、万把钩、牛菜,为菊科二年生草本植物。根肉质,茎粗壮,高 1 ~2 米,带紫色,有微毛,上部多分枝。基生叶丛生. 茎生叶互生,宽卵圆形或心形, 长 40 ~50 厘米,宽 30 ~40 厘米,上面绿色,无毛,下面密被灰白色绒毛,全缘,波状或有细锯齿,项端圆饨,基部心形,有柄,上部叶渐小。头状花序丛生或排成伞房状,有梗;总苞球形. 总苞片披针形;花全部筒状,淡紫色。瘦果椭圆形或倒卵形，灰黑色。花期6－9月，果期8～9月。牛蒡主要分布于中国东北、华北、西北地区，华东、华中，西南部分地区也有分布。牛蒡对土填要求不严，管理简便。园林中可作绿化材料应用于林间野外。牛蒡嫩叶、花、肉质根均可食用嫩茎叶者，于4－5月间采集，在沸水中焯下，换清水浸泡后炒食、做汽或盐渍。在0.5%的醋水中泡一下，可去掉涩昧，使期风味更佳：食用肉质根者，于秋末或冬季挖取。浸泡后多用于腌制咸菜，也可妙食。

牛蒡食用营养价值很高，每百克嫩叶中含蛋白质4.7克、脂肪0，日克碳水化合物3克、粗纤维2.4克、胡萝卜素3.9克、维生素B，0.02毫克、维生素B，0.29毫克、烟酸1.1毫克、维生家C 25毫克、磷61毫克、铁7.8毫克。

牛蒡的根、茎、果实均可入药，始见于南朝梁人隐弘景《名医别录》，中医认为其性寒味甘苦无毒，入手太阴经。能清热解毒、祛风湿、宜肺气，尤善清上，中二焦及头面部的热毒.对风毒面肿、咽喉肿痛、肺热咳嗽等症最为适宜。牛蒡可做外用药，捣敷或熬膏或煎水，洗治多种疾患均可。牛蒡根500克，捣汁入少许盐花，置银锅中熬膏，涂齿龈上，消费品热毒牙疼，齿龈肿痛。鲜牛蒡根捣烂取汁，每日滴耳数次，治急性中耳炎。牛蒡叶捣汁搽涂，治各种疮疡疔疖。牛蒡子炒熟，煎水含漱，治风龋牙痛。

4.壮丹：牡丹花作为“花中之王”，不仅是一种名贵的观赏花卉，同令人大饱眼福，也能使人一饱牡丹花馔的口福。

牡丹花入馔历史悠久，明代《二如亭群芳谱》既有记载：“煎牡丹花煎法与玉兰同，可食，可蜜钱”。“花瓣择，洗净，拖面、麻油煮食，至没”清代《养小录》中的《餐芳谱》，亦介绍了牡丹花配制食馔的20多种花卉食品的制作方法，且烹饪讲究，做工精细。牡丹花瓣和花粉，还可制作保健食品、饮料、用牡丹花制成的“牡丹酒”，色正味纯，清香爽口。

牡丹花食用方法多样，炸、烧、煎或做汤等，皆可成美味。如牡丹花和肉共烩制成“肉汁牡丹”，色泽鲜丽，味美适口，牡丹用面粉裹后油炸食用，鲜香诱人；用白糖浸渍又是上乘的蜜饯；做牡丹花银耳汤时，撒些牡丹花瓣，色鲜香瓣，色艳香浓，令人食欲大振，此外，牡丹熘鱼片，牡丹爆鸭脯等也是时令佳肴。

牡丹花瓣营养价值极高，含有丰富的蛋白质、脂肪、淀粉和糖类，此外还含有钙、磷、铁等矿物质及维生素A、D、C、E等，特别是所含的多种游离氨基酸，更易为人体所吸收。

牡丹花亦可入药，它性味平和、微苦，无毒，有调节活血的功效。明代医学家李时珍认为，牡丹花的颜色决定其功效：“赤花者利，白花者补。”实症患者宜选红牡丹花入药，能加清利效果；虚症患者选白牡丹花入药，有补益作用。

5. 食用仙人掌“米邦塔”:在墨西哥,仙人掌早就是人们餐桌上的美味佳肴了。但在我国,把仙人掌作为一种蔬菜,似乎才刚刚开始。“米邦塔”是墨西哥专家多年选育出的以食用为主、具有多种用途的仙人掌新品种。经过农业部优质产品开发中心,引种、示范推广应用和营养专家分析验证,它是一种适宜在我国种植推广的新型保健蔬菜良种。同时在药用、观赏果用、美容环保等方面也很有价值。“米邦塔”虽貌不惊人,却浑身是宝,营养丰富,富含钾、钙、铜、铁、锰、锶等多种矿物质及柠檬酸、苹果酸、磷脂、维生素、纤维素等,具有行气活血、清热解毒,促进新陈代谢,、降血糖、降血脐等功效。将别是它不含草酸,极利于人体对钙的吸收,是儿童及中老年人补钙的佳品。它的适口性好,如切成丝,通体碧绿透明,人见人爱,入口清香爽口,无异味,可炒菜,烧汤.也可凉拌,更是制作罐头、饮料、色拉的上等原料,它的果实是－种好看又好吃的新型水果,在思西哥大型超市及蔬菜水果市场上颇受欢迎。

“米邦塔”食用仙人掌粗生易长,抗性强,病虫害少,无须使用农药、化肥,是生产绿色无公害蔬菜和保健食品的首选品仲、它为多年生植物,可一次种机,连续采收10年以上,不像其它蔬菜,每年每季都要进行整地、施肥等一系列管理,大大减少了工序.降低了成本。且植株无刺或少刺.更于管理。种植后2个月左右即可采收.以后半个月至20天可再次采收,在气候适宜的情况下可周年上市。

“米邦塔 ”食用仙人掌;喜温暖 、干燥、通风的环境,喜强光,0℃以下易受冻害,作生产性栽培时要保5℃以上,一般情况下可不浇水,梅雨季节要注意排水。一年四季均不要遮阳,以保光照充足。仙人掌性喜砂质弱碱性土壤,如土壤黏重,可多施有机底肥,以增加通透性,它又是浅根性植物,如遇大风大雨要注意防倒伏,主要采用无限繁殖。当掌片呈半木质化时,用利刀沿某部割下,切口用多菌灵粉处理,稍晾干,扦插在疏松的基质中,插后不要浇水,生根后转科正常管理。

仙人掌也是一种极好的中药,其花,果,茎,肉以及肉浆的疑结物(玉芙蓉)均可供药用。祖国医学认为:仙人掌微苦性寒,有消肿止痛,行血活气,清热解毒,祛湿生机之功效,可治心胃气痛,痣血,咳嗽,喉痛,肺痛,疔疮、烫火伤等。

6. 费菜:(Sedum kamtschaticum Fisch)又名养心草、回生草、救心草,属景天科多年生草本植物。茎白根基部簇生,株高20－30厘米。叶豆片状,长2～3厘米,叶缘有细锯齿” 春季开花,伞状花序,小花黄色。黄花绿叶相衬,鲜艳夺目.有一定的观赏价值。费菜又是一种药用植物,全草人药。据《福建中草药》药理分析,其性味甘淡微酸平。有清热凉血、平肝:宁心功能。主治心脏病、癔病,吐血、咳血等,尤其对心脏病有特效,被称为“心脏病患者的救星”。口服费菜安全,未发现任何副作用,费菜口感好,不寒不温,适合任何胃口的病人服用;服用方法简单,可以煎汤,泡开水服用,也可榨汁服用,鲜嫩幼叶还可炒成菜肴食用" 防治高血压、心脏病一般有 5 种服法:①每日取鲜全草日 50 克(干草 15 克,1、同),洗净放入保温杯内,冲入 100℃的汗水,盖上盖,10 分钟即可饮用,可边饮用边加开水,十分方便。②取鲜草 50 克,丹参 15 克. 加水 500 克水煎,再加蜂蜜 20 克. 调匀服用,③鲜草 50 克加 1/3 只猪心炖服。④鲜草 50 克加 1/4 白鸽肉炖服,⑤鲜草 50 克加瘦猪肉炖服。费菜除可防治心脏病外,对老年人的血管硬化、高血脂等病症也有缓解作用,癔病患者可取鲜全草 100 克,猪心一个(去外部油脂、不加剖切)置瓦罐内,将药放在猪心周围,加蜂蜜酌量,倒入开水(以浸没猪心为度)放锅内炖至猪心熟,去药渣,分两次食尽。此外,鲜草 50 克切碎捣烂泡开水服用,可治惊悚、烦躁、胸闷、失眠、口臭、肝病、妇女更年期心烦失眠等,鲜根 50 克(或干根 20 克)水煎服,可治鼻出血、牙龈出血等出血性病症。

费菜繁殖快,易种易管。4 –9 月均可扦插,方法是剪取两节一段的行条,插于沙床半阴处养护。每日喷水 3 –4 次,10 天可成活。生根后移植到花盆中或地栽均可。繁殖率极高,一盆费菜一年可繁殖千余株。对种植土壤要求不严,但最好是砂质壤土,管理粗放,不怕强光,不怕大雨,只是冬天要注意防寒。3 月初,可施一次 1% 的复合肥作催芽肥,以后可每采摘一次鲜草,均要追施上述液肥,以促进分枝,提高草量,每年 4 –5 月为费菜生长旺期,可将鲜草采下晒干或烘干,以备冬季和春季服用。

费菜据有关资料记载,主要成分为生物碱、齐墩果酸、谷甾醇、黄酮类、景天庚糖、果糖、蔗糖、蛋白质和有机酸等,他的药理作用是养心宁心平肝,清热凉血活血止血。其中,谷甾醇能阻止人体对胆固醇的吸收,降血脂,防止血管硬化;黄酮类可扩张心脏血管,促进血液循环。因此,费菜是老年人一种很好地保健治病良药。研究表明人过 50 岁后容易发生高血脂、高胆固醇症,增加引发冠心病和中风的可能性。若在这些人中提倡种植,服作费菜,可防患于未然。此外,费菜中的齐墩果酸还可保护肝脏,起到延缓老年性肝组织纤维化的作用。即使没有明显疾患但时常感到胸闷、心悸、难眠的老人,经常服用,也能有益健康,延年益寿。

7. 清新益气话“花饮”:①辛夷花饮:辛夷,又名木笔,为木兰科植物,因初开时苞长半寸似笔头,故名。“辛”为味道,“夷”为幽远,是指辛夷的味道辛香而幽远。采初春含苞未开的辛夷花,晾晒后取 6 –9 克,加蜂蜜 20 克,同 6 克甘草在砂锅中煎煮 10 分钟,分数次代茶饮用,可祛风祛寒,宜通鼻窍。②金银花饮:5 –6 月间为采摘期,在晴天清晨露水未干时将花蕾采下,凉晒或阴干,有清热解毒的功效,可有效地抑制多种对人体有害的杆菌和球菌。取金银花、菊花各 10 克,用开水冲泡,代茶饮用,可消热止渴,去炎止痛,长期饮用还可治疗和预防冠心病。③厚朴花饮:厚朴载于《神农本草经》,为木兰科植物,花期4-5月,在春末夏初花蕾未开或稍开时采摘,置于蒸笼上蒸,上气后10分钟取出晒干或文火烘干,取厚朴花6克,佛手片12克(鲜品加倍),开水冲泡,浸10分钟后,温服代茶,可治疗肝郁气结,胱腹胀满,暖气频作等症状。

8.桃花:属蔷薇科李属落叶小乔木。"逃之夭夭,灼灼其华"这《诗经》上的名句描绘出在山上、水边、屋旁桃花盛开时,风中笑舞凝霞敷锦、点燃得春色分外美丽。桃花4月中旬开始绽放,花有桃红、粉红、深红色,也有白色和红白相间等颜色。不管是结果,还是专供观赏的桃花都可以食用,也强以入药。

桃花含山奈醋、香平精、袖皮素等特殊成分,性平,味微苦,无毒具有消食顺气、怯风镇静、美容润肤、养心活血、润肠通便、艳颜美容等功效,它既是良药又是极天然的美容品,还可烹调美味佳肴,香味柔和,怡人心神。桃原产我国,结果的桃树遍布各地。每年4、5月桃花盛开之际,果农们为了保证桃子的质量,要把过多、多密的桃花除去(蔬菜),以使桃树集中养料使果实匀称,提高色、香味,这就是提供了很多桃花制作的花馔,使餐桌上充溢桃花香气,也使生活充满了妩媚的桃红色。可将桃花制作成桃花蟹黄烩芙蓉、桃香烹牛蛙、桃花京鲜鱼等时令佳肴。

槐花:为槐树的鲜花。槐树为豆科落叶乔木,树形苍劲挺拔,枝繁叶茂,是良好的庭院绿化及四旁绿化树种。夏日花开,雪白晶莹,团团簇簇,甜香迷人。每到夏季当槐树的嫩叶长出或槐花盛开时,加入食盐、花椒、葱、姜蒜、小茴香等调味口,蒸熟食之,既可当饭,又可当菜。古时人们就知道采食槐树嫩叶,水焯后冷水浸泡除涩,在拌以姜、葱等调味品,做菜食用。古诗中也有食用槐叶的记载。杜甫诗云:"青青高槐叶,采掇付中厨。"槐花可制成槐菊茶,开水沏泡饮用,具有清肝降火,活血、止血之效。另外槐花可烹调成汤、粥及美味佳肴,款待宾客。

槐花营养丰富。据现代科学研究测定:每100克鲜品含水分78克,蛋白质3.1克,脂肪0.7,碳水化合物15克,钙8.3毫克,磷69毫克,铁3.6毫克,胡萝卜素0.04毫克,维生素B 0.04笔克,维生素B 20.l毫克,尼克酸6.6毫克,维生素C 66毫克。

槐树花蕾、花、果、根、叶、枝均可入药。祖国医学认为,槐花性味某凉,具有凉血止血、清肝降火的功效。主治肠风便血,痣血、尿血、血淋、崩漏、纽血、赤白痢、目赤、疮毒等。《本经逢源》载:"槐花苦凉,阴明厥阳血分药也,故大肠便血及目赤肿痛皆用之,且得血而能视,赤肿及血热之病"。

现代医学研究证实,槐花含芸香试,以及三萜皂甙.水解后得白桦脂醇、槐花二醇和葡萄糖、葡萄糖醛酸:另含槐米甲、乙、丙素,甲素为黄嗣类、乙、丙素为甾醇类。果实含9个黄酮类和异黄酮类化合物。药理实验证明芸香甙及槲皮素能保持毛细管正常的抵抗力,减少血管通透性,并能抗炎、抗菌解痉、抗溃疡作用;槲皮素可扩张冠状血管,改善心肌循环,且有降低肝、主动脉及血中胆甾醇量,可防治动脉硬化症。果实有升血糖、抗菌作用。

参考文献

[1]韦三立. 花卉组织培养[M]. 北京:中国林业出版社,2001.
[2]《中国花卉盆景》,1999 - 2002.
[3]《花木盆景》,1990 - 2002.
[4]《园艺学报》,1989 - 1998.
[5]《花卉商情》,1999 - 2000.

本文曾发表在2003年第3期,总第200期的甘肃农业上。

半夏疫病病原鉴定和防治研究

裴建文　孙新荣　呼丽萍　刘艳梅　王　鹏　裴国维*

为了明确甘肃清水半夏疫病的病原及其防治药剂，对该病病原菌进行了分离鉴定，并筛选防治该病的药剂。结果表明：半夏疫病的病原为寄生疫霉(*Phytophthora parasitica* Dast.)，寄生疫霉菌丝生长最适 pH 、温度分别为6.8 与 27.3℃；室内药效比较表明，58% 甲霜灵 · 锰锌可湿性粉剂和 70 % 甲基硫菌灵可湿性粉剂药效最好，1200 倍抑菌率均为 100 %；田间药效试验表明，58% 甲霜灵 · 锰锌可湿性粉剂 500 倍液药效最好，施药 2 次防效为 97.4%，70% 甲基硫菌灵可湿性粉剂 500 倍液施药 2 次后防效仅达 71.5%。

近年来，随着半夏野生资源的日益匮乏，栽培半夏已成为该药材的主要来源。然而随着半夏种植面积的不断扩大，病害问题已逐渐成为影响半夏产量和品质的重要因素之一。在甘肃清水半夏 GAP 生产基地进行病害调查分析发现，疫病是危害当地半夏生长后期的主要病害，发病率高达 20 % 以上。该病在连续的阴雨天气易发生，特别是在北方雨季 7 - 8 月份蔓延迅速，严重时发病率可达 100 %，产量损失达 15 % 以上。该病害主要危害半夏地上叶部，发病植株初期叶片出现暗绿色水渍状不规则形病斑，随后叶片皱缩扭曲、枯黄；严重时病斑布满全叶导致半夏倒苗。关于半夏叶部病害，曾令祥和李德友[1]报道了由半知菌亚门葡萄孢属真菌引起的叶斑灰霉病、病毒引起的病毒病以及生理性病变的白点斑病和紫斑病。申屠苏苏等[2]、陈集双等[3]研究表明侵染半夏的病毒主要为黄瓜花叶病毒 CMV(Cu - cumber mosaic virus)的天南星科株系、大豆花叶病毒 SMV(Soybean mosaic virus)的天南星科株系和芋花叶病毒 DsMV(Dasheen mosaic virus)。半夏疫病还未见报道，为此，本研究对该病害的病原菌进行鉴定并对该病的防治药剂进行

---

* 作者简介：裴建文(1957—　)，男，甘肃天水人，曾为天水师范学院生物工程与技术学院教授，学士，主要从事作物育种与中药材规范化种植研究。

筛选。

1 材料与方法

1.1 病原菌的分离和培养

发病的半夏植株于2007年7－8月采自甘肃清水半夏GAP生产基地。采用常规组织分离法分离病原菌,选取新发病叶片病健交界处的组织(5 mm × 5 mm),置于PDA平板上28 ℃恒温箱内培养。从分离到的菌落边缘挑取菌丝,继续转接纯化2 ～ 3次。

1.2 病原菌致病性测定

采用针刺接种法[4]测定组织分离物的致病性。将种植于灭菌土花盆中的半夏健康植株叶片用无菌水洗净,然后用70 %乙醇擦洗组织表面, 晾干后用灭菌针轻微刺伤表皮。然后用6 mm灭菌打孔器在培养5d的菌落边缘打取菌丝块,将菌丝块反贴到叶片上, 同时设立有伤对照和无伤对照(有伤对照是刺破表皮后接6 mm直径的培养基块,无伤对照是不刺破表皮贴菌丝块),并以无菌水浇湿的消毒棉团敷在菌丝块上, 保湿3 d, 观察、记录发病情况。每处理10株,3次重复。待接种植株表现出症状后对病斑进行组织分离并培养,根据柯赫氏法则验证分离物是否为病原。

1.3 病原菌的形态学鉴定

用6 mm灭菌打孔器在培养5d的菌落边缘打取菌丝块, 然后分别移植到PDA、菜豆粉琼脂培养基(KBA)、燕麦粉琼脂培养基(OMA)、胡萝卜琼脂培养基(CA)上,每种培养基3皿,置28 ℃恒温箱中,于2 、4 、6 d测量菌落直径,并观察孢子囊、厚垣孢子等的产生情况;配对培养在VA培养基上进行。上述培养基配方和孢子囊脱落性的测定均采用余永年等方法[5]。

1.4 菌丝生长温度和pH测定

采用2因子饱和D－最优设计,试验因子水平设计及编码值见表1。灭菌后的PDA培养基,用0.1%的NaOH和0.1%的HCl调节pH,使pH分别为4.0、6.2、7.5、9.0。PDA平板中央接种6 mm灭菌打孔器在培养5d的菌落边缘打取的菌丝块,分别置于5、18、25.9、35℃恒温箱中培养。每处理3次重复, 5 d后用十字交叉法测量菌落直径,按照下式计算平均值表示菌丝生长量。

菌丝生长量＝菌落直径－接种菌丝块直径。

表1 试验设计水平

编码值	pH(X1)	温度(X 2)/ ℃
1	9.0	35.0
0.3944(λ)	7.5	25.9
-0.1315(μ)	6.2	18.0
-1	4.0	5.0

1.5 室内药效比较

供试药剂名称及测试浓度如下:50 %多菌灵可溶粉剂(江苏省吴县市农药厂)、80 %代森锰锌可湿性粉剂(美国罗门哈斯公司)、53.8 %氢氧化铜水分散粒剂(郑州致信农化有限公司)、68.75 %恶唑菌酮·锰锌水分散粒剂(美国杜邦公司)、70 %甲基硫菌灵可湿性粉剂(日本曹达株式会社)、58%甲霜灵·锰锌可湿性粉剂(江苏宝灵化工股份有限公司),以上各种药剂测试浓度均为400、600、800、1 000、1 200倍。

室内药效比较采用生长速率法[4]。各种药剂采用梯度稀释法配成5级不同浓度的含药PDA培养基,以不加药剂为对照。含药PDA平板中央接种6 mm灭菌打孔器在培养5 d的菌落边缘打取的菌丝块,每处理3次重复,于28 ℃恒温箱中培养,5 d后采用十字交叉法测量菌落直径,并计算相对抑菌率。

$$抑菌率 = \frac{(对照菌落直径-0.6)-(处理菌落直径-0 6.)}{对照菌落直径-0 6.} \times 100\%$$

1.6 田间药效试验

试验地位于甘肃清水川郊水地半夏主栽区,海拔1378 m;土质中壤,肥力中等。半夏播种期为3月,播量(块茎)为300 kg/ 667 m^2,人工灌溉。选择长势整齐一致、发病均匀的半夏田块为试验田。

选择室内药效较好的2种药剂,每种药剂4级浓度。小区面积15 m^2。重复3次,清水为空白对照,共27个处理。试验药剂和空白对照的小区处理随机排列。

供试药剂及测试浓度为:58 %甲霜灵·锰锌可湿性粉剂、70 %甲基硫菌灵可湿性粉剂,测试浓度均为500、1 000、1 500、2 000倍。

7月6日,将供试药剂按照试验剂量及小区面积和重复计算各处理所需药量,用背负式压缩喷雾器均匀喷雾于植株地上各部位,从低浓度向高浓度顺序进行。第1次喷药后10d再喷1次,共喷2次,常规叶面喷雾。

分别于每次喷药前和第2次喷药后10d调查病害严重度。每处理5点取样,每点取样面积0.2 m^2(0.5 m ×0.4 m)。半夏疫病严重度分级标准如下。

0 级:不发病;

1 级;病斑面积占叶片面积的 1/5 以下;

2 级;病斑面积占叶片面积的 1/5 ~1/3 以下;

3 级:病斑面积占叶片面积的 1/3 ~1/2 以下;

4 级:病斑面积占叶片面积的 1/2(含)以上。

根据调查结果,逐叶记载严重度,计算病情指数和防治效果,公式如下:

$$病情指数 = \frac{\sum(各级病叶数 \times 各级代表值)}{调查总叶树 \times 最高一级代表值} \times 100\%$$

$$防治效果 = \frac{对照区病情指数增长值 - 处理区病情指数增长值}{对照区病情指数增长值} \times 100\%$$

1.7 统计方法

采用 Excel 处理原始数据,DPS 进行统计分析。

2 结果与分析

2.1 病原分离结果

对所采病株进行分离,结果表明,在分离的 150 个病样中,疫霉菌(Phytophthora sp.)53 个,占 35.3%,链格孢菌(Alternaria sp.)37 个,占 24.7 %,镰刀菌(Fusarium sp.)21 个,占 14 %;没有分离出任何菌的 11 个,占 7.3 %;其他合计 28 个,占 18.7 %。从病部分离出的主要是疫霉、链格孢、镰刀菌。

2.2 病原菌致病性测定

半夏疫病的病原菌能以菌丝从伤口侵入寄主,在 28 ℃条件下,无伤对照和有伤对照都没有出现病斑,有伤接菌后叶片卷曲且出现明显的暗绿色水渍状病斑,为不规则形,病健交界处波浪形,和大田发病基本一致。对接种发病株进行再分离,均可获得原菌株,说明该菌是引起 7 -8 月份田间半夏大量死亡的致病菌。而相应分离到的链格孢和镰刀菌都没有引起病斑。

2.3 病原菌形态特征

经培养和观察,该菌在 PDA 、KBA 、OMA 、CA 等人工培养基上均生长良好,菌落呈棉絮状,边缘明显。CA 培养基上气生菌丝茂盛,菌丝丝状,无隔,表面光滑,无膨大突起,直径为 3 ~7μm。孢子囊卵圆形、近球形,基部圆形,多数顶生,个别间生,大小(24.6 ~ 52.8)μm ×(18.1 ~ 46.6)μm,平均长宽比 1.3。乳突明显,平均乳突高 4.3 μm。孢子囊不脱落。厚垣孢子顶生或间生,球形,黄褐色,大小 20 ~43 μm,平均 32 μm,在以上固体培养基中都很容易大量产生。单独培养未见有性器官形成，在 VA 培养基上种内对生培养时形成，藏卵器穿雄生，蜜黄色，球形，大小 24.6 ~ 36.9 μm，平均 28.8 μm；雄器围生，近球形或鼓形，大小 (12.8 ~ 28.3) μm ×(12.8 ~ 14.6) μm，平均 18.7 μm ×13.1 μm。

2.4 菌丝生长温度和 pH 测定

试验结果见表 2。由表 2 资料得回归方程：$Y = 4.2617 + 0.6425X_1 + 0.6068X_2 - 2.9039X_1^2 - 0.6250X_2^2 + 0.1600X_1X_2$

（Y 为菌丝生长量，X_1 为 pH，X_2 为温度） （1）

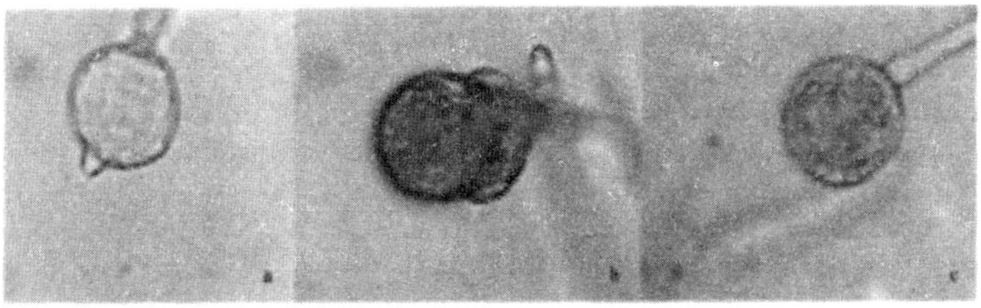

a: 孢子囊; b: 藏卵器、雄器; c: 厚垣孢子

图 1 半夏疫病病原菌形态特征（200 ×）

表 2 试验实施方案及结果

处理号	结构矩阵						菌丝生长量 /cm
	X_0	X_1	X_2	X_1^2	X_2^2	X_1X_2	
1	1	-1	-1	1	1	1	0.30
2	1	1	-1	1	1	-1	0.40
3	1	-1	1	1	1	-1	1.60
4	1	-0.1315	-0.1315	0.0173	0.0173	0.0173	2.00
5	1	1	0.394	1	0.1556	0.3944	4.60
6	1	0.3944	1	0.1556	1	0.39444	2.30

1）表中数据均为 3 次重复的平均值。

进行回归方程显著性检验，$F = 191.74 > F_{0.01(5,10)} = 5.64$，达极显著水平，表明菌丝生长量与 pH、温度间存在极显著回归关系。以 0.05 显著水平剔除方程中不显著项后，方程可简化为：

$Y = 4.2617 + 0.6425X_1 + 0.6068X_2 - 2.9039X_1^2 - 0.6250X_2^2$ （2）

对方程（2）求 X_1、X_2 的一阶偏导数并置 0，求出当菌丝生长最快时的 pH、温度编码值分别为：$X_1 = 0.1106$、$X_2 = 0.4854$，即最适 pH 和温度分别为 6.8 与 27.3 ℃。该回归方程本身已经过无量纲形编码代换，其偏回归系数已经标准化，故可直接由其绝对值大小来判断 pH 和温度对菌丝生长量的重要性,可见 pH 对该菌丝生长的影响较温度明显。

2.5 病原菌的鉴定

根据病原菌孢子囊的形态、长宽比、有明显的乳突，在固体培养基中可大量形成厚垣孢子，单独培养不易形成有性繁殖器官,不同培养基对菌落形态无影响,35 ℃以上可以生长,与 Waterhouse G M [6]、Ne - whook F J[7]、Ho H H[8]、魏景超[9]描述 Phytophthora

parasitica Dast. 基本一致,故鉴定为寄生疫霉。

2.6 室内药效比较

从表 3 看出,50 % 多菌灵可溶粉剂、80 % 代森锰锌可湿性粉剂、53.8 % 氢氧化铜水分散粒剂、68.75 % 恶唑菌酮・锰锌水分散粒剂、58 % 甲霜灵・ 锰锌可湿性粉剂、70 % 甲基硫菌灵可湿性粉剂对 P. parasitica Dast. 的菌丝生长均有抑制作用。不同药剂的抑制效果存在差异,其中 58% 甲霜灵・锰锌可湿性粉剂和 70% 甲基硫菌灵可湿性粉剂的效果最好,在设计浓度下抑菌率均为 100 %;80 % 代森锰锌可湿性粉剂和 68.75 % 恶唑菌酮・锰锌水分散粒剂次之, 400 倍抑菌率分别为 97.30 % 和 79.28 % 。

表 3 不同药剂对 *P. parasitica* Dast. 的抑菌率

药剂名称	抑菌率/%				
	400 倍	600 倍	800 倍	1000 倍	1200 倍
50% 多菌灵 SP	34.68	21.17	13.06	7.66	7.66
80% 代森锰锌 WP	97.30	95.05	75.68	61.71	50.90
53.8% 氢氧化铜 WG	73.87	56.31	42.79	38.74	7.39
68.75% 恶唑菌酮:锰锌 WG	79.28	75.23	71.17	69.82	68.47
58% 甲霜灵:锰锌 WP	100.00	100.00	100.00	100.00	100.00
70% 甲基硫菌灵 WP	100.00	100.00	100.00	100.00	100.00

2.7 田间药效试验

小区试验结果(表 4)表明, 第 1 次施药后 58 % 甲霜灵・锰锌可湿性粉剂(500 倍、1 000 倍)、70 % 甲基硫菌灵可湿性粉剂(500 倍、1 000 倍、1 500 倍)的防治效果均在 50 % 以上, 且 58 % 甲霜灵・锰锌可湿性粉剂 500 倍防治效果最好,其余均较低。经2次施药后58 %甲霜灵·锰锌可湿性粉剂500倍防治效果亦最好,防效为97.4%,其次防效达50 %以上的为58 %甲霜灵·锰锌可湿性粉剂1 000倍、70 %甲基硫菌灵可湿性粉剂500倍、70 %甲基硫菌灵可湿性粉剂1 000倍, 防治效果分别为74.6%、71.5 %、60.0 %。可见2次施药后除58 %甲霜灵·锰锌可湿性粉剂500倍有较好防效外,其余都较低。

表4　2种药剂防治半夏疫病的小区试验结果1)

药剂	稀释倍数倍	药前病情指数	第1次药后10d		第2次药后10d	
			病情指数	防效/%	病情指数	防效/%
58%甲霜灵*锰锌WP	500	11.0	11.3	98.6aA	12.2	97.4aA
	1000	10.0	16.7	68.7bB	21.7	74.6bAB
	1500	11.7	26.4	31.3cBC	34.9	49.6bcBCD
	2000	13.4	28.9	27.6cBC	51.7	16.7cCD
70%甲基硫菌灵WP	500	10.2	16.7	69.6bB	23.3	71.5bBC
	1000	13.3	23.3	53.3bcBC	31.7	60.0bBCD
	1500	11.3	21.6	51.9bcBC	43.6	29.8cCD
	2000	13.4	28.8	28.0cBC	53.3	13.4cD
CK(清水)	0	10.8	32.2		56.8	

1)表中同列数字后面小写字母或大写字母分别表示($p<0.05$)显著水平或($p<0.01$)极显著水平。

3　结论与讨论

本文通过病原菌致病性的测定和形态学观察,确定引起甘肃清水半夏GAP生产基地田间半夏生长后期大量死亡的病原菌为寄生疫霉(P. parasitica Dast.)。菌丝生长最适pH和温度分别为6.8与27.3℃,与成家壮等[10]分别从九重葛、忍冬及长春花上获得的寄生疫霉的生长温度28~30℃、24~30℃、28~30℃,朱建兰[11]从温室茄茎上获得的寄生疫霉的生长温度30℃左右都基本一致,初步表明二因子饱和D-最优设计在研究pH和温度对寄生疫菌菌丝生长影响的可行性,减少了以往试验的工作量。

通过室内药效比较并选择药效较好的2种药剂结合小区试验表明,70%甲基硫菌灵可湿性粉剂虽在室内药效较好,但大田试验中防治效果较低,70%甲基硫菌灵可湿性粉剂500倍液经2次施药后防效仅达71.5%。而58%甲霜灵·锰锌可湿性粉剂500倍液经2次施药后防效为97.4%,说明其可有效控制半夏疫病的发生发展。生产中为防止病菌抗药性产生,必须几种药剂交替使用,避免长期使用单一农药品种。

半夏植株矮小、茎秆纤细，一旦发病将很快死亡，因此，实施早期防病重于治病。预防半夏生长后期疫病的发生，可以有效延长半夏生长期，提高半夏产量。曾建红等[12]研究表明半夏单株平均生物碱产量以10月下旬至11月上旬为最高，对半夏疫病的有效预防还有利于提高半夏生物碱含量。

参考文献

[1]曾令祥，李德友. 旱半夏病虫害识别及防治[J]. 农技服务，2007，24(3)：73－76.

[2]申屠苏苏，王海丽，陈集双，等. 三叶半夏的2种病毒检测[J]. 中国中药杂志，2007，32(8)：664－667.

[3]陈集双，李德葆. 侵染半夏的两种病毒的分离纯化和初步鉴定[J]. 生物技术，1994，4(4)：24.

[4]方中达. 植病研究方法[M]. 北京：中国农业出版社，1998.

[5]余永年，李金亮，杨雄飞. 中国橡胶树疫霉种的研究[J]. 菌物学报，1986(4)：193－206.

[6]Waterhouse G M. Key to the species of P hy toph thora de Bary[J]. M ycologia Paper，1963，92：1－22.

[7]New hook F J. Tab ular key to the species of Ph ytop hthor a de Bary[J]. M ycologia Paper，1978，143：110.

[8]H o H H. Synoptic key s to the species of P hytophthora[J]. M ycologia，1981，73(4)：705－714.

[9]魏景超. 真菌鉴定手册[M]. 上海：上海科学技术出版社，1979：36－40.

[10]成家壮，韦小燕. 广州地区观赏植物疫霉种的鉴定及交配型研究[J]. 热带作物学报，1996，17(2)：100－104.

[11]朱建兰. 日光温室茄茎腐病病原鉴定[J]. 植物保护，2001，27(4)：6－9.

[12]曾建红，彭正松. 不同采收期半夏生物碱含量的变化规律[J]. 中南林学院学报，2004，24(4)：109－112.

本文曾发表在2010年第6期植物保护上。

施肥对西北地区各等级半夏总生物碱含量的影响

裴建文　孙新荣　王　鹏　裴国维
杨少平　刘红霞　裴建奎*

采用N、P、K肥三因素五水平二次通用旋转组合设计研究了半夏在人工栽培条件下施肥对总生物碱含量的影响。结果表明：影响半夏各等级总生物碱含量的主要因素是N肥；最可取的施肥方案为纯N 419.5 ~ 514.1kg·hm^{-2}、K_2O 150kg·hm^{-2}，不施P肥，此时可兼顾各等级半夏总生物碱含量都较高。

中药半夏为天南星科植物半夏(Pinellia ternata)的干燥块茎[1]，其中含有淀粉、半夏蛋白、生物碱类、苷类、酚类、甾醇类、氨基酸类、脂肪酸类和无机元素等多种成分[2]，生物碱是半夏主要药效成分之一。半夏所含的生物碱对慢性髓性白血病细胞(K 652)的生长有抑制作用，所含的季铵生物碱—葫芦巴碱对小鼠肝癌细胞亦有抑制作用[3]，以往的研究主要集中在不同粒径[4]、不同采收时期[5-6]、不同品种[7-9]对半夏总生物碱含量的影响，半夏总生物碱含量的提取方法[10-12]，南方施肥对半夏产量[13-15]的影响等方面。本文在西北地区半夏高产栽培施肥研究的基础上[16]，研究了施肥对半夏总生物碱含量的影响，旨在为半夏的高产优质栽培提供一定的科学依据。

1　材料与方法

1.1　试验地概况

试验于2007—2008年在甘肃省天水市清水县城郊川水地上进行，当地海拔1378m；土质中壤，肥力中等；播前土壤养分为全N 0.63g·kg^{-1}，全P_2O_5 1.29 g·kg^{-1}，全K_2O 16.85 g·kg^{-1}，碱解N 68.0mg·kg^{-1}，速效P_2O_5 22.7mg·kg^{-1}，速

* 作者简介：裴建文(1957—　)，男，甘肃天水人，曾为天水师范学院生物工程与技术学院教授，学士，主要从事作物育种与中药材规范化种植研究。

效 K_2O 173.5mg · kg^{-1}。前茬以小麦做匀地试验,试验区均匀一致。

1.2 试验材料

1.2.1 供试品种 采用甘肃省天水市清水县百家乡收集的野生半夏,种子块茎直径 1.0cm 左右,平均单粒重 0.82g。

1.2.2 供试化肥 氮肥用尿素(N 46%,兰州石化公司,磷肥用过磷酸钙(P_2O_5 12%,白银绿源磷复合肥有限公司),钾肥用氯化钾(K_2O 60%,青海利源化肥厂)。

1.3 田间试验方法

1.3.1 试验设计 采用三因素五水平二次通用旋转组合设计,因素水平设置见表1。

表1 试验因素及水平设置

水平	因素		
	N	P(P_2O_5)	K(K_2O)
	(x_1)	(x_2)	(x_3)
γ	600.0	300.0	300.0
1	478.4	239.3	239.3
0	300.0	150.0	150.0
-1	121.7	60.8	60.8
-γ	0.0	0.0	0.0

1.3.2 播种、田间管理及收获 为了便于田间管理,降低因化肥水平渗透作用对试验的影响,每小区做一畦,四周打埂,埂宽 10cm,高 15cm。小区净面积 12.25m^2(3.5m ×3.5 m),共20 个处理,分为2 个区组随机排列("0"点处理重复6次)。播种时先铲出畦内表土,整平畦底,将全部化肥混合后一次性施入畦底,耧入土中混匀后撒播种茎,最后均匀覆土 15 cm。整个试验期不施有机肥,生育期内也不追施化肥。2007 年 11 月 13 日播种,播种量(块茎)4200kg · hm^{-2},密度 510 万粒 · hm^{-2}。翌年出苗前浇水 1 次,并在地表覆盖麦草保持土壤松软湿润。全生育期视苗情、土壤和大气湿度随时浅灌,及时除去田间杂草,统一防治病虫害。待全田倒苗,叶片干枯后于 9 月 15 日一次性采挖。每小区四周除去 30cm 宽面积不计产(进一步消除肥料不同水平在小区间的横向渗透影响),剩余部分严格计产并取样检验。

1.3.3 等级划分 参考张贵君[17]的标准进行分级,统货为直径大于 0.5cm

的混合干块茎,一、二、三等品分别为直径大于1.5cm、直径1.0～1.5cm、直径0.5～1.0cm的干块茎。

1.3.4 土壤化验 土壤中的全氮用半微量开氏法测定;全磷用氢氧化钠熔融-钼锑抗比色法测定;全钾用氢氧化钠熔融-火焰光度法测定:碱解氮用碱解扩散法测定;速效磷用碳酸氢钠熔融-钼锑抗比色法测定;速效钾用乙酸铵提取-火焰光度法测定[18]。

1.4 总生物碱含量测定

1.4.1 仪器与试剂 TU-1810型紫外-可见分光光度计;盐酸麻黄碱对照品(东北制药集团公司沈阳第一制药厂,批号080404-1);0.05%溴麝香草酚蓝溶液;柠檬酸-柠檬酸钠缓冲液(pH=5.4)[19];其它试剂均为分析纯。

1.4.2 测定方法 采用紫外分光光度法[12-13],以盐酸麻黄碱为对照测定干燥至恒重的半夏块茎总生物碱含量。

1.5 统计方法

采用Excel 2003和DPS 7.05软件进行统计分析。

2 结果与分析

2.1 半夏总生物碱含量与施肥间的相关性

经三因素五水平二次通用旋转组合设计的各等级半夏总生物碱含量的结果见表2.建立各等级半夏总生物碱含量与施肥间的回归方程如表3所示:各方程均拟合良好,除$\hat{y}1$方程达0.1显著水平外($F=2.6>F_{0.1}=2.3$)水平外,其余均达0.01极显著水平,剔除不显著项后得各等级半夏总生物碱与施肥的回归方程为:

一等品总生物碱含量:$y_1=0.0497+0.0063_{x1}-0.0047{x_1}^2$

二等品总生物碱含量:$y_2=0.0395+0.0072_{x1}-0.0030{x_1}^2$

三等品总生物碱含量:$y_3=0.0319+0.0064_{x1}-0.0025{x_3}^2$

统货总生物碱含量:$y_4=0.0403+0.0067_{x2}-0.0011{x_1}^2$

可见,前二者仅为N肥的二次方程,后二者为N、K或N、P肥的二次方程。N、P、K三因素两两间的交互项均不显著。

表 2　试验结果

处理	编码值			总生物碱含量//%								干重比例（一等:二等:三等）
				一等品		二等品		三等品		统货		
	$x1$	$x2$	$x3$	$y1$	$\hat{y}1$	$y2$	$\hat{y}2$	$y3$	$\hat{y}3$	$y4$	$\hat{y}4$	
1	-1	-1	-1	0.0388	0.0386	0.0311	0.0293	0.0268	0.0231	0.033 2	0.031 3	1: 7.59: 2.91
2	-1	-1	1	0.0427	0.0386	0.0293	0.0293	0.0233	0.0231	0.0308	0.031 3	1: 5.49: 1.29
3	-1	1	-1	0.0407	0.0386	0.0262	0.0293	0.0234	0.0231	0.0271	0.029 0	1: 3.56: 3.57
4	-1	1	1	0.0277	0.0386	0.0263	0.0293	0.0240	0.0231	0.0261	0.029 0	1: 4.33: 3.70
5	1	-1	-1	0.0482	0.0513	0.0440	0.0437	0.0412	0.0358	0.0445	0.044 6	1: 3.64: 0.64
6	1	-1	1	0.0492	0.0513	0.0482	0.0437	0.0342	0.0358	0.0447	0.044 6	1: 3.38: 0.43
7	1	1	-1	0.0450	0.0513	0.0423	0.0437	0.0394	0.0358	0.0422	0.042 3	1: 3.41: 0.98
8	1	1	1	0.0461	0.0513	0.0434	0.0437	0.0380	0.0358	0.0436	0.042 3	1: 2.54: 0.43
9	-1.682	0	0	0.0239	0.0257	0.0228	0.0188	0.0181	0.0212	0.0205	0.019 2	1: 2.82: 2.46
10	1.682	0	0	0.0523	0.0470	0.0425	0.0430	0.0369	0.0426	0.0400	0.041 6	1: 1.77: 1.19
11	0	-1.682	0	0.0410	0.0497	0.0372	0.0395	0.0363	0.0319	0.0402	0.042 2	1: 6.31: 1.81
12	0	1.682	0	0.0535	0.0497	0.0407	0.0395	0.0298	0.0319	0.039 6	0.038 4	1: 2.75: 0.61
13	0	0	-1.682	0.0572	0.0497	0.0482	0.0395	0.0252	0.0249	0.0419	0.040 3	1: 4.18: 1.42
14	0	0	1.682	0.0468	0.0497	0.0451	0.0395	0.0204	0.0249	0.0365	0.040 3	1: 2.79: 0.68
15	0	0	0	0.0451	0.0497	0.0427	0.0395	0.0302	0.0319	0.0391	0.040 3	1: 1.74: 1.33

续表

处理	编码值			总生物碱含量//%								干重比例(一等:二等:三等)
				一等品		二等品		三等品		统货		
	$x1$	$x2$	$x3$	$y1$	$\hat{y}1$	$y2$	$\hat{y}2$	$y3$	$\hat{y}3$	$y4$	$\hat{y}4$	
16	0	0	0	0.0442	0.0497	0.0353	0.0395	0.0328	0.0319	0.0418	0.040 3	1: 2.54: 0.73
17	0	0	0	0.0500	0.0497	0.0390	0.0395	0.0366	0.0319	0.0395	0.040 3	1: 3.05: 0.89
18	0	0	0	0.0496	0.0497	0.0402	0.0395	0.0352	0.0319	0.0402	0.040 3	1: 2.26: 0.97
19	0	0	0	0.0604	0.0497	0.0379	0.0395	0.0329	0.0319	0.0397	0.040 3	1: 1.96: 1.58
20	0	0	0	0.0479	0.0497	0.0415	0.0395	0.0248	0.0319	0.0415	0.040 3	1: 2.37: 1.63

注:y_i为实际值;y_1为依方程计算出的理论值.

表 3　回归方程及其显著性

$\hat{y}_1$	b_0	b_1x_1	b_2x_2	b_3x_3	$b_{12}x_{12}$	$b_{13}x_{13}$	$b_{23}x_{23}$	$b_{11}x_{12}$	$b_{22}x_{22}$	$b_{33}x_{33}$
$\hat{y}_1$	0.0497**	0.0063**	0.0001	-0.0018	0.0009	0.0014	-0.0021	-0.0047*	-0.0015	0.0002
$\hat{y}_2$**	0.0395**	0.0072**	-0.0006	-0.0001	0.0002	0.0009	-0.0002	-0.0030**	-0.0008	0.0019
$\hat{y}_3$**	0.0319**	0.0064	-0.0009	-0.0014	0.0006	-0.0007	0.0012	-0.0008	0.0011	-0.0025*
$\hat{y}_4$**	0.0403**	0.0067	-0.0011*	-0.0008*	0.0009	0.0006	0.0003	-0.0035**	-0.0001	-0.0003

注:*表示达到显著水平,**表示达到极显著水平.

2.2 半夏各等级总生物碱含量最大时N、P、K肥的最佳施用量

对$\hat{y}_1$、$\hat{y}_2$两方程求解知：当$x_1=0.670$（N 419.5 kg·hm^{-2}）时，一等品总生物碱含量（$\hat{y}_1$）达极值0.0518%；当$x_1=1.200$（N514.1kg·hm^{-2}）时，二等品总生物碱含量（$\hat{y}_2$）达极值0.0438%。对$\hat{y}_3$方程降维求解知：当$x_1=1.68\ 2$（N施量600.0 kg·hm^{-2}，x_1在本试验设计范围内未找到佳点施量，从图1看出施量还可增加）、$x_3=0$（K_2O施量150.0kg·hm^{-2}）时，三等品总生物碱含量（$\hat{y}_3$）达极值0.0427%。对$\hat{y}_4$方程降维求解知：当$x_1=0.957$（N施量470.7kg·hm^{-2}）、$x^2=-1.682$（不施P_2O_5）时，统货总生物碱含量（$\hat{y}_4$）达极值0.0454%。可见，在本试验土壤条件下，要使各等级半夏总生物碱含量均较高时，应以施N、K肥为好，不施P肥，施P反倒会使统货总生物碱含量下降。

2.3 N、P、K肥对各等级总生物碱含量的单施效应

2.3.1 N肥的单施效应对4个方程中的三因子分别取-1.682（不施），使方程降维得到单因子总生物碱含量的方程，求解方程并绘图（表4、图1）. 从图1可看出，N肥对一、二等品和统货总生物碱含量的影响均呈开口向下的抛物线；N肥对三等品总生物碱含量的影响呈线型，为正效应。在最佳施量之前，单独增施N肥可显著提高各等级半夏总生物碱的含量。其中对一等品的效果最好，总生物碱含量极值0.0518%（当$x_1=0.670$，即施N419.5 kg·hm^{-2}时）比不施肥（0.0257%）提高0.0261%；其次为统货，总生物碱含量极值0.0454%（当$x1=0.957$，即施N470.7kg·hm^{-2}时）比不施肥（0.0211%）提高0.0243%；再次为二等品，总生物碱含量极值0.0438%（当$x1=1.200$，即施N514.1kg·hm^{-2}时）比不施肥（0.0188%）提高0.0250%；N肥对三等品总生物碱含量的影响呈线性关系，在设计范围内，随N肥施量的增加而增加. 施N量平均每增加1kg·hm^{-2}，一等品、二等品、三等品、统货的总生物碱含量分别增加6.2204×10^{-7}、4.8972×10^{-7}、3.5667×10^{-7}、5.1625×10^{-7}。在试验设计范围内，除三等品不存在施N过量的问题外，当N肥超过佳点施量时，其余等级总生物碱含量均有不同程度的下降。施N量平均每增加1kg·hm^{-2}，一等品、二等品、统货的总生物碱含量分别下降2.6606×10^{-7}、8.9385×10^{-8}、$1.469\ 8\ \times10-7$。可见，单位N肥在半夏各等级总生物碱含量上的效果（农学效率）不同：一等品>统货>二等品>三等品。

表4　N、P、K 肥单施对半夏各等级总生物碱含量的影响

水平	总生物碱含量/%					
	一等品	二等品	三等品		四等品	
(xi)	$\hat{y}1(x1)$	$\hat{y}2(x1)$	$\hat{y}3(x1)$	$\hat{y}3(x3)$	$\hat{y}4(x1)$	$\hat{y}4(x2)$
γ 1.200	0.0470	0.0430 0.0438 *	0.0356	0.0142	0.0435	0.0173
1	0.0513	0.0437	0.0313	0.0188	0.0453	0.0181
0.957					0.0454 *	
0.670	0.0518 *					
0	0.0497	0.0395	0.0249	0.0212 *	0.0422	0.0192
-1	0.0386	0.0293	0.0186	0.0188	0.0320	0.0203
-γ	0.0257	0.0188	0.0142	0.0142	0.0211	0.0211

注：$\hat{y}_i(x_i)$为单施x_i时$\hat{y}_i$的理论值；* 为$\hat{y}_i(x_i)$的极值

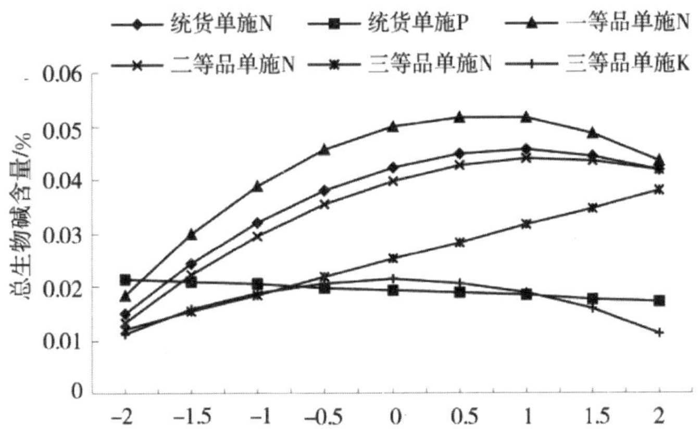

图1　N、P、K 肥单施对半夏各等级总生物碱含量的影响

2.3.2　P 的肥单施效应由 4 个方程和表 4、图 1 看出，P 肥对统货总生物碱含量的影响呈线型，P 肥对一、二、三等品均无明显影响（不显著项已剔除），但对统货总生物碱含量有显著负效应，这是由于统货是各等级归并后的和所致。单施 P 肥时，统货总生物碱含量随施量增加呈线性平缓下降趋势，施量每增加 1kg · hm^{-2}，总生物碱含量则降低 1.2667×10^{-7}%。

2.3.3　K 肥的单施效应由 4 个方程及表 4、图 1 看出，K 肥只对三等品总生物碱含量有影响。单施 K 肥当 $x_3=0$(K_2O150 kg · hm^{-2})时,三等品总生物碱含量达极值 0.0212 %,平均每增施 K_2O1kg · hm,施量佳点前三等品总生物碱含量提高 4.6667×10^{-7},越过最佳点则降低 4.6667×10^{-7}。各等级总生物碱含量受 N 肥的影响最大,一、二等品只施 N 肥有效;单位肥料效果在三等品上 K > N,P 无效;统货上 N > P,K 无效。

结论

在半夏栽培中,要利用施肥措施实现各等级总生物碱含量都达最大,是无一共同施量佳点可循的,但 N 肥却有一个可取的施量范围:x_1取 0.670 ~ 1.200,即施 N419.5 ~ 514.1 kg · hm^{-2}、K 肥取 x3 = 0,即施 K_2O150.0kg · hm^{-2}时为好,试验地 P 肥不是太缺时,最好不施,这样即可实现各等级生物碱含量均较高。不过,在这一施 N 区间, N 施量越大,一等品总生物碱含量会有所降低,而二、三等品总生物碱含量则会有所升高。陈中坚[9]等研究认为,影响半夏总生物碱含量的主要因素是 N 肥,其次为 P 肥(P 肥为正效应),K 肥最小。本文结论与之不同的是 P 肥为负效应。初步认为这可能是由南北方土壤及气候环境差异造成的,可能与供试品种的不同也有关,尚待深入研究。

参考文献

[1]中华人民共和国卫生部药典委员会. 中国药典: Ⅰ 部[M]. 北京:化学工业出版社,2005.

[2]关虎吕. 中药现代化研究与应用:第 2 卷[M]. 北京:学苑出版社,1997.

[3]侯家玉. 中药药理学[M]. 北京:中国中医药出版社,2002.

[4]曾建红,彭正松,陈旭,等. 半夏块茎不同粒径总生物碱含量的研究[J]. 时珍国医国药,2008,19(4):829 – 830.

[5]李西文,马小军,宋经元,等. 半夏不同生长发育时期总生物碱含量动态变化的研究[J]. 中国中药杂志,2006,31(8):687 – 688.

[6]曾建红,彭正松. 不同采收期半夏生物碱含量的变化规律[J]. 中南林学院学报,2004,24(4):109 – 112.

[7]曾建红,彭正松. 半夏不同克隆株生物碱含量的比较研究[J]. 安徽农业科学,2008,36(16):6797 – 6798.

[8]魏淑红,彭正松,王祖秀. 半夏种内各变异类型总生物碱含量的变异规律[J]. 中国中药杂志,2008,33(2):191 – 193.

[9]于超,张明,王宇,等. 栽培、野生及不同产地半夏总生物碱测定[J]. 中国中药杂志,2004,29(6):583 – 584.

[10]谢一辉,肖宏浩,曹凯,等.比色法测定半夏中总生物含量[J].江西中医学院学报,1993,5(2):22.

[11]曾建红,彭正松,魏淑红,等.半夏生物碱最佳提取条件的研究[J].中药材,2003,26(5):361－363.

[12]于超,张明,王宇,等.紫外分分光度法测定不同产地半夏药材中总生物碱的含量[J].时珍国医国药,2002,13(2):73－74.

[13]陈中坚,孙玉琴,赵雄廷,等.施肥水平对半夏产量和质量影响的研究[J].中药材.2006,29(8):757－759.

[14]蒋燕,翟玉钤,王惠,等.半夏配方施肥模型研究[J].安徽农业科技,2007,35(25):7887－7888.

[15]卢立兴.半夏施肥技术的探讨[J].中国中药杂志,1992,17(3):142－143.

[16]王鹏,裴建文,孙万仓,等.半夏高产高效栽培最佳施肥数学模型研究[J].中国中药杂志,2009,34(06):669－673.

[17]张贵君.中药商品学[M].北京:人民卫生出版社,2005.

[18]中国科学院南京土壤研究所.土壤理化分析[M].上海:上海科学技术出版社,1978.

[19]中国科学院上海药物研究所.中草药有效成分提取与分离[M].1版.上海:上海科学技术出版社,1972.

本文曾发表在2010年第2期甘肃农业大学学报上。

泡菜生产的微生物区系分析

张宗舟　王玉洁　石宝珍*

通过对泡菜水的微生物区系分析，证明了泡菜发酵是一种特殊的混菌共酵。发酵初期，各类微生物相互竞争，大量繁殖，数量增加较快，渐而乳酸菌、醋酸菌成为优势菌群，泡菜液的 pH 值开始明显下降，其他微生物受到抑制。在发酵 2－3d，微生物数量达到高峰，之后所有微生物数量开始下降，并逐渐平稳。整个泡菜发酵周期确定为 7d，在泡菜发酵第 7 天，乳酸菌群数量 $245.0\times10^5\sim404.6\times10^5$CFU/mL；醋酸菌群数 $13.1\times10^4\sim14.0\times10^4$CFU/mL；丁酸菌群数量 $2.5\times10^3\sim3.2\times10^3$CFU/mL；酵母菌群数量 $4.0\times10^2\sim5.1\times10^2$CFU/mL。微生物的数量变化说明了泡菜的发酵过程，在生产中注意把握微生物的变化动态，有利于控制产品质量。

泡菜，古称葅，1400 年以前已有记载。泡菜是以甘蓝、胡萝卜等为原料，经洗涤、切分处理后放入泡菜坛内，加入一定浓度的食盐溶液，再经乳酸发酵等制成的一种腌制蔬菜制品[1]。泡菜味道宜人，组织脆嫩，盛产于四川，流行于全国。凡含纤维丰富的蔬菜、水果，都可以用来腌制泡菜。泡菜含有丰富的维生素和钙、磷等无机物，既能为人体提供充足的营养，又能预防动脉硬化等疾病[2]。有人认为泡菜是乳酸发酵，乳酸菌越纯，泡菜质量越好，越具有吸引力。实际上泡菜发酵和乳酸发酵从发酵机理到发酵工艺都有很大的差异[3-4]。从生物化学角度来讲，泡菜是酿造而不是发酵，只是人们习惯称泡菜发酵。泡菜在发酵过程中参与的微生物比较复杂，其是多种微生物共同发酵的结果[5-6]。发酵初期，多种微生物对环境因素以及营养物质进行竞争，形成一种动态平衡。发酵稳定后，泡菜成熟，营养物质比较恒定，泡菜内部条件基本不变，蔬菜得以较长时间保存[7-10]。微生物决定

---

* 作者简介：张宗舟（1957—　）男，甘甘肃礼县人，曾为天水师范学院生物工程与技术学院教授，博士，主要从事应用微生物学研究。

着泡菜的风味、质量，直接影响着泡菜的发酵进程和泡菜的保存期。发酵过程中微生物的种类与数量的变化可以解释泡菜的发酵原理，泡菜的发酵工艺[11-12]。本研究采用4种常用泡菜原料—苹果、白萝卜、黄萝卜、大白菜，分别在同一种工艺、同一种泡菜母液、同样的发酵条件、同样的发酵时间下，定期测定主要微生物数量[13-14]。因乳酸菌群、醋酸菌群、丁酸菌群和酵母菌群是主体菌群，所以只测这4类菌群，测定结果供有关科技工作人员参考。

1 材料与方法

1.1 材料与试剂

1.1.1 原料

苹果、白萝卜、黄萝卜和大白菜均为天水当地生产。食盐：天水市盐业公司，满足国标GB5461—2000《食用盐》中二级标准。花椒、白酒由陇南市生产。

1.1.2 培养基[15-16]

醋酸菌培养基：葡萄糖10g、酵母膏10g、$CaCO_3$ 10g、琼脂20g、蒸馏水1000mL，pH6.8；

乳酸菌培养基：牛肉膏10g、蛋白胨10g、酵母膏10g、葡萄糖10g、吐温80 0.1mL、K_2HPO_4 2g、$MgSO_4 \cdot 7H_2O$ 0.6g、$MnSO_4 \cdot 4H_2O$ 0.3g、琼脂20g、蒸馏水1000mL；丁酸细菌用牛肉膏蛋白胨培养基：牛肉膏10g、蛋白胨10g、酵母膏10g、葡萄糖10g、吐温80 0.1mL、$KHPO_2$ g、$MgSO_4 \cdot 7H2O$ 0.6g、$MnSO_4 \cdot 4H2O$ 0.3g、琼脂20g、蒸馏水1000mL；酵母菌用麦芽汁琼脂培养基：干麦芽与水比例为1∶4，在65℃水浴中糖化3-4h，糖化程度可用碘滴定至不变色，说明全部糖化。加水约20mL，过滤。将滤液稀释到糖度5-6°Bé，pH值约6.4，加入2%琼脂。

1.2 仪器与设备

DZKW-4型电子恒温水浴锅：黄骅市渤海电器厂；722型分光光度计：上海欣茂仪器有限公司；PHS-2C型精密酸度计：上海虹益仪器仪表有限公司。

1.3 试验方法

1.3.1 泡菜发酵工艺

原料加工→修整、洗涤、晾晒、切分成条状 }
加盐→盐水冷却→泡菜盐水 } →加入调味料装坛→发酵→成品

操作说明：将4种原料苹果、白萝卜、黄萝卜、大白菜分别称2kg，洗净，切片。都用3 000mL 6%食盐水为发酵液，加入母液50mL、花椒20g和52度白酒30mL，入泡菜坛，用双层保鲜膜封口，置25℃发酵。

1.3.2 试验设计

发酵时间从封口开始设为1d、2d、3d、4d、5d、6d、7d。每隔1d取样，每次取样20mL。测pH值，并稀释、涂布，在25℃条件下培养3d。测定泡菜发酵液中微生物数量(CFU/mL)[17-19]。微生物数量用稀释平板法测定，接种时细菌用103、104、105稀释倍数计数；酵母菌用101、102、103稀释倍数计数。计数时取3个稀释度中菌落数较接近的平皿进行统计，重复测定3次取平均值。

2 结果与分析

2.1 泡菜发酵液中乳酸菌群数量变化[20-21]

如图1所示，乳酸菌群的数量大约在16.2× 105～440.8×105CFU/mL。乳酸菌在前2d数量较低，第2～3天数量增长迅速，在第3天达到高峰，产生的乳酸相对较多，乳酸菌属于广谱抗菌物，在达到一定量时，其他微生物被乳酸菌产生的乳酸抑制，泡菜微生物形成一种平衡。3～4d乳酸菌数量缓慢下降，5d后趋于稳定。各种原料发酵液的乳酸菌数量差异不太大，达到一种平衡。

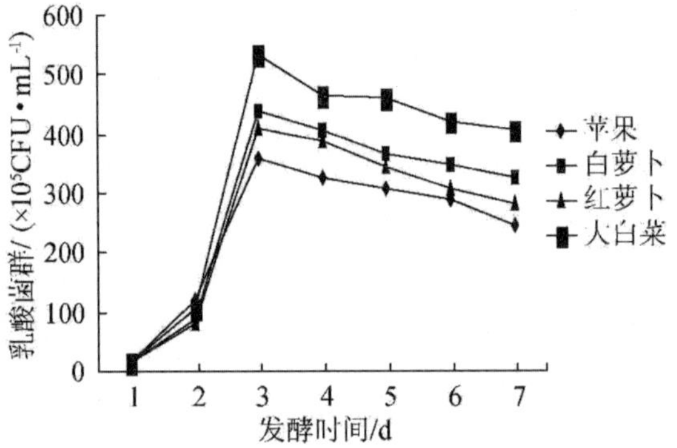

图1 发酵时间对发酵液中乳酸菌群数量的影响

2.2 泡菜发酵液中醋酸菌群数量变化

如图2所示，醋酸菌群的数量9.0×104～15.8×104CFU/mL，各种原料的发酵液中，醋酸菌变化趋势相同，都是在第1天数量较低，在2～3d达到高峰，4d后基本稳定。醋酸菌群在4种发酵液中的数量差异很小，且全部低于乳酸菌。泡菜产品的特色就是乳酸中有醋酸，醋酸中有乳酸，使得成品口感清爽宜人。

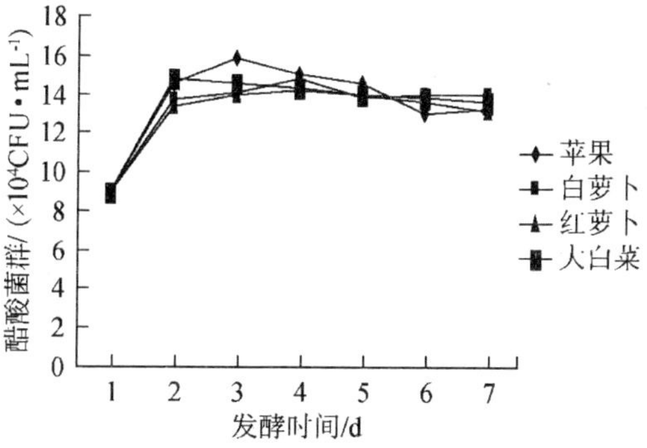

图 2　发酵时间对发酵液中醋酸菌群数量的影响

2.3　泡菜发酵液中丁酸菌群数量变化

如图 3 所示，丁酸细菌群的数量 2.0×103～8.6×103CFU/mL，各种原料发酵液中的丁酸菌群数量变化趋势相同，都是在第 2 天达到高峰，以后逐渐下降，一直到第 7 天丁酸菌群数仍呈下降趋势，且数量均小于乳酸菌和醋酸菌群，说明丁酸菌的代谢可能受到乳酸菌和醋酸菌的抑制。丁酸菌群代谢产物是丁酸，丁酸的味道沉闷，如果丁酸多了，会给泡菜带来不愉悦的气味，如果泡菜中没有丁酸，会让人感到味淡薄，特色不明显。

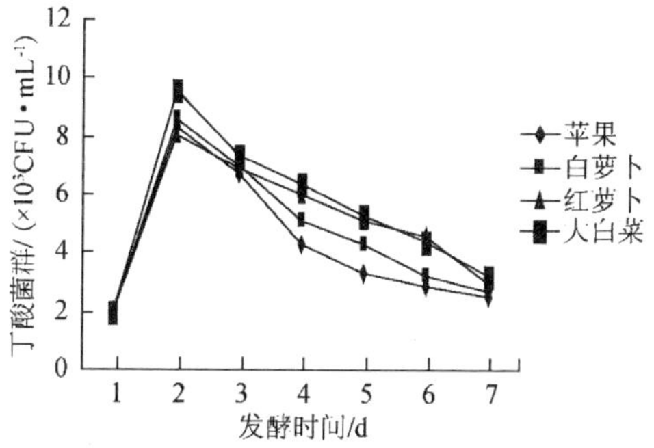

图 3　发酵时间对发酵液中丁酸菌群数量的影响

2.4　泡菜发酵液中酵母菌群数量变化

如图 4 所示，泡菜中酵母菌群 数量为 3.5× 102～36.4×102CFU/mL。各种原料发酵液中酵母菌都在第 2 天达到高峰，以后逐渐下降，但下降速度有差异，5d 后数量基本稳定。酵母菌的作用是将葡萄糖在缺氧的条件下转化为乙醇，乙醇的存在使泡菜更香，醇酸酯化，酯类物质含量会增加，但酵母菌增加（特别是野生耐盐酵母）会使发酵液中出现白花，影响泡菜的质量，缩短泡菜的保存期。

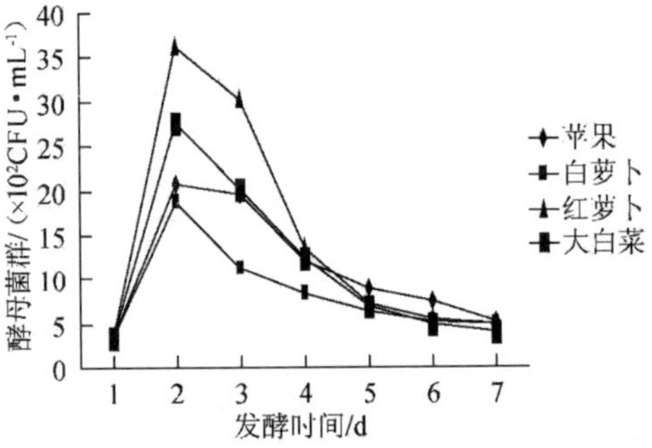

图 4　发酵时间对发酵液中酵母菌数量的影响

2.5　泡菜发酵过程中发酵液的 pH 值变化

4 种泡菜发酵过程中的发酵液 pH 值变化如图 5 所示，在发酵的前 3d，pH 值迅速下降，第 3 天后下降速度缓慢，4d 之后 pH 值稳定，pH 值约为 3。由于在发酵 48～72h，微生物数量达到高峰，乳酸菌、醋酸菌为优势菌群，其发酵产生的乳酸和醋酸使泡菜液的 pH 值开始明显下降，其他微生物受到抑制。之后随着代谢产物积累，所有微生物数量开始下降，并渐渐衡定，pH 值也基本达到稳定，使泡菜能够较长期保存.

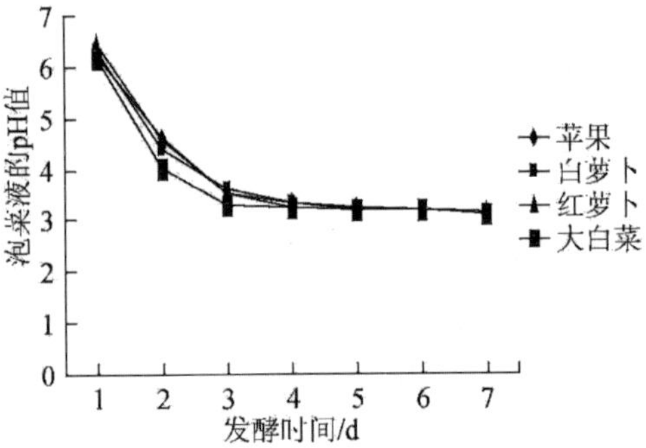

图 5　发酵时间对发酵液 pH 的影响

2.6　泡菜中微生物的平衡关系

从以上微生物数量分析中，还能看到泡菜的发酵是一种动态平衡。刚加入原料和发酵母液、盐水时，各种微生物得到营养，且条件适宜于各种微生物生长。母液中、环境带来的、原料表面的各种微生物大量繁殖，数量大增，24h 数量是原来的 5～20 倍。工艺的特色带来产品风格的特色，不同原料，营养成分不同，营养物质的可给性不一样，各种微生物的数量高峰出现的时间和增加的倍数也不一样。在生长繁殖过程中，随着代谢产物的积累，有些微生物被抑制，有些微生物死亡，使得微生物数量降低;平衡期时,泡菜味道已初步形成,pH 和菌种数量也基本达到稳定,渐而进入稳定状态。

结论

泡菜发酵是一种特殊的混菌共酵,其中乳酸菌、醋酸菌起主要作用,丁酸菌、酵母菌起辅助作用。发酵的第 1 天,各类微生物大量繁殖,数量增加很快,是一个竞争过程。渐而乳酸菌、醋酸菌成为优势菌群,发酵产生的乳酸和醋酸使泡菜液的 pH 值开始明显下降,其他微生物受到抑制,pH 值在第 4 天基本稳定。在发酵 2 ~ 3d,微生物数量达到高峰,之后所有微生物数量开始下降,并渐渐恒定。整个泡菜发酵周期确定为 7d,在第 7 天,4 种原料发酵液中乳酸菌群 $245.0 \times 10^5 \sim 404.6 \times 10^5$ CFU/mL;醋酸菌群 $13.1 \times 10^4 \sim 14.0 \times 10^4$ CFU/mL;丁酸菌群 $2.5 \times 10^3 \sim 3.2 \times 10^3$ CFU/mL;酵母菌群 $4.0 \times 10^2 \sim 5.1 \times 10^2$ CFU/mL。说明在泡菜发酵中乳酸菌、醋酸菌起主要作用,这 2 种菌也是泡菜口感丰厚,酸而不浮的原因。酵母菌群也会因为带来的酒精发酵而使泡菜的香味更浓。这种微生物的数量变化反映了泡菜的发酵过程,在生产中注意把握微生物的变化动态,有利于控制产品质量。

参考文献

[1]郑其良,赵喜茹.影响泡菜质量的因素及其质量控制[J].中国酿造,2005,24(2):29-31.

[2]黄业传,曾凡坤.自然发酵与人工发酵泡菜的品质对比[J].食品工业,2005(3):41-43.

[3]苏扬,陈云川.泡菜的风味化学及呈味机理的探讨[J].中国调味品,2001(4):28-31.

[4]周晓媛,夏延斌.蔬菜腌制品的风味研究进展[J].食品与发酵工业,2004,4(30):104-107.

[5]吕育新.蔬菜品种对发酵蔬菜质量及纯乳酸菌发酵过程的影响[J].中国调味品,2009(2):59-61.

[6]孙力军,李正伟,孙德坤.纯种接种和促菌物质的添加对苔菜泡菜发酵过程及其品质的影响[J].食品与发酵工业,2003,29(8):103-105.

[7]赵小蓉.纤维素分解菌对不同纤维素类物质的分解作用[J].微生物学杂志,2000,20(3):12-14.

[8]OLSSON L,HAHN-HAGERDAL B. Fermentation of lignocellulosic hy-drolysates for ethanol production[J]. Enzyme Microb Tech,1996(18):312-331.

[9]孔 健. 农业微生物技术[M]. 北京:化学工业出版社,2005:112 - 136.

[10]赵 斌,何绍江. 微生物学试验[M]. 北京:科学出版社,2002.

[11]叶姜瑜. 快速识别纤维素分解菌的新方法[J]. 生物学通报,1997,32(12):34.

[12]曲音波,高培基,王祖农. 斜卧青霉纤维素酶的研究[J]. 山东大学学报,1987,22(3):97 - 104.

[13]邬敏辰,李江华,邬显章. 黑曲霉固态培养生产纤维素酶的研究[J]. 酿酒,1997(6):5 - 9.

[14]李 琴,杜风刚. 在酱油生产中应用双菌种制曲的探索[J]. 中国酿造,2004,23(4):282 - 291

[15]赵德安. 混合发酵与纯种发酵[J]. 中国调味品,2005(3):3 - 8.

[16]NAKAJIMA A,EBIHARA K. Effect of prolonged vinegar feeding on postprandial blood glucose response in rats[J]. J Jpn Soc Nutr Food Sci,1988(41):487 - 489.

[17]张国春. 多菌种酿造酱油的特点[J]. 中国酿造,1998,17(5):302 - 321.

[18]顾立众,翟玮玮. 发酵食品工艺学[M]. 北京:中国轻工业出版社,1998.

[19]欧阳平凯,曹竹安,马宏建,等. 发酵工程关键技术及其应用[M]. 北京:化学工业出版社,2005:15 - 21

[20]熊 俐,杨跃寰,胡 洋. 物理诱变技术在食品工业微生物育种上的应用进展[J]. 江苏农业科学,2010(5):191 - 23.

[21]张红岩,申乃坤,周 兴. 基因敲除技术及其在微生物育种中的应用[J]. 酿酒科技,2010(4):21 - 25.

注:本文曾发表在2014年第3期中国酿造期刊上

棘托竹荪深层发酵胞外酶活性的研究

李仲芳　李冬琳*

本实验以棘托竹荪菌株为试材,采用液体培养的方法研究了棘托竹荪深层发酵过程中6种胞外酶,即羧甲基纤维素酶、淀粉酶、蛋白酶、漆酶、多酚氧化酶、过氧化物酶的酶活性变化,并对菌丝生物量进行了测定。结果表明:酶活性与菌丝生物量增长有密切关系。菌丝生物量增加呈S曲线,第4~6天增长最快。棘托竹荪菌丝生物量达到最大值之前,羧甲基纤维素酶和淀粉酶活性出现峰值。过氧化物酶、漆酶、多酚氧化酶和蛋白酶的活性与菌丝生物量同步增加。说明发酵过程中棘托竹荪菌丝首先分解利用淀粉和纤维素,然后利用木质素和蛋白质作为碳源和氮源。要提高棘托竹荪深层发酵效率,缩短培养周期,就必须在其菌丝达到最大生物量之前保证碳源和氮源的均衡供给。可根据不同酶的分泌高峰期,确定菌丝的营养利用情况和发酵周期,以收获最大菌丝生物量。

棘托竹荪(*Dictyophora echinovolvata*)是担子菌亚门腹菌纲鬼笔科竹荪属名贵的食药用真菌,不仅香气浓郁,脆嫩爽口,还有降血脂,降胆固醇,延缓食品腐败等功效,主要分布于四川、湖南、贵州等南方地区[1]。在自然条件下其生长发育所利用的主要原料是腐竹、腐木等不溶于水或难溶于水的纤维素、半纤维素、木质素等高分子物质,这些物质必须被竹荪产生的胞外酶降解才能吸收利用。所以,在竹荪人工培养过程中可向培养基中分泌多种胞外酶[2],胞外酶活性大小与食用菌种类、生长发育状态和培养基质等因素密切相关。

目前,对于竹荪的研究大多集中在竹荪液体发酵所需的C源、N源、无机盐等营养条件和理化条件的探索及多糖提取方面[3-5],其中对培养产物的抑菌作用也有研究报道[6]。与香菇、侧耳等相比,竹荪的液体发酵培养相对较难,对竹荪液体

---

* 作者简介:李仲芳(1963—),女,甘肃陇南人,现为天水师范学院生物工程与技术学院教授,学士,主要从事林果和花卉的生物技术应用研究。

发酵过程生理生化方面的基础研究较少，尚未见到对竹荪液体培养胞外酶活性变化规律的研究。

本试验采用摇瓶对棘托竹荪进行深层发酵培养，研究培养过程其羧甲基纤维素酶、淀粉酶、蛋白酶、漆酶、多酚氧化酶、过氧化物酶活性大小和动态变化规律，及其与竹荪菌丝体生物量的关系，为提高竹荪深层液体发酵效率，缩短培养周期提供理论依据。

1 材料与方法

1.1 供试菌株：棘托竹荪菌株从四川绵阳食用菌研究所引进。

1.2 培养基

1.2.1 活化培养基：牛肉膏蛋白胨培养基。

1.2.2 种子培养基：

竹屑（过 40 目筛）97%、白糖 1.2%、K_2HPO_4 0.2%、$MgSO_4$ 0.1%、石膏粉 1.5%，含水量 55% ~60%，pH 自然。

1.2.3 深层发酵培养基：

葡萄糖 20g、玉米粉（过 100 目筛）30g、KH_2PO_4 2g、$MgSO_4$ 1g、VB_1 10 mg、水 1000ml。pH 自然。

1.3 供试菌的培养：

1.3.1 母种活化培养

将棘托竹荪菌接种于上述活化培养基试管斜面上，28℃恒温黑暗培养 8 天菌丝长满培养基表面后待用。

1.3.2 种子培养基制作

将上述种子培养基按配方配制，装入 250mL 三角瓶至 2/3 处，高压灭菌后接种已活化的竹荪母种，长满三角瓶待用。

1.3.3 深层发酵培养

按上述发酵培养基配方配制液体培养基，注入 250ml 的三角烧瓶，每瓶装 100ml，高压灭菌后，将长满菌丝的种子培养基挖块，充分压细碎，每个培养瓶接种 1 小勺（约 10mg），接种后摇匀，在 28℃恒温黑暗静置培养 24h 后，置摇床 28℃恒温黑暗培养，振荡频率为 140r/min。

深层培养过程中，每天测定菌丝体干重和菌球直径、发酵液 pH 值，每 12h 做相关酶活性测定。菌丝体干重的测定，即每天随机抽取 1 瓶将菌丝体过滤，清水洗涤数次，80℃烘干至恒重，称量；菌球直径测定[8]，培养第 1 – 2 天每天从摇瓶中吸取发酵液，用测微尺随机测定 1 个视野的菌球大小，求平均值。从第 3 天开始随机抽取发酵液中10个菌球,在培养皿中排成一列,测总长度,求平均值。

1.4 酶活性的测定:

每12小时,取发酵液10mL,4℃,4000rpm离心10min,取上清液,即为粗酶液。

1.4.1 淀粉酶活性测定[7]:

向试管中加入0.5%的可溶性淀粉溶液0.5mL,样品管加粗酶液0.5mL,混匀,38℃恒温水浴保温30min,取出立即加入DNS试剂0.75mL(对照管加入DNS后再加入0.5mL酶液),沸水浴5min,取出立即冷却,加蒸馏水10mL,摇匀,于520nm测OD值,依据葡萄糖标准曲线($y=406.47x+13.58$, $R^2=0.95$)求得还原糖量。每组3个重复,求平均值。

定义每30分钟内底物被水解生成1微克葡萄糖所需的酶量为一个酶活力单位(U)。

1.4.2 羧甲基纤维素酶(CMC酶)活力测定[8-9]:

向试管中加入0.5%的CMC-Na溶液0.8mL,样品管加粗酶液0.2mL,50℃恒温水浴保温30min,取出立即加入DNS试剂0.75mL(对照管加入DNS后再加入0.2mL酶液),沸水浴5min,取出立即冷却,加蒸馏水10mL,摇匀,于520nm测OD值,依据葡萄糖标准曲线($y=406.47x+13.587$, $R^2=0.95$),求得还原糖量。每组3个重复,求平均值。

定义每30分钟内CMC酶水解生成1微克葡萄糖的所需的酶量为一个酶活力单位(U)。(1U=葡萄糖1ug/30min·mL发酵液)

1.4.3 过氧化物酶活性的测定[10]:

向样品管中加入粗酶液1.0mL,对照管中加入煮沸的粗酶液1.0mL,再分别加入过氧化物反应混合液3.0mL,倒入比色皿,立即开启秒表计时,测定470nm的OD值。以煮沸的酶液作对照,每组3个重复,求平均值。

1.4.4 多酚氧化酶活力测定[11]:

取0.5mL,10mM邻苯二酚作为底物,加人3.4mL pH6.0 0.05M的磷酸缓冲溶液,加人0.1mL粗酶液,28℃水浴准确保温30min后,410nm处测OD值,以煮沸的酶液作对照,每组3个重复,求平均值。

1.4.5 漆酶活性的测定[12]:

向试管中加入3.36mM邻联甲苯胺0.5mL,pH4.6、0.1M乙酸缓冲液3.0mL,样品管中加入粗酶液0.5mL,对照管中加入煮沸的粗酶液0.5mL,28℃恒温水浴保温30min,取出后立即于600nm测OD值。每组3个重复,求平均值。

过氧化物酶、多酚氧化酶和漆酶活力以每分钟引起0.001个OD值改变所需的酶量为一个酶活力单位(U)。

1.4.6 蛋白酶活性的测定：

蛋白酶活性的测定参照文献[10]的方法，Tyr 标准曲线回归方程为：$y = 122.42x - 2.1105$，$R^2 = 0.99$。取 4 支试管，分别加入 1ml 稀释酶液，其中一支为空白管，三支为平行实验管。置入 40℃水浴中预热 3～5min，在三支平行试管中分别加入 1ml 2% 酪蛋白溶液，准确计时保温 10min。立即加入 2mL 0.4M 三氯乙酸溶液，摇匀并静置过滤。分别吸取 1mL 上清液，加 5mL 0.4M 碳酸钠溶液，最后加入 1mL Folin－酚试剂，摇匀，于 40℃水浴中显色 20min 于 660nm 测 OD 值。对照试管测定方法同上，在加酪蛋白之前先加 0.4M 三氯乙酸 2mL，使酶失活，再加入酪蛋白。

定义每 10 分钟催化蛋白酶水解生成 1 微克酪氨酸的酶量为一个酶活力单位（1U = 酪氨酸 1μg/10min · mL 发酵液）。

2 结果与分析

棘托竹荪栽培过程及深层发酵过程中的形态变化见图 1，球形菌蕾破裂后，伸出笔状菌伞，然后菌裙落下，似穿着白纱裙的王后，故有“菌中皇后”之美称。深层发酵培养初期菌球表面致密光滑，后期菌球表面呈星芒状。

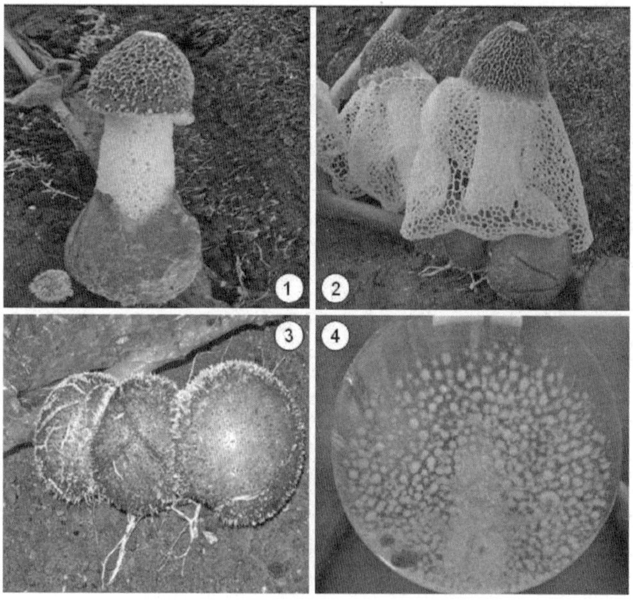

图 1 棘托竹荪发育及深层发酵过程中的形态变化

说明：①突破菌蕾的棘托竹荪；②已撒裙的棘托竹荪；③棘托竹荪的菌蕾；④深层发酵过程中的棘托竹荪菌丝球

2.1 棘托竹荪深层发酵过程生物量的变化

为了解棘托竹荪菌丝体在深层发酵过程中的生长变化情况,连续测定7天内菌球直径、菌丝体干重及发酵液的PH值,结果见表1和图2。

表1 深层发酵过程生物量及PH的变化

培养天数(d)	1	2	3	4	5	6	7
菌球直径(mm)	0.35	0.87	2.27	3.46	4.59	5.60	6.12
菌丝体干重(g)	0.0282	0.0439	0.1252	0.2641	0.3971	0.4748	0.5145
发酵液PH值	5.6	5.4	5.1	4.7	4.6	4.4	4.3

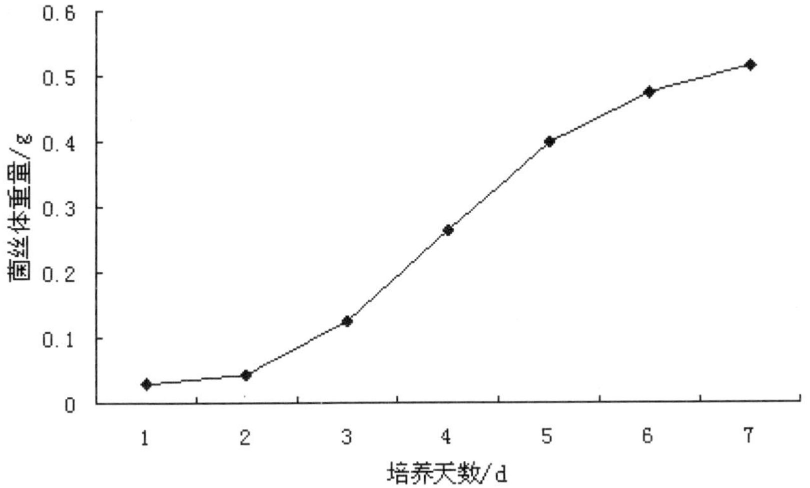

图2 深层培养过程菌丝生物量的变化

由表1和图2可见,在深层发酵过程中,棘托竹荪的菌球直径和干重随培养天数的增加而增加,发酵液的PH值则随着培养天数的增加而下降。菌丝体的干重增加呈S曲线,表现出典型的适应期、对数增长期和稳定期。在初始培养的1-3天增长缓慢,第4-6天快速增长,第7天趋于平缓,但尚未开始衰亡。培养后期,发酵液pH值的不断下降对菌丝生物量的增加有抑制作用,这可能是由于较低的PH影响棘托竹荪所分泌的各种酶的活性,加之培养基营养物质消耗而影响了棘托竹荪自身物质的合成,使得菌丝干重达一定值时不再快速增加。

2.2 CMC酶和淀粉酶活性变化

纤维素和淀粉都是作为棘托竹荪的碳源的大分子物质。CMC酶属于纤维素酶系中的一种,可以将纤维素分解为葡萄糖供菌体利用。同样,淀粉在淀粉酶等一系列酶的作用下可水解为葡萄糖等小分子物质供菌体利用。在液体发酵过程中 CMC 酶和淀粉酶活性变化与棘托竹荪菌丝生物量的增加密切相关(见图 3)。从总体酶活性趋势来看,CMC 酶和淀粉酶活性变化基本一致,CMC 酶活性相对于淀粉酶变化幅度小,淀粉酶总体活性略高于 CMC 酶。棘托竹荪菌丝生物量的增加随 CMC 酶、淀粉酶的活性增强而增加,即在生物量达到最大值之前,CMC 酶和淀粉酶的活性出现峰值。自培养的第 3 天活性明显增强,第 4 天酶活性达到最大,之后逐渐下降。这说明 CMC 酶和淀粉酶的大量分泌,将纤维素和淀粉分解为葡萄糖,为菌丝快速生长提供足够的碳水化合物。

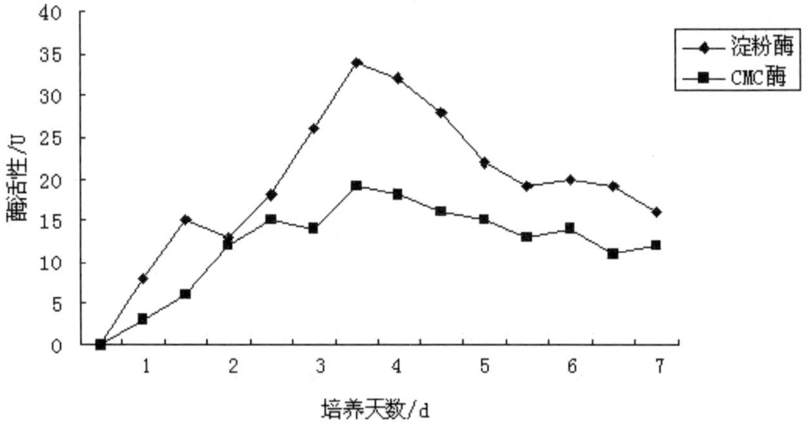

图 3 淀粉酶和 CMC 酶活性与培养天数的关系

2.3 过氧化物酶、漆酶和多酚氧化酶活性变化

过氧化物酶、漆酶、多酚氧化酶是木质素分解酶系的主要成员,其活性大小可以用来衡量菌株对木质素的降解能力。在棘托竹荪深层发酵过程中,3 种木质素分解酶的活性变化见图 4,这三种酶的活性变化趋势较为一致,从培养的第 2 天起活性增加明显,第 4 – 6 天活性很强,比 CMC 酶和淀粉酶的峰值出现稍晚,这说明棘托竹荪菌丝首先利用的是淀粉和纤维素作为碳源,然后才利用木质素。当培养第 5 – 6 天,淀粉和纤维素已被大量利用,淀粉酶和 CMC 酶活性下降时,漆酶等大量形成,棘托竹荪开始利用液体培养基中的木质素作为碳源。

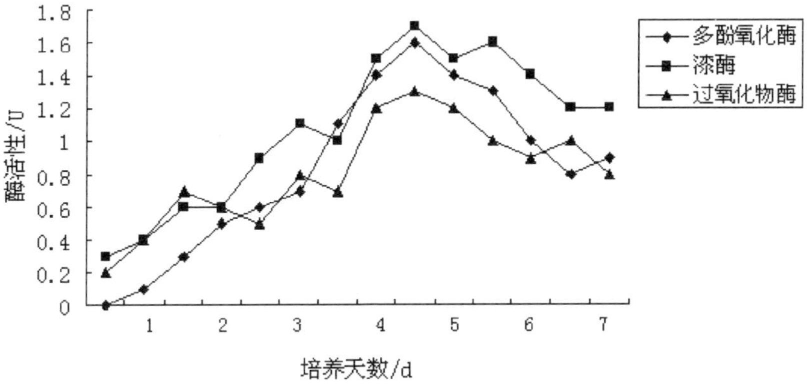

图 4　漆酶、过氧化物酶和多酚氧化酶的活性

2.4　蛋白酶活性变化

蛋白酶的作用是分解蛋白质的肽键,使其分解为小分子的蛋白胨、小肽或氨基酸。当培养基中氮源减少的情况下,食用菌为获得蛋白质中的氮素而分泌蛋白酶。随着氮源不断减少,蛋白酶会被大量诱导出来,故在菌丝生长阶段蛋白酶活性会呈现逐步升高的趋势[13]。从图 5 可见,棘托竹荪在深层发酵培养的第 2 天就显示出较强活性,随着菌丝生物量的增加逐步增强,第 5 天达到最大值,第 7 天仍然保持相对较高的水平,这就为菌丝的快速生长提供充足的氮源。

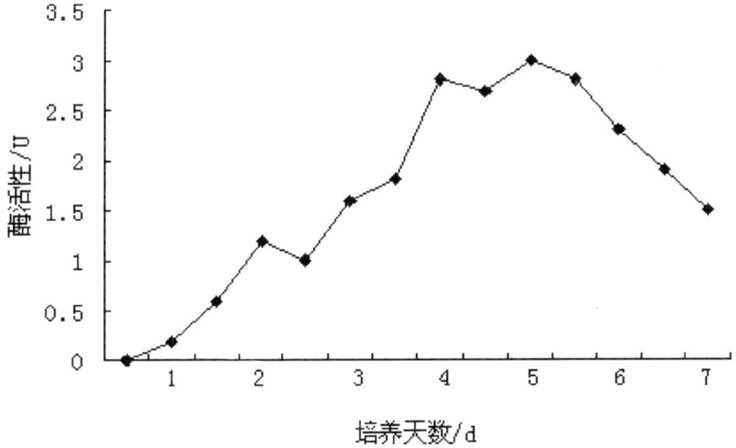

图 5　蛋白酶活性与培养天数的关系

3 结论

棘托竹荪深层发酵过程中,6 种胞外酶即羧甲基纤维素酶、淀粉酶、蛋白酶、漆酶、多酚氧化酶、过氧化物酶活性变化与菌丝生物量增长有密切关系。生物量增加呈 S 曲线,第 4 – 6 天增长最快。棘托竹荪菌丝生物量的增加随 CMC 酶、淀粉酶的活性增强而增加,在菌丝生物量达到最大值之前,羧甲基纤维素酶和淀粉酶活性出现峰值。过氧化物酶、漆酶、多酚氧化酶和蛋白酶的活性与菌丝生物量同步增加。这说明发酵过程中棘托竹荪菌丝首先分解利用淀粉和纤维素,CMC 酶和淀粉酶的大量分泌,将纤维素和淀粉分解为葡萄糖,为菌丝快速生长提供足够的碳源。当培养第 5 – 6 天,淀粉酶和 CMC 酶活性下降时,漆酶、过氧化物酶、多酚氧化酶和蛋白酶大量形成,棘托竹荪开始利用液体培养基中的木质素和蛋白质作为碳源和氮源。因此,在棘托竹荪深层发酵过程中,要提高棘托竹荪菌丝生物量,就必须在液体培养的第 4 – 6 天菌丝达到最大生物量之前保证碳源和氮源的均衡供给。这样才能提高深层液体发酵效率,缩短培养周期。还可以根据不同酶的分泌高峰期,确定菌丝的营养利用情况和发酵周期,以收获最大菌丝生物量。

4 讨论

本试验在棘托竹荪深层培养过程中,虽未在液体培养基中加入纤维素、木质素和蛋白质,但均能测出胞外羧甲基纤维素酶、蛋白酶、漆酶、多酚氧化酶、过氧化物酶的活性,这可能是由于液体培养基中添加玉米粉的缘故。

在本试验中,培养到第 7 天时生物量仍持续增加,这可能是由于深层发酵用培养基接种的竹屑菌种分散性好,可以多点萌发,形成的菌球体积小,数量多,众多的菌丝球能增加外围生长区域体积,充分的利用培养液中的营养物质,获得更大的菌丝生长量,且菌丝球内部不易积累过多的代谢产物或氧浓度过低而自溶。棘托竹荪深层发酵的最大效率期与发酵终点有待进一步研究。

发酵液的 PH 值随培养天数的增加而降低,菌丝干重增加减缓,这与竹荪菌丝体在培养过程中酸性代谢产物积累,发酵液中营养物质减少,PH 值下降,胞外酶活性下降,培养环境恶化使菌丝体自身异化作用增强有关。

胞外酶是菌丝体在发酵过程分泌到液体培养基中的活性物质,它的变化一定程度上可以表明菌丝体分解利用营养物质的能力大小及其生长状况。羧甲基纤维素酶、淀粉酶、漆酶、多酚氧化酶、过氧化物酶的活性变化反应棘托竹荪菌丝体对碳源的利用状况。蛋白酶活性的变化可以反应菌丝体对氮源的利用状况。这些酶的活性与棘托竹荪菌株深层发酵的高产性能有关。

本次试验初步研究棘托竹荪深层培养胞外酶的活性,对了解其胞外羧甲基纤维素酶、淀粉酶、蛋白酶、漆酶、多酚氧化酶、过氧化物酶分泌特点、活性大小及动态变化趋势,推测棘托竹荪在液体培养不同生长发育阶段对培养基中天然木质素、纤维素、淀粉等大分子营养成分的降解动态,提高深层培养效率有重要的意义。对于液体培养基中的营养物质组成、温度、PH 值等环境条件对棘托竹荪的胞外酶活性的影响有待进一步研究。

参考文献

[1]袁德培. 竹荪的研究进展. 湖北民族学院学报. 医学版[J],2006,4(23):39.

[2]刘祖洞,罗信昌. 食用蕈菌生物技术及应用[M]. 北京:清华大学出版社. 2002.

[3]卢惠妮,潘迎捷,赵勇,等. 长裙竹荪和棘托竹荪碳源、氮源、无机盐的筛选[J]. 中药材,2010,33(1):10 – 12.

[4]杜昱光,卜宗式. 竹荪深层培养的理化因子研究[J]. 中国食用菌,1999,15(2):37 – 39.

[5]檀东飞,吴若菁. 棘托竹荪菌丝深层培养的研究[J]. 福建师范大学学报,2002,18(1):63 – 66.

[6]暴增海,周超,夏振强,等. 棘托竹荪发酵液的抑菌作用研究[J]. 北方园艺,2010(3):171 – 173.

[7]王玉万,王云,构菌栽培过程中对木质纤维素的降解和几种多糖分解酶的活性变化[J]. 微生物学通报,1989,16(3):137 – 189.

[8]张凤芹. 不同糖类对鸡腿菇胞外酶活性及多糖分泌的影响[J]. 食用菌. 1999,21(4):9.

[9]Mandels M,Hontz L,Nystrom J et al. Enzymatic hydrolysis of waste cellulase[J]. Biotechnol Bioeng,1974,16:1471 – 1493.

[10]张志良主编. 植物生理实验指导(第三版)[M]. 北京:高等教育出版社. 2003.

[11]李忠光,龚明. 植物多酚氧化酶活性测定方法的改进[J]. 云南师范大学学报,2005,25(1):44 – 49.

[12]潘迎捷,陈明杰,郑海歌. 香菇和平菇生长发育中漆酶、酪氨酸酶和纤维素酶活性的变化[J]. 上海农业学报,1991,7(2):21 – 26.

[13]Lee,S. Y. Characterization and production of antitumor polysacchari deform besi diomycetes[J]. Biotech. News,2005,(3):399 – 403.

注:本文曾发表在《北方园艺》2011 年第 9 期

苹果霉心病病原研究

呼丽萍　马春红　杨光明　谭维军*

1989—1993年,在甘肃苹果产区采集红星或元帅品种病果和无症状果实,逐果取果心组织分离,获得30余个真菌分离物,以链格孢(*Alter naria alternata*)出现率最高,占51.3%;其次为粉红单端孢(*Trichothecium roseum*)、棒盘孢(*Corynewn* sp.)、节孢状镰刀菌(*Fusarium arthrosporioides*)、狭截盘多毛孢(*Truncatella angustata*)等,依次占11.8%、12.0%、8.9%、7.9%。每个单果大多是只能分离出一种菌,少数出现2—3种菌。不同症状的果实,出现的真菌种类不同,霉心果以链格孢出现率最高,占60%—80%;心腐果中链格饱出现率显著较少,占10%—30%;而粉红单端孢、节孢状镰刀菌、棒盘抱、狭截盘多抱等种菌的出现率较高,占7.8%—25.5%。用果心不带菌的果实人工接种,对12种分离物进行致病性测定看出,不同菌株之间致病力有明显差异,致心腐的病菌主要有5种,分别是粉红单端孢、棒盘孢、节孢状镰刀菌、狭截盘多毛孢和一种不产孢的浅色丝状菌。链格孢主要致霉心症状。混合接种试验表明,如有几种真菌进入果心,则致病性强的一种首先占据优势,使其它真菌在再分离中不易出现。

苹果霉心病是元帅系苹果的重要病害,导致果实心室生霉或心室以外果肉腐烂,呈现霉心和心腐两种类型的症状。此病在苹果产区均有不同程度发生,在甘肃陇东南元帅系苹果集中产区,病果率一般为40%—50%,重的达70%,苹果贮运期因果心腐烂所致的经济损失常年在10%上下,严重年份可达30%。

苹果霉心病是由多种病原真菌侵染引起的病害,据前人研究,曾从病果果心分离出20多属真菌[1-5],其中以链格孢[*Alternaria alternata*(Fr.)Keissl.]出现率最高,大多认为链格孢和粉红单端孢[*Trichothecium roseum*(Bull.)Link]是主要致

* 作者简介:呼丽萍(1962—　),女,甘肃通渭人,现为天水师范学院生物工程与技术学院研究员,学士,主要从事果树栽培及病虫害研究。

病菌[1-7]。为了明确本地区主要病原种类,有效地进行防治,我们于1987—1993年进行了系统研究,本文报道病原研究的结果。

1 材料和方法

1.1 病原菌分离与鉴定

1989—1993年,从甘肃各地采集贮藏期的红星和元帅果实,用70%酒精擦洗表面3遍,无菌操作,剖果,取每个单果的心室壁的全部组织或部分组织,对心腐病取其病健交界处果肉组织培养。逐果分离,记录症状类型,病情级别。PSA平板,PSA均按常用量加入乳酸,25℃培养7d后,检查菌落出现情况,纯化,编号,登记,鉴定。

1.2 致病性测定

1.2.1 单一分离物接种试验　取心室不带菌的国光或金冠果实,用70%酒精擦洗,以无菌注射器将待测菌株的孢子悬液(10×16,每视野20－30个孢子),不产孢的菌株则用菌丝段悬液,从果实胴部注射接种到心室中,以注射无菌水为对照。在20℃室温条件下,经15d后剖果检查发病情况,并进行再分离。病情分5级,大致分为霉心和心腐两种类型。

1.2.2 不同菌株混合或先后接种试验　选致病力不同或相近的致心腐或致霉心的菌株按相同或不同比例两两组合,混合接种或先后接种。

2 结果

2.1 果心区真菌种类及出现频率

前后5年从甘肃省15个县(区)采集元帅系果实共2757个(其中心腐型病果838个、霉心型病果762个,无症状健果1157个),逐果分离,共获得30余个分离物,已初步鉴定出了其中15个,分别是链格孢[*Alternaria alternata*(Fr.)Keissl.]、粉红单端孢[*Trichothecium roseum*(Bull.)Link]、棒盘孢(*Coryneum* sp.)、狭截盘多毛孢[*Truncatella angustata*(*pers. ex LK*) *Hughes*]、节孢状镰刀菌(*Fusarium arthrosporioides sherb.*)、串珠镰刀菌(*F. moniliforme* Sheld.)、砖红镰刀菌(*F. lateritium* Nees)、头孢霉(*Cephalosporium* sp.)、枝孢霉(*Cladosporium* sp.)、蒂地顿壳霉(*Coniothyrium tirolensis* Bub.)、仁果盾壳霉[C. pirinum(Sacc.)Sheld.]、青霉(Penicillium sp.)、拟青霉(Paecilomuces sp.),还有一些不产孢的丝状菌。分离出现率以链格孢为最高,占51.3%;其次为棒盘孢、分红单端孢、节孢状镰刀菌、狭截盘多毛孢和一种浅色丝状菌,出现率分别为12.0%、11.8%、8.9%、7.9%和6.9%。各地均以链格孢占绝对优势,其它菌的出现率在不同年份及地点有一定变动。

不同症状类型的果实,分离出的真菌种类相差不多,但各种真菌所占比率有明显差异。以1992年分离结果(表1)为例,在健果中,链格孢占60.7%;霉心果中也是以链格孢出现率为最高,达79.9%;其它各类菌的出现率为0—8.1%;在心腐果中,链格孢的出现率为29.4%,而粉红单端孢、棒盘孢、盘多毛孢、镰刀菌等5种菌的出现率明显增高,分别为25.5%、11.8%、7.8%、13.7%、7.8%。说明症状类型与病原种类有密切关系。

从病果、健果群体中,分离出的真菌有多种,但从每个单果中,则大多是只能分离出一种菌,少数出现2—3种菌。从单个健果中分离出2、3种菌的较多,而从病果中,绝大多数只能分离出一种菌。由此可见,在健果心室内起初可能有2—3种菌定殖,但最终扩展致病多是其中一种菌,如果某一种菌占居了优势,另外的菌就不易分离出来。

表1 健果、霉心果、心腐果果心区带菌情况(1992)

分离材料	分离果数(个)	果实带菌率(%)	分离组织块数(块)	组织块带病率(%)	几种主要真菌的出现频率(%)							一果中出现两种真菌(%)	一果中出现三种真菌(%)
					链格孢	单端孢	棒盘孢	盘多毛孢	镰刀菌	SP* -3	其他		
健果	95	88.4	4572	21.4	60.7	1.2	3.6	10.7	16.7	10.7	54.8	32.1	26.1
霉心果	134	100.0	6532	88.4	79.9	0.0	4.5	1.5	8.1	2.2	7.5	3.7	0.0
心腐果	51	100.0	2433	98.1	29.4	25.5	11.8	7.8	13.7	7.8	5.9	2.0	0.0

注:1 每个果实均取整个心室壁组织块培养。2. 出现率(%)= 出现某种菌或两种菌的果数/带菌果数×100。

*一种浅色丝状菌

2.2 致病性测定

2.2.1 单一菌株接种试验 供试12个菌株,发病情况和症状类型从表2可以看出,不同菌株致病力强弱有明显差异,致心腐的病菌主要有5种,分别是粉红单端孢、节孢状镰刀菌、狭截盘多毛孢和一种浅色丝状菌(SP-3),其中以粉红单端孢致病力最强;棒盘孢接种大多出现心腐,少数出现霉心;链格孢接种大多呈现霉心,少数出现心腐;其他呈现不同程度的霉心症状。

表 2　果心真菌致病力测定结果(1992)

测定菌种(或代号)	接种果数(水)	发病			再分离*	
		发病果数	霉心果率(%)	心腐果率(%)	分离果数	接种菌出现率(%)
粉红单端孢	24	24	0.0	100.0	24	100.0
狭截盘多毛孢	36	36	0.0	100.0	36	100.0
节孢状镰刀菌	20	20	0.0	100.0	20	100.0
棒盘孢	20	27	15.0	85.0	20	85.0
蒂地盾壳霉	20	20	100.0	0.0	20	80.0
链格孢	30	30	86.7	13.3	30	100.0
仁果盾壳霉	20	20	100.0	0.0	20	80.0
头孢霉	20	20	100.0	0.0	20	100.0
头孢霉	20	20	100.0	0.0	20	95.0
拟青霉	20	20	100.0	0.0	20	100.0
串珠镰刀菌	20	20	100.0	0.0	20	95.0
SP-3	30	30	0.0	100.0	30	100.0
无菌水(CK)	38	0	0.0	0.0	38	0

*心腐果取病健交界处果肉培养,霉心果取心室壁组织,0.25% 次氯酸钠液灭菌 20min 培养。

接种菌出现率(%)=接种菌出现果数/分离果数×100。

表 3　不同菌种混合接种试验结果(1992)

处理组合		接种果数	发病			再分离	
			霉心果率(%)	心腐果率(%)	病情指数	第一种菌出现率(%)	第二种菌出现率(%)
单接	粉红单端孢	24	0.0	100.0	87.5	100.0	—
	狭截盘多毛孢	25	0.0	100.0	91.3	100.0	—
	节孢状镰刀菌	20	0.0	100.0	66.0	90.0	—
	SP-3	19	0.0	100.0	63.1	100.0	—
	棒盘孢	18	40.0	60.0	52.0	85.0	—
	链格孢	24	87.5	12.5	38.3	100.0	—
	蒂地盾壳霉	20	100.0	0.0	22.7	81.7	—
	头孢霉	15	100.0	0.0	21.3	100.0	—

续表

处理组合		接种果数	发病			再分离	
			霉心果率(%)	心腐果率(%)	病情指数	第一种菌出现率(%)	第二种菌出现率(%)
心腐(菌)+心腐(菌)	粉红单端孢 + 狭截盘多毛孢	14	0.0	100.0	91.4	100.0	0.0
	粉红单端孢 + 节孢状镰刀菌	13	0.0	100.0	67.7	0.0	100.0
	粉红单端孢 + SP - 3	14	0.0	100.0	81.4	100.0	0.0
	粉红单端孢 + 棒盘孢	10	0.0	100.0	82.2	100.0	0.0
	狭截盘多毛孢 + 节孢状镰刀菌	10	30.0	70.0	62.0	0.0	100.0
	狭截盘多毛孢 + SP - 3	10	0.0	100.0	82.0	100.0	0.0
	狭截盘多毛孢 + 棒盘孢	10	60.0	40.0	42.0	100.0	0.0
	节孢状镰刀菌 + SP - 3	10	0.0	100.0	62.0	100.0	0.0
	节孢状镰刀菌 + 棒盘孢	10	40.0	60.0	52.0	100.0	0.0
	SP - 3 + 棒盘孢	10	50.0	50.0	56.0	100.0	0.0
霉心(菌)+心腐(菌)	链格孢 + 粉红单端孢	14	0.0	100.0	81.6	0.0	100.0
	链格孢 + 狭截盘多毛孢	13	0.0	100.0	89.2	0.0	100.0
	链格孢 + 节孢状镰刀菌	13	15.4	84.6	46.2	0.0	100.0
	链格孢 + SP - 3	13	7.7	92.3	56.9	100.0	0.0
	链格孢 + 棒盘孢	13	15.4	84.6	52.3	100.0	0.0
霉心+霉心	链格孢 + 蒂地盾壳霉	10	100.0	0.0	22.0	30.0	0.0
	链格孢 + 头孢霉	10	100.0	0.0	24.0	0.0	50.0

2.2.2 不同菌株混合或先后接种试验 将两种致病力相同或不同的菌株按相同比例混合接种后，发病情况与前项菌株单独接种的无明显差异，发病后从病组织中只能分离到其中一种菌，一般为致病力强的一种（表3）。将两种致病力不同或相同的菌株按不同比例混合接种后，仍表现致病力较强的一种首先占居优势并致病（表4）。将两种致病力不同的菌株按不同间隔时间先后接种，致病力强的病菌占居优势后，可完全抑制致病力弱的病菌扩展；而致病力弱的病菌先占居优势后，对致病力强的病菌的扩展仅有一定的抑制作用（表5）。

表 4　不同接种量与致病的关系(1992)

处理组合		接种比例	发病			再分离(%)			
			接种果数	霉心率	心腐率	心室		果肉	
			个	%	%	第一种菌	第二种菌	第一种菌	第二种菌
心腐+心腐	粉红单端孢与不等量的狭截盘多毛孢	1:08	10	0.0	100.0	60.0	40.0	100.0	0.0
		1:16	10	0.0	10.0	30.0	70.0	80.0	20.0
心腐+霉心	粉红单端孢与不等量的链格孢	1:02	20	0.0	100.0	—	—	100.0	0.0
		1:04	20	0.0	100.0	—	—	100.0	0.0
		1:08	20	0.0	100.0	40.0	60.0	70.0	30.0
		1:16	20	0.0	100.0	30.0	70.0	60.0	40.0
霉心+霉心	链格孢与不等量的头孢霉	1:02	20	100.0	0.0	50.0	50.0	—	—
		1:04	20	100.0	0.0	25.0	75.0	—	—
		1:08	20	100.0	0.0	15.0	85.0	—	—
		1:16	20	100.0	0.0	5.0	95.0	—	—

表 5　病菌进入心室先后与致病的关系(1992)

处理	接种间隔天数(d)	发病		再分离			
		接种果数(个)	心腐率(%)	心室		果肉	
				链格孢	单端孢	链格孢	单端孢
先接种链格孢	2	20	100.0	90.0	10.0	60.0	40.0
后接种单端孢	4	20	100.0	100.0	0.0	90.0	10.0
	8	20	100.0	100.0	0.0	100.0	0.0
先接种单端孢	2	20	100.0	0.0	100.0	0.0	100.0
后接种链格孢	4	20	100.0	0.0	100.0	0.0	100.0
	8	20	100.0	0.0	100.0	0.0	100.0

3　结论与讨论

苹果霉心病是多种非专化性弱寄生真菌侵染所致的病害,具潜伏侵染特征,元帅系品种果实普遍带菌。在研究病原中,前人曾从苹果果心中分离出 20 多个属的真菌,多是从分离物的出现频率来认定病原,未做深入研究。有人进行过花期接种试验,但由于试验材料、方法上的困难,也无法作出确切结论。本项研究对呈现不同症状类型的各个单果逐果进行病菌分离，对照症状特征，检查病原物出现种类；选用心室不带菌果实为接种试材，改变接种方法，遵循柯赫氏法则进行致病性测定，明确了苹果霉心病的主要致病菌种类及其致病特点。

研究结果表明，元帅系苹果的果心普遍带菌，5 年间从甘肃苹果果心中共获得 30 余个分离物。无病健康果实心室的带菌率高达 80% 以上，一半左右可以分离出 1 种真菌，另一半左右可能出现 2—3 种。病果中绝大多数只能分离到一种菌，分离出两种真菌的仅占 2.0%—3.7%，心腐果的腐烂果肉组织，则只能分离出一种真菌。霉心果心室带菌以链格孢所占比率最高，达 60%—80%，而心腐果心室内的病菌，粉红单端孢、棒盘孢、镰刀菌等病菌的出现率明显增多。因年份及地区的不同，病果中优势菌种类及出现频率不尽相同。

接种试验结果表明，从果心区分离出的各个菌株致病力有明显差异，粉红单端孢、节孢状镰刀菌、狭截盘多毛孢、棒盘孢和一种浅色丝状菌主要致心病，链格孢主要致霉心，是此病的主要致病菌；头孢霉、盾壳霉等菌株仅致轻微霉心症状或不致病，不是此病的主要致病菌。这一结果和病原物分离的出现频率表现一致。将两种致病菌同时混合接种或先后接种于心室后，最终引起发病的是其中一种，即致病力强的一种致病。关于各种真菌之间的关系及寄主组织上真菌种群的演替有待继续研究。

参考文献

[1] Ellis MA, Barrat JG. Colonization of Delicious apple fruits by *Alternaria* sp. and effect of fungicidal sprays on mouldy core. Plant Disease, 1983, 67: 150 – 152.

[2] 陈延熙，梅汝鸿，赫玉强. 苹果霉心病的发生与防治[J]. 植物保护, 1984(4): 9 – 11.

[3] 高桥俊作(何国重译). 苹果霉心病发生原因及其防治方法[J]. 天津农业科学, 1980 (1): 47 – 51.

[4] Combrink JC, Kotze JM, Wehner FC. Fungi associated with core of Starking apples in South Africa[J]. Phytophylactica, 1985, 17(2): 81 – 83.

[5] 冷怀琼，刘勇. 苹果霉心病的病原四川果品, 1989(4): 9 – 11.

[6] 魏康年，韩金声，李红等. 苹果霉心病侵染规律研究[J]. 中国果树, 1980(3): 32 – 34.

[7] 陈策. 苹果霉心病的发生与防治[J]. 山西果树, 1990, (3): 11 – 12.

注：本文曾在 1996 年 3 卷 13 期果树科学期刊上

羊传染性脓疱病毒42K囊膜蛋白基因克隆及表达

王廷璞　赵菲佚　安建平　孙春香　党　岩*

用特异性引物扩增羊传染性脓疱病毒42K囊膜蛋白基因,与不同的表达载体重组后用DIG标记的特异性DNA探针进行Dot－blotting和Southern blotting杂交检测,挑选出能稳定传代的重组菌9株。经IPTG诱导后,用光敏生物素标记抗囊膜蛋白IgG进行Western blotting分析,检测到能高水平表达的重组菌pGEL－4A,每升培养液中所表达的病毒囊膜蛋白达560mg。

羊传染性脓疱病毒也称Orf病毒(Contagious Ecthyma,Orf Virus,CE),属痘病毒科,副痘病毒属[1],该病毒主要引起羊和人的口疮,对羔羊危害极为严重,本病在世界各国均有流行。Orf病毒属高度嗜上皮毒,在血液中产生的抗体水平低下,用常规方法均无法检测,主要以细胞免疫和局部免疫为主[2,3]。由于本病在免疫上的特殊性使得对本病的诊断和免疫还没有完整的,特异的方法和良好的疫苗。用SDS和NP40裂解纯化的病毒后,可分离出18－33个病毒囊膜亚单位[7],经western－blotting分析,其中的54、42、39、37、35KU等囊膜亚单位是主要的保护性抗原成分,其中42KU的保护性较强,可刺激家兔产生高滴度的IgG。从新西兰NZ2毒株和我国HCE毒株中提取的基因组DNA中,用BamHⅠ和HindⅢ消化的A片断中含有42KU的囊膜蛋白基因。我们将Orf病毒HCE毒株的主要抗原42K囊膜蛋白(42Kep)基因进行了重组克隆,得到了较好的表达,为基因工程苗的研制奠定了基础,报道于下。

---

* 作者简介:王廷璞(1965—　),男,甘肃秦安人,现为天水师范学院生物工程与技术学院研究员,学士,主要从事病毒分子生物学,细胞生物学、中草药有效成份提取及鉴定、天然药物开发等方面的研究。

1 材料和方法

1.1 毒株

HCE 毒株:甘肃肃南皇城毒株,用其痂块毒进行病毒 DNA 的提取与纯化,供本研究用。

1.2 PCR 引物[4]:

根据文献[4]所提供的新西兰 NZ2 毒株的参考序列,合成 42Kep 基因特异性引物,由华美生物工程公司合成。5′端:5′ AAGTCGACTATGGATGAAAAT 3′,3′端:5′ CCTGCAGGATCCTATCTAATATCTGCTGTA 3′。

1.3 药品、试剂

核酸内切酶、T4DNA 连接酶,核酸分子量标准品、蛋白分子量标准品、光敏生物素标记试剂盒、硝酸纤维素膜(nitrocelluose Membrane)、PCR 扩增试剂盒均购自华美生物工程公司。地高辛标记试剂盒为 Boehringer Mannheim 产品,丙烯酰胺、BCIP、X－gal、IPTG 为 Sigma 产品。蛋白酶 K,溶菌酶,2 －巯基乙醇(2－ME)为 MERCK 公司产品,琼脂糖 4B(Sepharose 4B)为 pharmacia 产品,其他试剂均为国产或进口分析纯试剂。光敏生物素标记抗囊膜蛋白 IgG(1.55mg/ml)、DIG 标记 DNA 探针(0.24ug/ul),均由本室制备[5,6]。

1.4 菌株、质粒

pGEM－3Z、pGEM－4Z、pGEMEX－1、JM109、JM109(DE3)均购自华美生物工程公司。

1.5 病毒 DNA 之提取与纯化

按参考文献[7]进行。

1.6 病毒 DNA 之酶切和片断回收

取纯化的病毒 DNA 7ug,用 Hind Ⅲ进行完全消化,在 20ul 反应体系中,于 37℃保温 30min 后,加入 2ul 0.5M EDTA 终止反应,于 0.8% 琼脂糖凝胶中 30V 电泳 18h,紫外检测仪检测结果,并用离心法回收 A 片断[8],用 TE 稀释至 0.12ug/ul 作模板。

1.7 PCR:

扩增 A 片断中 42Kep 基因:取 10 × Reaction Buffer ($MgCl_2$ － Free)5ul,$MgCl_2$(25mm)3ul,HCE DNA Hind Ⅲ/A 片断(0.12 ug/ul,Primer1(12.5pmoles/ul)2ul,Primer2 (12.5pmoles/ul)2ul,Taq DNA polymerase 1ul,4 × dNTP Mixture(each 2.5mM)4ul,Nuclease－Free Water 29ul。将反应组分混合后,加入 50ul 石蜡油,离心数秒。按下列参数进行 PCR 扩增:预变性 94℃ 10min;变性 94℃ 60S,退火 55℃

45S,延伸72℃ 60S,30个循环后,于72℃ 5min。于0.8%琼脂糖凝胶中30V电泳,回收扩增片断。

1.8 42Kep基因的重组克隆

按参考文献[9]方法,取HindⅢ酶切质粒DNA 6ul,加入扩增DNA 10ul,指弹混匀。置45℃ 5 min后冷却至0℃退火。加入T4 DNA Ligase 1ul,10×连接buffer 2ul 5mmol/lATP 1ul指弹混匀。16℃4h后4 ℃ 过夜或于15℃水浴过夜。用等量酚:氯仿抽提,乙醇沉淀后溶于10ul TE中转化感受态大肠杆菌。

1.9 重组DNA转化、筛选和重组菌质粒DNA之快速提取

用$CaCl_2$法制备感受态大肠杆菌。每管0.3ml感受态细胞中加1ul质粒连接物,混匀,4℃冰浴30min,移至37℃ 150r/min,摇振培养1h,吸取培养物0.1ml接种选择培养基(含Amp和X-gal与IPTG),同时作未转化的细菌作为对照,置37℃培养24h。挑选白色菌落,接种含Amp的LB平板传代,能够传下去的重组菌株再接种含Amp,X-gal与IPTG的选择培养基应为白色,不应出现蓝色菌落。按参考文献[10]方法提取重组质粒DNA。

1.10 重组质粒DNA之酶切分析与杂交检测

将快速提取的质粒DNA用前述0.8%的琼脂糖凝胶30V电泳8-16h,紫外灯下切取所需条带,置核酸回收管内,离心回收,酚氯仿抽提后,无水乙醇沉淀,15 000r/min离心10min沉淀质粒DNA,沥干液体,溶于适量TE中。按前述方法用相同的酶对重组DNA进行切割,同标准分子量一起进行琼脂糖凝胶电泳,检测所重组的条带,并按文献[5]方法进行Dot-blotting和Southern blotting杂交检测。

1.11 42Kep基因的诱导及重组菌的SDS-PAGE

取单克隆重组菌PGRL-4A 37℃培养过夜,以1%转接于20ml含100ug/ml Amp的LB培养液中,150r/min培养至吸光度A_{600}为0.4-0.6,加入IPTG至终浓度0.4nmol/L,继续诱导培养3小时,收集菌液,10 000r/min离心30s,沉淀菌体,吸出上层培养液,用0.5ml预冷的50mM/L Tris·Cl重悬,同上离心30s回收菌体,弃上清,加入25ul水振荡重悬,立即加入25ul 2×SDS凝胶加样缓冲液,继续振荡20s,超声波乳化后以10 000r/min离心10min,取上清,弃沉淀物,取样品25ul进行SDS-PAGE[7]。

1.12 转移电泳与Western blotting检测[6]

SDS-PAGE后,在转移电泳槽中,80mA电泳6h,将样品从凝胶转移到硝酸纤维素滤膜,进行Western blotting杂交。同时,凝胶用考马斯亮兰R250染色6h后,用含25%甲醇和10%冰乙酸的洗脱液脱色至背景为白色。硝酸纤维素滤膜用PBS洗涤3×15min后,加5%牛血清白蛋白封闭液20ml,4℃封闭过夜。加入杂交

液（犊牛血清 2ml，PBS 20ml，光敏生物素标记 IgG 1ml < 1.55mg/ml > Tween20 5ul）于 4℃杂交 2h，用 PBS 洗涤 3×10min，将膜移入新袋中，加入新配的 AV－AP 液（AV－AP 15uL，小牛血清 2ml PBS 18ml Tween－20 5ul）于室温作用 15min。取出膜用 PBS 洗涤 3×15min，加显色液，于暗室中显色 1.5h 至条带清晰后用 TE 缓冲液洗涤中止反应，保存于塑料袋内。

2 结果

2.1 羊传染性脓疱病毒 42Kep 基因的重组克隆

用特异性引物扩增 Orf 病毒 42Kep 基因结果见图 1。以 HCE HindⅢ/A 片断为模板，PCR 扩增出病毒 DNA 42Kep 基因，经电泳分离后，用离心法回收，紫外测定含量为 0.12ug/ul。将此片断与 HindⅢ切割的 pGEM－3Z、pGEM－4Z、pGEMEX－1 表达载体进行连接重组后，转化感受态大肠杆菌 JM109 和 JM109（DE_3），共挑选出重组菌 56 个克隆株，筛选出能稳定传代的 12 个克隆株，对这 12 株重组菌进行质粒的快速提取，用 HindⅢ消化后，与扩增片断同时电泳，检测到 9 株重组正确的菌株，这 9 株细菌中，重组到 pGEM－4Z 中的有 3 株，pGEM－3Z 中的有 5 株，pGEMEX－1 中的有 1 株。对这些重组菌株用特异性 DIG 标记的 42Kep DNA 探针进行 Dot－blotting 和 Southern blotting 杂交检测。均显示极强的阳性，见图 2。

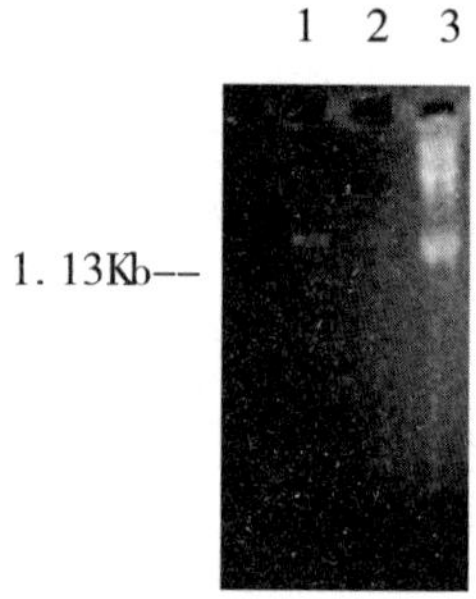

图 1：羊传染性脓疱病毒 42Kep 基因扩增结果 1，2 为扩增的片断，3 为 λHindⅢ

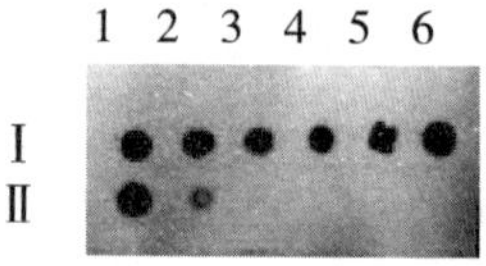

图 2：DIG 标记 DNA 探针检测重组菌结果

2.2 重组 42Kep 基因的表达检测

将所筛选的各克隆菌株，用 IPTG 诱导后裂解，提取融合蛋白，与病毒裂解液一起进行 SDS－PAGE，电泳后，用考马斯亮蓝 R－250 染色，结果见图 3。同时，进行转移电泳，用光敏生物素标记的特异性 IgG 进行 Western－blotting 杂交分析。结果显示，在重组菌中，pGRL－4A 的表达较强，薄层扫描分析，每升培养液中所表达的病毒囊膜蛋白达560mg,pGRH－3A的表达较弱,其它重组菌表达很少或不表达。在表达的重组菌中,pGRL－4A传代稳定,重复性好,表达产物较高,是筛选出的表达较好的重组菌。

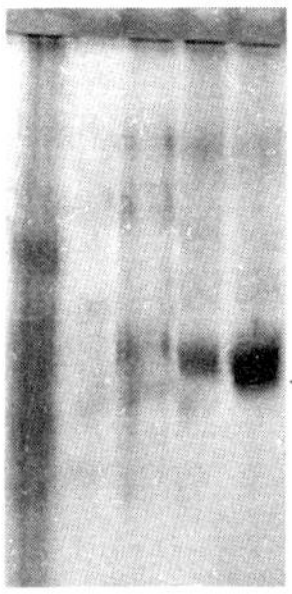

图3:重组菌经IPTG诱导后的SDS—PAGE

1 病毒裂解物 2. 低分子量蛋白标准品

3. PGRH—4A 4. PGRH—3A 5. PGRL—4A

Fig2:Dig labeling DNA probe detected recombinant bacteria.

1	2	3	4	5	6	
Ⅰ	PGRL—4A	PGRL—3A	PGRL—4A	PGRL—3C	PGRL—4D	HCE
Ⅱ	PGEMK—A	PGRL—4C	鱼精DNA			

3 讨论

羊传染性脓疱病毒是引起人和动物传染性接触性皮炎的病毒,传播迅速,传染范围广,危害较为严重,特别对养羊业造成的损失尤为严重。而且,该病毒具有高度的嗜上皮特性,其所介导的是细胞免疫,体液免疫水平低下,并且主要以局部免疫为主。羊传染性脓疱病毒在免疫上的这些特性,使得对其所致疾病的诊断和免疫还没有完整的特异的方法和良好的疫苗。目前,世界各国特别是养羊业发达的国家,用分子生物学的方法进行疾病诊断,并加紧进行病毒分子生物学研究和基因工程苗的研究工作[11－18]。在基因工程苗的研究工作中,主要以病毒囊膜蛋白亚单位中的42KU、39KU、37KU和35KU的基因作为保护性抗原进行基因重组,但由于这些囊膜蛋白所介导的免疫反应较弱,一些科学家正在试图建立以Orf病毒为载体的重组病毒疫苗,如在Orf病毒中引入了细粒棘球绦虫的抗原基因,以求获得对Orf病毒和对细粒棘球绦虫的双重保护[19]。我们将本病毒的主要抗原基因42K囊膜蛋白基因通过PCR扩增,重组到表达载体pGEM-4Z中,筛选出具有较高表达水平且能稳定传代的重组菌pGRL-4A,为基因工程苗的研制奠定了基础。

参考文献

[1]殷震,刘景华主编.动物病毒学[M].科学出版社,1985:786-800.

[2]王廷璞,薛双虎,吕武夫.羊传染性脓疱病毒的免疫特性[J],中国兽医科技,1990(2):15-17.

[3]Torfason,EG;Gunadottir,S. Polymerase chain reaction for laboratory diagnosis of orf virus infections[J]. Journal of Clinical Virology,2002,vol. 24,no. ER1-2,79-84.

[4]John T. Sullivan,Andrew A. Mercer,Stephen B. Fleming,and Anthony J Robinson. Identification and characterization of an Orf Virus Homologue of the Vaccinia Virus Gene Encoding the Major Envelope Antigen P37K [J]. Virology. 1994,202:968-973.

[5]王廷璞,赵菲佚,孙春香等.光敏生物素和DIG标记Orf病毒DNA探针的研究[J].生物技术2002,12(6):21-23.

[6]王廷璞,赵菲佚,党岩等.光敏生物素标记IgG检测羊接触传染性脓疱性皮炎病毒囊膜蛋白及重组表达蛋白[J].中国兽医科技2003,33(5):28-30.

[7]王廷璞,吕武夫,薛双虎等.羊传染性脓疱性皮炎病毒核酸及蛋白类型分析[J].中国兽医学报1998,18(4)328-333.

[8]王廷璞,吕武夫,薛双虎等.一种新的DNA回收方法[J].中国兽医学报1996,16(3):284.

[9]Andrew A. Mercer,Kate Fraser,Gillian Barns,and Anthony J. Robinson. The structure and cloning of orf virus DNA[J]. Virology 1987,157:1-12.

[10]Maxur. C;Rangel Fiho. F B;Galler. R,Molecular analysis of contagious pustular dermatitis virus:a simplified methed for viral DNA extraction from scab material[J]. Journal of virological Methods. 1991,35(3)265-272.

[11]Anderson,IE;Reid,HW;Nettleton,PF et al. Detection of cellular cytokine mRNA expression during orf virus infection in sheep:differential interferon-g mRNA expression by cells in primary versus reinfection skin lesions[J]. Veterinary Immunology and Immunopathology,2001,vol. 83,no. ER3-4,161-176.

[12]McInnes,CJ;Wood,AR;Nettleton,PF;Gilray,JA. Genomic Comparison of an Avirulent Strain of Orf Virus with that of a Virulent Wild Type Isolate Reveals that the Orf Virus G2L Gene is Non-Essential for Replication[J]. Virus Genes,2001,vol. 22,(2)141-150.

[13]Mazur,C;Ferreira,II;Galler,R. Molecular characterization of Brazilian isolates of orf virus[J]. Veterinary Microbiology,2000,vol. 73,(4)253.

[14] Haig, DM; Fleming, S. Immunomodulation by virulence proteins of the parapoxvirus orf virus[J]. Veterinary immunology and immunopathology, 1999, vol. 72, (1/2) 81.

[15] Housawi, FMT; Roberts, GM; Mercer, AA. The reactivity of monoclonal antibodies against orf virus with other parapoxviruses and the identification of a 39 kDa immunodominant protein[J]. Archives of virology, 1998, vol. 143, (12) 2289.

[16] McInnes, CJ; Wood, AR; Mercer, AA. Orf Virus Encodes a Homolog of the Vaccinia Virus Interferon – Resistance Gene E3L[J]. Virus genes, 1998, vol. 17, (2) 107.

[17] Gilray, JA. Restrictionendonuclease profiles of orf virus isolates from the British Isles[J] The Veterinary record, 1998, vol. 143, (9) 237.

[18] Haig, DM; Thomson, J; McInnes, C; McCaughan et al. Orf virus immuno – modulation and the host immune response[J]. Veterinary Immunology and Immunopathology, 2002, vol. 87, (3 – 4). 395 – 399.

[19] B. J. Marsland, D. J. Tisdall, D. D. Heath, A. A. Mercer. Construction of a recombinant orf virus that expresses an Echinococcus granulosus vaccine antigen from a novel genomic insertion site [J]. Arch Virol. 2003 Mar; 148(3): 555 – 562.

注：该文发表于《中国预防兽医学报》2006 年 1 月第 28 卷第 1 期 29 – 32

镉诱导黄瓜金属硫蛋白抗体的制备及纯化

王廷璞　安建平　邹亚丽　马　腾　张春成*

用镉诱导黄瓜叶片组织产生金属硫蛋白(MT),经分离纯化,血清白蛋白偶联后作为抗原。采用背部皮内、皮下和耳缘静脉注射免疫家兔,制备出效价为1∶16的免疫血清。用饱和硫酸铵盐析法和Sephadex G-200柱层析分离纯化IgG。经测定纯化后的IgG蛋白含量为0.9364 mg/ml。用纯化的IgG建立黄瓜Cd—MT的斑点免疫试验检测方法可检测黄瓜Cd—MT的最低含量为200pg/ml。

随着现代工业化的飞速发展,环境问题 越来越来受人们的普遍关注,尤其重金属污染。如Ag^{+}、Pt^{2+}、Hg^{2+}、Cd^{2+}等的污染。金属硫蛋白[1](Metallothionein,MT)是一类广泛存在于生物界的低分子量蛋白,缺少芳香氨基酸[2]的金属结合蛋白[3]。MT具有参与机体微量元素贮存、运输、代谢、拮抗电离辐射[4]、清除自由基和重金属解毒[5]等多种作用,有着广泛的生物学功能。金属硫蛋白具有与其结合金属相对应的金属光吸收峰[6],不同的金属结合蛋白有不同的光吸收值。如镉金属硫蛋白在254nm处有最大光吸收值[7]。MT主要有三种形式:MT—Ⅰ,MT—Ⅱ,MT—Ⅲ,植物MT大多属于第三类,非基因编码的金属硫肽[8]。

我们以黄瓜为材料,用镉诱导后产生MT,经液氮研磨,Tris-HCl提取,Sephadex G-50纯化,PEG-6000浓缩后作为抗原,用戊二醛与BSA偶联后采用背部皮内,皮下,耳缘静脉多种途径免疫家兔,取得较好效果,在较短时间内制备出抗血清,经饱和硫酸铵盐析法和Sephadex G—200柱层析纯化后建立Dot Immunobinding Assay检测方法,为我们检测植物机体金属硫蛋白奠定了基础,可作为环境重金属污染的检测指标,具有重要的生物学意义。

---

* 作者简介:王廷璞(1965—　),男,甘肃秦安人,现为天水师范学院生物工程与技术学院研究员,学士,主要从事病毒分子生物学,细胞生物学、中草药有效成份提取及鉴定、天然药物开发等方面的研究。

1 材料与方法

1.1 材料与仪器

津杂四号黄瓜种子，购于天水市农业科学研究所。家兔，购于天水市郁友养兔场。

弗氏完全佐剂、弗氏不完全佐剂均购于 Sigma 公司。Tris、SDS 、丙烯酰氨(Acr)、甲叉双丙烯酰氨(Bis)、碱性磷酸酶标记的山羊抗兔 IgG、硝基蓝四氮唑(NBT)、5－溴－4－氯－3 吲哚－磷酸盐(BCIP)、Tween—20 均购于华美生物工程公司。其他均为国产分析纯试剂。SDS—PAGE 凝胶电泳低分子量标准蛋白为中科院上海生物化学研究所产品，882SB—紫外检测系统为实达英创有限公司产品，低温冷冻离心机为上海安亭科学仪器厂制造。

1.2 方法

1.2.1 镉诱导黄瓜金属硫蛋白的制备

取子粒饱满的黄瓜种子，洗去包衣，在 25℃ 水中浸泡 4h 后，用滤纸发芽法使其生根，待根长到 1cm 时，移栽到石英砂基中，每日浇以适量的 Hongland 营养液，在无菌室 24 小时不间断光照，25—28℃ 条件下培养，直到第三片真叶开始长出。将真叶较大的黄瓜苗从石英砂中连根拔出来，在去离子水中将石英砂洗掉，用吸水纸吸掉黄瓜根上的水，然后移栽于盛 有 40ml，1000ppm 氯化镉溶液的小烧杯中，每个烧杯栽三株黄瓜苗，在 25℃ 左右的无菌室内全天候光照下，胁迫 48～72 h。称取胁迫后的黄瓜叶，用液氮研磨，分次加入 10 倍体积预冷的 0.01mol/L Tris－HCl 缓冲液(pH8.0)充分研磨，4℃，5000r/min 离心 10min；取上清液加入等量无水乙醇，4℃ 静置 12h，除去易变性蛋白质，4℃，5000r/min 离心 10min；取上清液于 80℃ 水浴加热 5min，自然冷却至室温后，4℃，5000r/min 离心 10min；取上清液，装透析袋用 PEG－6000 透析浓缩，得到粗提 MT。

1.2.2 MT 的分离纯化

取适量浓缩好的 MT 进行 Sephadex G－50 凝胶过滤层析，层析柱事先用 0.01mol/L Tris－HCl(pH7.4)缓冲液平衡，上样后控制流速 1ml/min，在 254nm 波长处收集最大吸收峰，将收集到的组分浓缩后进行 SDS－PAGE 凝胶电泳，测定分子量。

1.2.3 抗原的偶联

取 7ml 方法 1.2.2 纯化浓缩的 MT(0.3mg/ml)与 2ml BSA(1mg/ml)相混合，在 25℃，磁力搅拌下逐滴加入 1% 的戊二醛 35 ul。然后，再持续搅拌 1h，4℃ 冰箱静置 12h 后，用 0.01mol/L 的 PBS(pH7.4)透析除去未交联的戊二醛。

1.2.4 抗 MT 免疫血清的制备

首次免疫取 5ml 纯化浓缩后的 MT 原液，加等体积的弗氏完全佐剂，用超声波乳化器制成乳剂，于家兔背部脊柱两侧皮下进行多点注射，每点 0.2 - 0.4ml，共 2ml。第二周用 MT 与等体积的弗氏不完全佐剂制成乳剂后，皮内多点注射 1ml，同时用与 BSA 偶联的 MT 静脉注射 0.5ml。第三周仍用 BSA 偶联的 MT，背部脊柱两侧皮内多点免疫注射 2ml。第四周只用戊二醛把 MT 进行分子间偶联，仍为皮内多点注射共 2ml。最后一次免疫后的第六天，在家兔耳缘动脉采血，分离血清。

1.2.5 免疫血清效价检测（免疫琼脂双向扩散法）

用巴比妥缓冲液（BBS pH7.4）溶解琼脂糖，并加入 1% 的 NaCl 制成凝胶按模型打孔。中央孔中加入 1∶8 稀释的抗原 MT，边缘孔加入倍比稀释的免疫血清和阴性血清。在 37℃湿盒中自由扩散 48 ~ 72h，观察结果，确定抗体效价。

1.2.6 抗体 IgG 的分离与纯化

取抗血清 5ml 加入等量的生理盐水，在磁力搅拌下逐滴加入与稀释血清等体积的饱和硫酸铵溶液使得硫酸铵饱和度达到 50%，在 4℃下静置 3h 以上充分沉淀，3000r/min，离心 20min，弃上清 用 10 ml 生理盐水溶解沉淀。在不间断搅拌下，逐滴加入 5ml 的饱和硫酸铵，使得硫酸铵的饱和度达到 33%，4℃静置 3h 以上使充分沉淀，3000 r/min，离心 20min，弃上清。将最终所得的沉淀用 0.02mol / LpH7.4 的 PBS 缓冲液溶解到 5ml，装入透析袋，在 4℃对 PBS 透析，除盐换液 3 次。透析后的 IgG 进行 SephadexG - 200 柱层析，先用 0.02mol/L PBS 缓冲液（pH7.4）平衡层析柱，上样后用 0.02mol/L PBS 缓冲液（pH7.4）洗脱，控制流速 1ml/min，收集 280nm 波长处最大吸收峰的组分，装入透析袋，用 PEG - 6000 浓缩到 5ml。紫外分光法测定 IgG 蛋白含量。

1.2.7 Dot Immunobinding Assay 定量检测

将纯化的 MT 稀释为 100ug/ml、10ug/ml、1ug/ml、100ng/ml 、10ng/ml、1ng/ml、800pg/ml. 400pg/ml、200pg/ml、100pg/ml、50pg/ml 后各取 1ul 点于硝酸纤维素膜上（NC）在 37℃烘烤 30min。将 NC 膜放入杂交袋中，用封闭液在 37^0 C 封闭 30min，再用洗涤液（含 0.5% Tween—20）洗涤 3 次，每次 15min，用吸水纸吸干 NC 膜后装入新的杂交袋。加入按 1∶8 稀释的抗体 IgG 后在 25℃静置 30min 后，洗涤 3 次，每次 15min，吸干 NC 膜放入新的杂交袋。加入碱性磷酸酶标记的山羊抗兔 IgG，在 37℃静置 30min，按上述洗涤后吸干放入新的杂交袋中，加入显色液（NBT + BCIP）后，在黑暗中反应 10—30min，观察显色效果最佳时立即终止反应，以免本底过高。

2 结果分析

2.1 MT 的分离纯化

在方法 1.2.2 中对粗提物 Cd－MT 的 Sephadex G－50 柱层析中，空白对照组(未用氯化镉胁迫的黄瓜苗)在 254nm 处的最大值为 39.06，这与生物机体本身含有的金属硫蛋白的量大致相当，，而经过用 1000ppm 的 $CdCl_2$ 溶液胁迫的实验组，在 254nm 处的最大吸收峰可达 82.10，在 280nm 处没有明显吸收峰(图 1)。

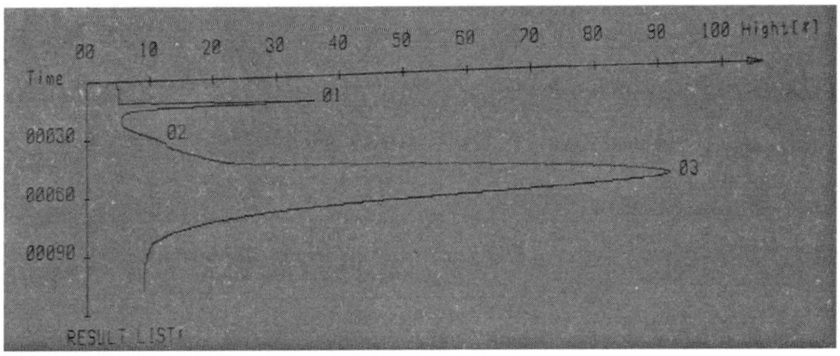

图 1 黄瓜叶片组织 Cd－MT 粗提液的 Sephadex G－50 层析图：
Column Size：1.0cm × 50cm，eluted with 0.001mol/L Tris－HCl(pH7.4)
At a flow rate of 1ml/min. The samples were monitored by measuring the absorbances at 254nm

2.2 MT 分子量的测定

对分离纯化过的 MT 进行 SDS－PAGE 凝胶电泳测的分子量较小 9100Da，同时在 27,000Da 处还有一条带，我们推测为 Cd－MT 三聚体。金属硫蛋白在有 O_2 存在下会形成多聚体[9]。

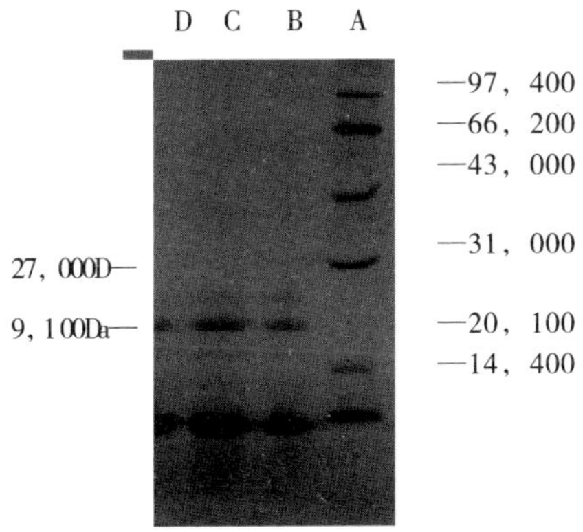

图 2 纯化后黄瓜 Cd—MT SDS 聚丙烯酰胺凝胶电泳图谱
A：Marker proteins；B、C、D：purified Cd－MT.

2.3 免疫血清效价检测

琼脂糖双扩散法检测后可看到,沉淀线由粗变细,免疫血清效价为 1∶16 且只出现一条沉淀线(图 3),证明免疫效果很好,所提取的 MT 较纯。

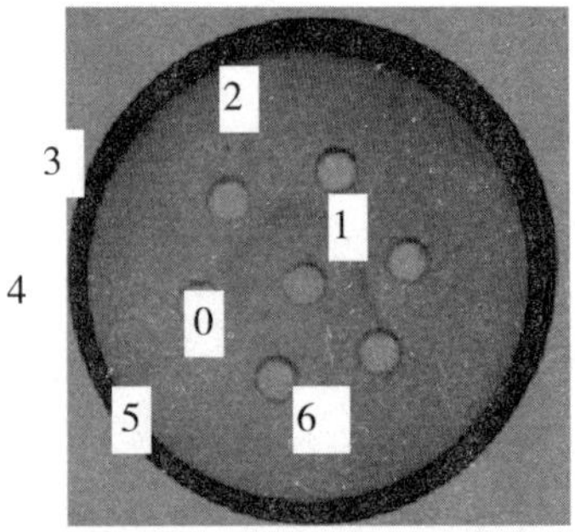

图 3 琼脂糖免疫双扩散试验

0:Purified cucumber Cd－MT,1～6:Rabbit antiserum diluted to 1∶2,1∶4,1∶8,1∶16,1∶32,1∶64,respectively.

2.4 IgG 的纯化

在纯化抗血清时,我们采取盐析法与 Sephadex G－200 相结合的方法,盐析过程中硫酸铵饱和度为 50% 时,有少量的球蛋白及大量的拟蛋白析出;硫酸铵饱和度为 33% 时 γ 球蛋白析出,达到对血清的纯化目的,再经过 PBS 透析除铵后用 Sephadex G－200 层析分离后,得到了纯度较高的抗体 IgG。紫外分光光度法测定 IgG 蛋白含量为 0.9364mg/ml。

2.5 黄瓜 Cd－MT 的 Dot Immunobinding Assay

我们用间接法封闭不同浓度的黄瓜 Cd－MT,建立了斑点免疫试验检测方法,可检测到最低的黄瓜 Cd－MT 含量为 200pg/ml,IgG 的稀释倍数为 1∶8。(图 4.1)以 Dot Immunobinding Assay 方法鉴定了此抗体的免疫交叉反应。结果显示此抗体与百合 Cd—MT、牛蛙 Cd—MT 有完全交叉反应,表明镉金属硫蛋白在免疫反应上有高度保守性(图 4.2)。

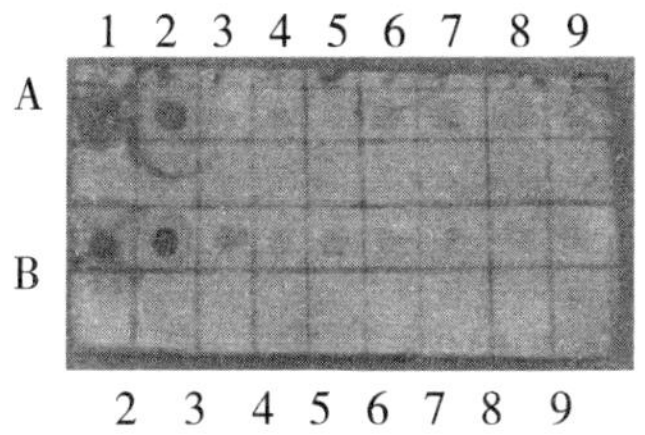

图 4.1 斑点免疫杂交试验

A 为黄瓜叶中 Cd－MT，A_{1-9}表示黄瓜叶中 Cd－MT 的浓度分别为 10ug/ml，1ug/ml，100ng/ml，10ng/ml，1ng/ml，800pg/ml，100ng/ml，500pg/ml，250pg/ml，200pg/ml.

B 为黄瓜叶中 Cd－MT，B_{1-9}表示黄瓜叶中 Cd－MT 的浓度分别为 100ug/ml，10ug/ml，1ug/ml，100ng/ml，10ng/ml，1ng/ml，800pg/ml，400pg/ml，200pg/ml

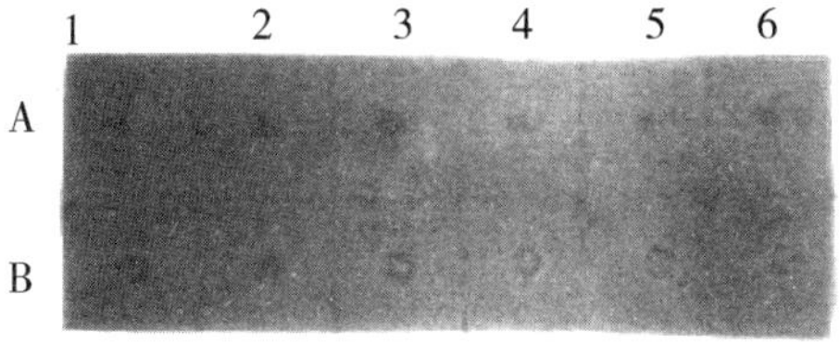

图 4.2 斑点免疫杂交试验

A 为牛蛙 Cd—MT，A_1-_6表示 Cd－MT 的浓度分别为 10ug/ml，1ug/ml，10ng/ml，1ng/ml，800pg/ml.

B 为百合 Cd—MT，B_1-_6表示 Cd－MT 的浓度分别为 10ug/ml，1ug/ml，100ng/ml，10ng/ml，1ng/ml，800pg/ml.

3 结论

3.1 目前用免疫学方法检测动物金属硫蛋白的文章较多，而用于检测植物金属硫蛋白的文章还未见报到。我们通过免疫学方法对黄瓜 Cd—MT 的检测，取得了较好的效果并建立 Dot Immunobinding Assay，为以后进一步检测环境中植物体金属硫蛋白奠定了良好的基础。

3.2 鉴于 MT 抗原性较弱[10]，我们在免疫过程中不仅使用了佐剂，还采用牛血清白蛋白与 MT 分子偶联的方法，以及 MT 分子间偶联的方法进行免疫，并且注重免疫途径和抗原用量，结果在较短的时间内获得了较好的免疫效果。

3.3 由于 MT 有较好的同源性[11]，我们的结果也证明 MT 抗体与百合 Cd—MT、牛蛙 Cd—MT 有完全交叉反应，说明镉金属硫蛋白在免疫反应上有高度保守性，可用 MT 抗血清检测黄瓜以外的其它生物体的 MT 含量，为环境中镉污染的检测建立免疫分子水平的快速检测方法奠定了基础。

参考文献

[1] Kagi J H R, Kojima Y. Chemistry and Biochemistry of Experimentum[M], 1989(52): 25－61.

[2] 茹炳根，潘爱华，黄秉乾，等. 金属硫蛋白生物化学与生物物理进展[M], 1991, 18(4): 254－259.

[3] Kagi J H, chaffer A. Biochemistry of metallothinein [J]. Biochemistry, 1988, 27

(23):8509.

[4] Cherian MG, chan HM. In: Suzuki. Imura M. etal. Metallothionein Biological Roles and Medical Implications[M]. Birkhauser Verlag: Basel, 1993. 87 – 109.

[5] Sato M. Bremner I. Oxygen Free Redicals and Metallothionein [J]. Free Redical BiolMed. 1993. 14325 – 337.

[6]张博润,蔡向荣.金属硫蛋白的研究进展及应用前景[J].微生物学报, 1999. 26(5): 355 – 357.

[7] Serra R, Isani G, Tramontano G, etal. Seaonal Dependence of Cadmium Accumlation and Cd – binding proteins in mytilus Galloprovincialis Exposed to Cadmium. comparative biochemistry and physiology c – pharmacology toxicology & endocrinology[J]. 1999, 123(2): 165 – 174.

[8]张晓钰 茹炳根 植物类 MT 和植物络合肽[J]. 生命科学 12(4): 170 – 122.

[9] Douglas MT, Cherian MG. Chemical modification of metallothionein. Biochem [J] 1984: 221(3): 569.

[10] Diponkar B, Satomi O. Cherian MG. Immunohistochemical localization of Mtin cell nucleus and cytoplasm of rat liver and kidney. Toxicology[J], 1982; 24(1): 95.

[11] Mehra P K, Tarbet E B, Gray W R etal. Biochem[J], 1990, 265(11)6369 – 6375.

注:该文发表于《生态学杂志》2008 年第 27 卷第 9 期 1592 – 1595

拟南芥 AtJ3 与 PKS5 相互作用参与植物 ABA 响应

赵菲佚　焦成瑾　陈　荃　贾　贞　王太术　周　辉*

脱落酸(abscisic acid,ABA)在植物生长、发育及环境胁迫响应中有着广泛的作用。前期研究已鉴定了诸多参与植物 ABA 信号转导的元件。本研究以拟南芥(*Arabidopsis thaliana*)蛋白激酶 PKS5(SOS2 - like protein kinase 5)为诱饵蛋白,使用酵母双杂交筛选到与 PKS5 相互作用的蛋白分子伴侣 AtJ3(Arabidopsis thaliana DnaJ homolog 3)。*AtJ3* T - DNA 突变体 *atj3* - 1 与 *atj3* - 2 表现出萌发期 ABA 表型。在外源 ABA 处理下,*atj3* - 1 与 *atj3* - 2 种子发芽率降低、幼苗黄化、生长矮小。*atj3* - 1 与 *PKS5* 双突变体 *atj3* - 1*pks5* - 1、*atj3* - 1*pks5* - 3 和 *atj3* - 1*pks5* - 4 表现出与 *AtJ3* 或 *PKS5* 突变体相同的 ABA 表型。亚细胞定位与转基因研究显示 AtJ3 与 PKS5 有相同的表达模式。免疫共沉淀与体外磷酸化测试确认 AtJ3 与 PKS5 存在相互作用并抑制 PKS5 激酶活性。研究结果表明:AtJ3 与 PKS5 相互作用,通过抑制 PKS5 激酶活性共同参与植物 ABA 信号响应过程。

植物激素脱落酸(abscisic acid,ABA)在植物的整个生活周期中起着关键性的作用。已有研究表明:ABA 参与植物胚胎发育、种子成熟与休眠、根与茎的生长及叶面蒸腾[1-3]。此外,ABA 还介导了植物对外界各种胁迫的响应,如干旱或渗透诱导的气孔关闭及盐、低氧、冷、伤害或病理等信号诱导的抗性[4]。对植物 ABA 响应信号途径中作用元件的鉴定与作用机理解析已成为植物激素研究的重要内容。

PKSes(SOS2 - like protein kinases)蛋白激酶属于拟南芥 SnRK3(SNF1 - relat-

---

* 作者简介:赵菲佚(1972—　),男,江苏镇江人,现为天水师范学院生物工程与技术学院教授,博士,主要从事植物分子生物学研究。

ed kinase3)家族[5,6]。该家族中,PKS12 在调节 ABA 与冷信号中起作用[7]。在植物萌发和幼苗期,PKS18 参与 ABA 响应。PKS18T/D(PKS18 组成型表达形式)转基因植物显示出对 ABA 的超敏感性,而该激酶 RNAi 转基因植物对 ABA 不敏感[8]。PKS3(SOS2 - like protein kinase 3)与 SCaBP5(SOS3 - like calcium binding protein 5)存在相互作用。*pks*3 或 *scabp*5 在种子萌发、幼苗生长和气孔关闭上表现出对 ABA 的超敏感性[9]。PKS24 不仅在 PHYA(phytochrome A)介导的抑制拟南芥远红光黄化苗转绿过程中起作用[10],也在盐及 ABA 的胁迫应答中具有功能[11]。

PKS5 为 PKS 家族 26 个成员之一。PKS5 通过磷酸化 AHA2(质膜 ATPase 的等位形式之一)第 931 位丝氨酸而阻止其与 14 - 3 - 3 蛋白相互作用,对 AHA2 活性进行负向调控,导致植物对外源高 pH 具有抗性[12]。近期发现,PKS5 可与 SCaBP8(SOS3 - like calcium binding protein 8)相互作用,通过磷酸化 AHA2 参与植物对外界盐碱的响应过程[13]。此外,PKS5 可对植物抗病途径中一个主要共激活子 NPR1(Nonexpressor of Pathogenesis - Related gene 1)的磷酸化介异 WRKY38(WRKY DNA - binding protein 38)与 WRKY62(WRKY DNA - binding protein 62)的表达,参与植物抗病过程[14]。

植物存在由高温、高盐、渗透等胁迫因素诱导产生的含 J 结构域的 DnaJ 蛋白[15-17]。大多数 DnaJ 蛋白在一级结构上包含 1 个由 70 个氨基酸组成的 J 结构域、相邻 G/F 结构域和末端锌指(CxxCxGxG)结构域[18,19]。在二级结构上,J 结构域含有四个螺旋和一个位于第 2 和第 3 螺旋间的由 His、Pro 和 Asp 三肽组成的高度保守环形区[20]。J 结构域与热激蛋白 Hsp70(Heat shock protein 70)的结合可稳定 Hsp70 与底物的相互作用[21]。DnaJ 蛋白在进化上具有保守性。除了在细胞蛋白的翻译、折叠、解折叠、移位及降解过程中作为蛋白分子伴侣起着重要作用外,还可直接结合于转录因子对转录激活进行调控。此外,DnaJ 在核内体形成和类胡萝卜素积累中也具有功能[15-17]。拟南芥中存在 89 个含 J 结构域的蛋白[22],AtJ3 分子伴侣为其中之一。拟南芥 AtJ3 介导开花信号的整合,对植物开化时间进行调控[23,24]。AtJ3 也可作为正向调节子对质膜上质子泵活性进行调节,从而参与植物对外源盐碱胁迫的响应[25]。已发现部分 PKS 家族成员参与植物 ABA 响应,PKS5 及与其有相互作用的 AtJ3 是否也在 ABA 信号途径中起作用目前未见报道。

本研究使用 *AtJ*3 与 *PKS*5 突变体,对两者在外源 ABA 处理下的表型进行了考察,并对两者在 ABA 响应中可能的作用机理进行了探讨。研究结果为 ABA 信号途径鉴定了新的参与元件,并为 AtJ3 与 PKS5 调控 ABA 信号机理提供依据。

1 材料与方法

1.1 实验材料

拟南芥野生型(Col－0)、T－DNA 突变体 *pks*5－1(SALK_108074)、*atj*3－1(SALK_131923)和 *atj*3－2(SALK_141625)订购自 SALK。*pks*5－3 与 *pks*5－4 为 TILLING 点突变体(http://tilling.fhcrc.org:9366),订购自 ABRC(Arabidopsis Biological Resource Center)。大肠杆菌 DH5α、BL21(DE3)及研究中所用质粒载体均为本实验室保存。

1.2 实验方法

1.2.1 引物设计

本研究中除用于植物点突变鉴定引物外,其余引物使用 Primer Premier 5.0 软件进行设计,在上海生物工程公司合成。引物序列见表 1。

表 1 引物序列

Tab. 1 Sequences of primers used in this study

引物名称	引物序列(5'－3')	限制性酶
A31TF	GCTGTTGACGGCTTAGGTAG	
A32TF	TTCGACTCGATCTTGCGTTT	
P5TF	ATGCCAGAGATCGAGATTGCC	
Lba1	TGGTTCACGTAGTGGGCCATCG	
LBb1	GCGTGGACCGCTTGCTGCAACT	
P53CF	GCGTTTGATTTGATTTCTTACTCCT	
P53CR	CACCACAAGCAAATCATTCAA	
P54CF	GTTTCGGATTTCGGTCTAAACG	
P54CR	CTCTCCTTTATAAATCTTC	
Act2F	GAAGATTAAGGTCGTTGCACCACCTG	
Act2R	ATTAACATTGCAAAGAGTTTCAAGGT	
P5H	CGGGATCCATGCCAGAGATCGAGATTGCC	*Bam*HI
P5R	GGAATTCTTAAATAGCCGCGTTTGTTG	*Eco*RI
A3F	CGGGATCCATGTTCGGTAGAGGACCCTC	*Bam*HI
A3R	GCGTCGACTTACTGCTGGGCACATTGCA	*Sal*I
AGF	CGGGATCCATGTTCGGTAGAGGACCCTCGA	*Bam*HI

续表

引物名称	引物序列(5'-3')	限制性酶
AGR	ACGCGTCGACTTACTGCTGGGCACATTGCACCC	*Sal*I
PYF	CCGCTCGAGATGCCAGAGATCGAGATTGCC	*Xho*I
PYR	CCGCTCGAGAATAGCCGCGTTTGTTGAC	*Xho*I
P5YF	GGAATTCATGCCAGAGATCGAGATTGCC	*Eco*RI
P5YR	CGGGATCCTTAAATAGCCGCGTTTGTTG	*Bam*HI
A3YF	CCGCTCGAGATGTTCGGTAGAGGACCCTC	*Xho*I
A3YR	CGGGATCCTTACTGCTGGGCACATTGCA	*Bam*HI
T1YF	CCGCTCGAGATGGATAATTCAGCTCCAGATTC	*Xho*I
T1YR	GGAATTCTCAAACTCTAAGGAGCTGCATTTTG	*Eco*RI
S1YF	CCGCTCGAGATGTCGCAGTGCGTTGACGGTATC	*Xho*I
S1YR	CGGGATCCTCAGGTATCTTCAACCTGAGAATG	*Bam*HI
P5PF	CGGGATCCATGCCAGAGATCGAGATTGCC	*Bam*HI
P5PR	GGAATTCAATAGCCGCGTTTGTTGACG	*Eco*RI
T1PF	CATGACTAGTATGGATAATTCAGCTCCAGATTC	*Spe*I
T1PR	GGAATTCAACTCTAAGGAGCTGCATTTTGTTAG	*Eco*RI
A3PF	CGGGATCCATGTTCGGTAGAGGACCCTC	*Bam*HI
A3PR	GGAATTCCTGCTGGGCACATTGCACCC	*Eco*RI
S1PF	CGGGATCCATGTCGCAGTGCGTTGACGGTATC	*Bam*HI
S1PR	GGAATTCGGTATCTTCAACCTGAGAATGGAA	*Eco*RI

序列中下划线为限制性酶作用位点碱基。

1.2.2 植物突变纯合体鉴定

*atj*3-1、*atj*3-2 与 *pks*5-1 突变纯合体鉴定正向引物分别为 A31TF、A32TF 及 P5TF,反向引物使用 T-DNA 左边界引物 Lba1 与 LBb1。*pks*5-3 与 *pks*5-4 点突变重合体鉴定使用衍生型酶切扩增多态性序列(Derived Cleaved Amplified Polymorphic Sequence,dCAPS)分子标记技术。点突变鉴定引物使用 dCAPS Finder 2.0(http://helix.wustl.edu/dcaps/dcaps.html)设计。*pks*5-3 使用 P53CF 与 P53CR 引物对,PCR 扩增 263bp 片段,*Ava*I 内切酶切野生型扩增产物。*pks*5-4 使用 P54CF 与 P54CR 引物对,PCR 扩增 236bp 片段,*Mlu*I 内切酶切野生型。双突变纯合体 *atj*3-1*atj*3-2、*atj*3-1*pks*5-1、*atj*3-1*pks*5-3、*atj*3-1*pks*5-4 和 *pks*5-

3*pks*5 -4 分别使用各单突变纯合体进行杂交,并经 PCR 与测序鉴定获得。

1.2.3 植物培养条件及 ABA 表型观察

试验种子用消毒液(20% 次氯酸钠 +0.1% Triton X -100)在 EP 管中灭菌 10 min,灭菌 ddH_2O 洗 5 次后点种于 MS 或含 0.3μmol · L^{-1} ABA 的处理 MS 固体平板上。处理 MS 平板在 4 ℃下春化 2 d 后置于 16 h 光(22 ℃)/8 h 暗(20 ℃)光周期条件下进行培养。培养一定时间(9 或 12 d)后,对平板上种子萌发率进行统计,并观察幼苗 ABA 表型。

1.2.4 RNA 分离及 RT - PCR 分析

拟南芥野生型(Col -0);*atj*3 -1、*atj*3 -2、*pks*5 -1 突变体总 RNA 提取使用柱式植物总 RNA 抽提纯化试剂盒(上海生物工程公司,SK8661)进行。提取总 RNA 使用 Promega 公司 M - MLV 反转录酶(Promega,M1705)在 42 ℃下进行反转录,获得 cDNA 第一链。植物总 RNA 提取及 cDNA 第一链合成操作按各自试剂盒推荐方法进行。为确认 *atj*3 -1 及 *atj*3 -2 中 *AtJ*3 的表达,以 Col -0、*atj*3 -1、*atj*3 -2、*pks*5 -1 的 cDNA 第一链为模板,使用 *AtJ*3 的 CDS 全长引物 A3F 与 A3R 进行 RT - PCR,同时以 *ACTIN*2 为内参,Act2F 与 Act2R 为内参引物,PCR 为 25 个循环。

1.2.5 AtJ3 及 PKS5 点突变序列克隆与蛋白原核表达

*PKS*5 序列克隆以野生型(Col -0)及 *pks*5 -3、*pks*5 -4 各点突变体基因组 DNA 为模板,以 P5H 与 P5R 为引物对,PCR 扩增得到野生型及点突变 *PKS*5 编码序列。*Bam*HI 与 *Eco*RI 酶切 PCR 扩增产物与 pET28a(Novagen,69864 -3)原核表达载体,胶回收纯化后使用连接酶(NEB,M0202S)进行连接。*AtJ*3 编码序列使用引物 A3F 和 A3R,以 *AtJ*3 cDNA 为模板。PCR 扩增产物与 pET28a 使用 *Bam*HI 与 *Sal*I 酶切,胶回收纯化后使用连接酶(NEB,M0202S)进行连接。构建载体转化大肠杆菌 DH5α。阳性候选克隆送上海生物工程公司测序验证。原核表达载体经测序验证后从 DH5α 提取质粒以电击转化方法转入 BL21(DE3)表达菌株,再次经 PCR 筛选得到阳性克隆后保存备用。

含目标蛋白表达载体的 BL21(DE3)菌株按 1:1000 的比例接入含 30 μg · L^{-1} 氯霉素和 50μg · L^{-1}的卡那霉素的 60 mL 液体 LB 培养基中,并于 37℃过夜小量培养。次日将 10 mL 过夜培养物转入 1 L 含 50μg · mL^{-1} Kan 的 LB 培养基中继续培养,当菌液 $O.D._{600}$达 0.7 时收菌,离心后沉淀加入 30 mL 裂解液(50 mmol · L^{-1} NaH_2PO4,300 mmol · L^{-1} NaCl,10 mmol · L^{-1} imidazole,pH 8.0)重悬,在超声破碎仪(VCX130,SONICS)上破碎(250 W,超声 3 s,间隔 3 s,1 min 循环,共 5 min)后离心,上清液中加入 100 μL Ni - NTA 树脂(Qiagen,No. 30210)进行蛋白结合,最后使用洗脱缓冲液(50 mmol · L^{-1} NaH_2PO4,300 mmol · L^{-1} NaCl,250 mmol ·

L^{-1} imidazole,pH 8.0)洗脱得到目标蛋白。后续融合蛋白纯化方法按 Qiagen 公司的树脂提取方法进行。结合于树脂上的重组蛋白保存于 4 ℃冰箱中备用。

1.2.6 PKS5 点突变重组蛋白体外激酶活性测试

激酶体外磷酸化试验按照以下方法进行。冰浴 EP 管中混合激酶与底物蛋白 MBP(Myelin Basic Protein),反应体积为 20 μL,计算激酶反应缓冲液(20 mmol · L^{-1} Tris – HCl,pH 7.2,5 mmol · L^{-1} $MgCl_2$,0.5 mmol · L^{-1} $CaCl_2$,10 μmol · L^{-1} ATP,2 mmol · L^{-1} DTT)的用量,每个反应使用 2 μCi($\gamma-32^{P}$) ATP。将反应所需体积的重组蛋白加入激酶反应缓冲液中,30℃水浴中反应 30 min 后加入 6 × SDS 蛋白上样缓冲液,并在干浴器(SBH130D/3,Stuart)上 95 ℃加热 5 min。离心后于 12% 的 SDS 蛋白胶上进行变性电泳,考马斯亮蓝染色后在脱色液中脱色,凝胶成像仪上(GelDoc – It2 310,UVP)照相后使用保鲜膜包好蛋白凝胶,压于磷屏上,次日于磷屏成像仪(STORM 860,Amersham Biosciences)上进行成像。

1.2.7 AtJ3 与 PKS5 荧光融合蛋白在原生质体与转基因植物中的亚细胞定位观察

使用引物 AGF 与 AGR,以野生型(Col – 0)cDNA 为模板进行 PCR 扩增。扩增产物克隆至 pCAMBIA1205 – GFP 植物转化载体上,构建植物 GFP – AtJ3 绿色荧光融合蛋白表达载体。使用引物 PYF 与 PYR,以野生型(Col – 0)基因组为模板进行 PCR 扩增,扩增产物克隆至 pTA7002 植物表达载体上,构建植物 PKS5 – YFP 黄色荧光融合蛋白表达载体。测序正确的植物荧光表达载体经质粒纯化试剂盒 Maxprep Kit(Vigorous Biotechnololgy)纯化后,分别转化拟南芥野生型(Col – 0)原生质体与植物中。原生质体转化使用 Jen 方法进行[26]。*pCAMBIA*1205 – *GFP* – *AtJ*3 转化的原生质体及生长 5 d 的 T2 代转基因植物用于 AtJ3 亚细胞定位。*PTA*7002 – *PKS*5 – *YFP* 转化的原生质体及生长 5 d 的 T2 代转基因植物使用 10 mmol · L^{-1} DEX(dexamethasone)处理 24 小时后在 confocal 显微镜(Zeiss 510 META)下进行 GFP – AtJ3 与 PKS5 – YFP 荧光观察。

1.2.8 酵母转化及双杂交试验

以 P5YF 和 P5YR 为引物对,拟南芥野生型(Col – 0)基因组为模板,PCR 扩增 *PKS*5 蛋白编码序列并克隆至酵母载体 pGBT9 上作为诱饵蛋白,对定购自 ABRC 的 50μg 酵母质粒猎物库进行筛选。为验证 AtJ3 与 PKS5 相互作用,使用引物 A3YF 和 A3YR;T1YF 和 T1YR;S1YF 和 S1YR,以拟南芥野生型(Col – 0)cDNA 为模板,分别扩增 *AtJ*3、*TTG*1(*TRANSPARENT TESTA GLABRA*1)与 *SCaBP*1(*SOS*3 – *like calcium binding protein* 5)蛋白编码序列并克隆至酵母表达载体 pGAD10 上。酵母表达载体以试验组合共转化至酵母菌株 PJ69 – 4A 中。酶母转化、筛库操作与相互作用试验按 Clontech 公司酵母操作手册进行。

1.2.9 蛋白原生质体表达与免疫共沉淀(Coimmunoprecipitation, Co - IP)试验

使用引物 P5PF 和 P5PR; A3PF 和 A3PR; S1PF 和 S1PR,以拟南芥野生型(Col - 0)基因组或 cDNA 为模板,PCR 分别扩增 PKS5、TTG1、AtJ3 与 SCaBP1 蛋白编码序列并克隆至 pCAMBIA1307 - 3xFLAG 植物转化载体上与 FLAG 标签 N 端转译融合。使用 BamHI 与 SalI 双酶切从 pET28a - PKS5 或 pET28a - AtJ3 切下 PKS5 或 AtJ3 蛋白编码序列,亚克隆至 pCAMBIA1307 - 6xMyc 植物转化载体上与 Myc 标签 C 端转译融合。构建载体以不同组合转化拟南芥野生型(Col - 0)原生质体中进行蛋白表达。原生质体转化使用 Jen 方法进行[26]。Co - IP 按 Quan 的方法进行[27]。

2 结果与讨论

2.1 *AtJ3*、*PKS5* 基因与蛋白结构及突变位置

AtJ3 定位于拟南芥基因组第三条染色体上(At3g44110),基因全长 1.95 Kb,包含 6 个外显子和 5 个内含子(图 1:A),从 SALK 订购得到 AtJ3 的两个等位突变体,被命名为 atj3 - 1(SALK_132923)和 atj3 - 2(SALK_141625),T - DNA 分别插入第 4 与第 6 个外显子上。AtJ3 的 CDS 长度为 1 263bp,编码含 420 个氨基酸的蛋白质,预测大小为 46.31kD,蛋白含 J(14 - 72)、G/F(82 - 107)、Zn(135 - 221)及 C(234 - 358)4 个结构域(图 1:B)。

PKS5 定位于拟南芥第二条染色体上(At2g30360),其 CDS 共有 1308 个碱基,不含内含子。编码蛋白含有 435 个氨基酸,为一丝氨酸/苏氨酸蛋白激酶。整个蛋白由位于 N 端的激酶结构域和 C 端的激酶调控域[28]组成。激酶结构域内含有一个推测的激活环(KDAL),调控结构域内含有两个子结构域,FISL 基序和 PPI 结构域。FISL 结构域是 PKS 家族中保守的结构域,是与 SCaBP(SOS3 - like calcium binding protein)相互作用所必须的结构域[8]。在激酶结构域和 FISL 基序之间由连接结构域(JK)进行连接[29]。从 SALK 与 ABRC 获得 *PKS5* 不同类型突变体。*pks5 - 1* 为 T - DNA 插入突变体,*pks*5 - 3 发生了 c - t 的转换,蛋白第 168 位氨基酸由丙氨酸变为缬氨酸(A168V); *pks5 - 3* 突变位于已知的激酶激活环内,*pks5 - 4* 发生了 c - t 的突变,第 317 位氨基酸由丝氨酸变为亮氨酸(S317L); *pks*5 - 4 的突变位于 FISL 基序内(图 1:C)。

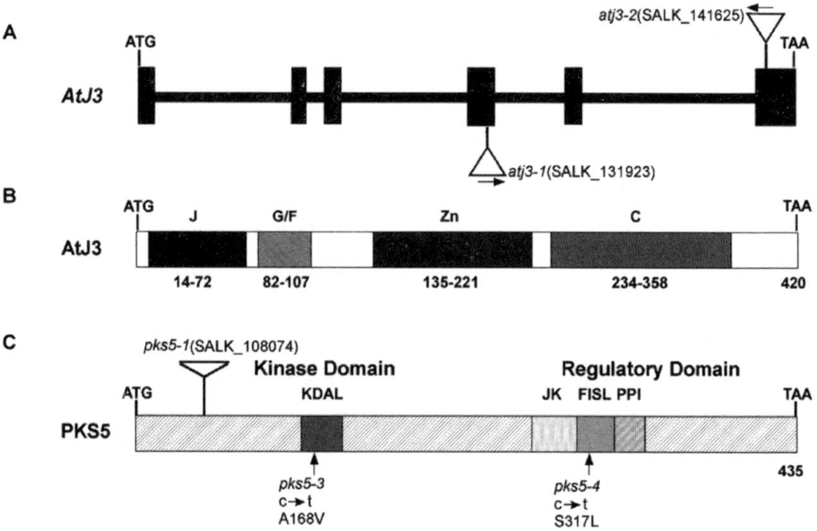

图1　AtJ3、PKS5 基因与蛋白结构

（A）AtJ3 基因结构（方框表示外显子；黑色棒表示内含子；三角形指示 *AtJ*3T－DNA 插入位置）；（B）AtJ3 蛋白结构（方框表示不同结构域；J：J 结构域；G/F：G/F；结构域；Zn：锌指结构域；C：C 末端结构域；数字表示从转译起始点计各结构域氨基酸位置）；（C）PKS5 蛋白结构与突变位置（KDAL：激酶催化结构域激活环；JK：连接区；FISL：FISL 基序；PPI：与 PP2C 蛋白磷酸酶作用结构域；箭头指示各点突变位置；三角形指示 *PKS*5 T－DNA 插入位置）

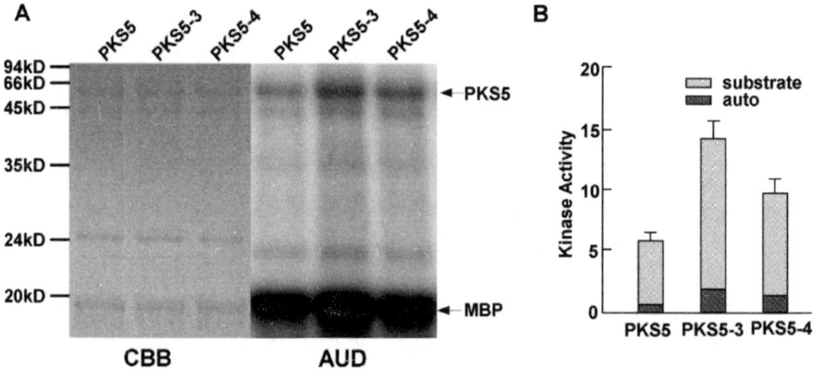

图2　PKS5 点突变磷酸化活性比较

（A）PKS5－3 和 PKS5－4 较 PKS5 有较高的磷酸化活性（MBP：麦芽糖结合蛋白；CBB：考马斯亮蓝染色；AUD：放射自显影）；（B）PKS5，PKS5－3 和 PKS5－4 相对磷酸化活性（auto：激酶自磷酸化活性；substrate：激酶 MBP 底物磷酸化活性）

2.2 PKS5 不同结构域点突变体外激酶活性测试

为考察 PKS5 不同结构域发生点突变后各突变激酶活性变化,对 PKS5 不同的点突变 CDS 进行克隆并在原核中表达。不同 PKS5 点突变重组蛋白的激酶活性比较如图 2A 所示。结果显示,不同的 PKS5 点突变激酶活性发生改变,与 PKS5 对照相比较,PKS5 - 3 与 PKS5 - 4 活性增高,其中 PKS5 - 3 活性最高。PKS5 点突变激酶底物磷酸化(MBP)活性与自磷酸化活性变化一致(图 2:B)。此结果说明:PKS5 不同结构域发生突变对 PKS5 激酶活性的影响不同。PKS5 - 3 突变发生于激酶激活环内(KDAL),PKS5 - 4 突变发生于 FISL 基序内。KDAL 突变对激酶活性影响要大于 FISL 内突变。从 PKS5 点突变激酶活性可得知,与 PKS5 野生型相比较,拟南芥 *PKS5* 突变体 *pks*5 - 3 与 *pks*5 - 4 为功能获得性突变体。

2.3 *pks*5 - 3 与 *pks*5 - 4 萌发期 ABA 表型

为考察 *PKS*5 突变体是否具有 ABA 表型,对 *PKS*5 点突变体 *pks*5 - 3、*pks*5 - 4 和 *PKS*5 的 T - DNA 突变体 *pks*5 - 1 的 ABA 表型进行测试,结果如图 3。结果显示:*pks*5 - 3 和 *pks*5 - 4 在含 0.3μmol · L^{-1} ABA 的 MS 平板上表现出 ABA 敏感表型,萌发率与野生型相比较降低,幼苗生长矮小,黄化(图 3:A)。而 *pks*5 - 1 在此条件下与野生型比较不显示 ABA 敏感表型。将 *pks*5 - 3 与 *pks*5 - 4 进行杂交,F1 子代在含 0.3 μmol · L^{-1} ABA 的 MS 平板上表现出与亲代相似的 ABA 敏感表型,但其对 ABA 的敏感性增加(图 3:B)。*pks*5 - 3 与 *pks*5 - 4 在不同 ABA 浓度下萌发率曲线表明:*pks*5 - 3 与 *pks*5 - 4 在不同 ABA 浓度下的萌发率随着 ABA 浓度的升高而降低,在 0.15 - 0.75μmol · L^{-1} ABA 浓度区间内,*pks*5 - 4 较 *pks*5 - 3 具有较高的萌发率(图 3:C)。*pks*5 功能获得性点突变体 ABA 表型说明:PKS5 参与 ABA 响应途径,为 ABA 信号转导的元件之一。

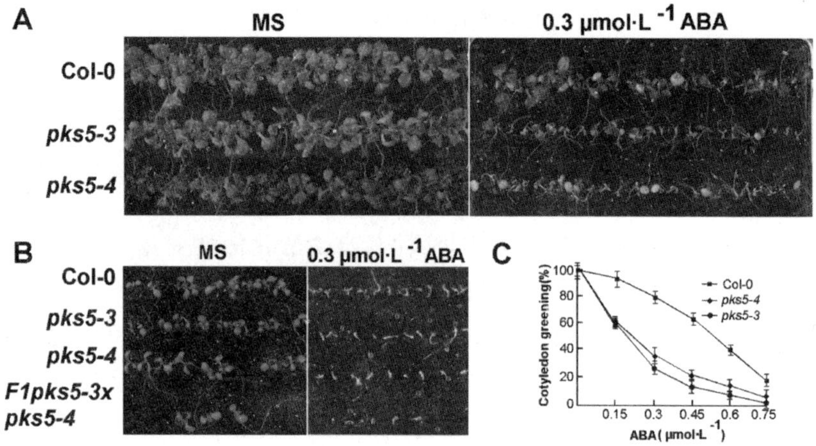

图3 *pks5* 点突变体 ABA 表型

(A)*pks5* 点突变体 *pks5* -3 与 *pks5* -4 生长 12 d 的 ABA 敏感表型;(B) *pks5* -3 与 *pks5* -4 杂交一代生长 9 d 表现出与亲代相似的 ABA 敏感表型;(C) 不同 ABA 浓度下 *pks5* 点突变体萌发率曲线

2.4 AtJ3 与 PKS5 的相互作用

为获得与 PKS5 相互作用元件,将 *PKS5* 的 CDS 克隆至 pGBT9 酵母表达载体上,以 PKS5 为诱饵蛋白筛选拟南芥 cDNA 文库。筛选到编码 DnaJ 类似热激蛋白 AtJ3。为确认 PKS5 与 AtJ3 间的相互作用,将 AtJ3 的 CDS 克隆至 pGAD10 酵母载体上,与 *PKS5* 酵母载体共转入酵母菌株 PJ69 -4A 中,在筛选培养基上验证 PKS5 与 AtJ3 间的相互作用。图 4A 结果表明 PKS5 与 AtJ3 在酵母中的确存在相互作用。为进一步确认 PKS5 与 AtJ3 在植物体内的相互作用,构建 PKS5 与 AtJ3 不同标签蛋白植物表达载体,共转入拟南芥野生型(Col -0)原生质体中。Co -IP 试验表明:在植物体内,PKS5 与 AtJ3 也存在相互作用(图 4:B)。在原生质体与植物中表达 AtJ3 绿色荧光与 PKS5 黄色荧光融合蛋白,使用 Confocal 观察两者的亚细胞定位,图 4C 表明 AtJ3 与 PKS5 在细胞膜、细胞质与细胞核中均有表达。PKS5 与 AtJ3 表达在细胞中有重叠。此支持 AtJ3 与 PKS5 在酵母与原生质体中相互作用结果。上述结果说明体内 PKS5 与 AtJ3 存在相互作用,共同参与特定的信号过程。

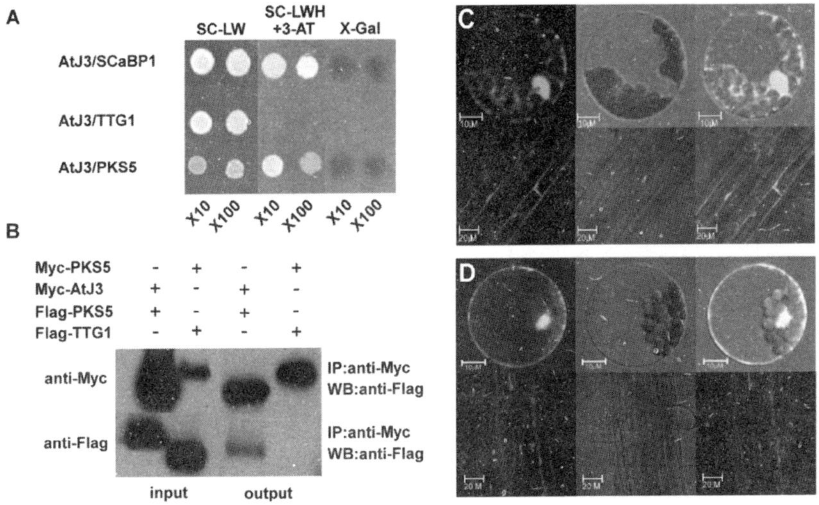

图4 AtJ3 与 PKS5 存在相互作用

(A) AtJ3 与 PKS5 在酵母中相互作用验证(SC－LW:缺乏亮氨酸与色氨酸的合成完全培养基;SC－LWH＋3－AT:缺乏亮氨酸、色氨酸和组氨酸并含 20 $mmol \cdot L^{-1}$ 3－氨基－1,2,4－三唑的合成完全培养基;X－Gal:SC－LWH＋3－AT 平板上酵母 β－半乳糖苷酶活性试验;SCaBP1 与 TTG1 为正负对照;X10、X100 表示酵母培养液稀释倍数);(B)体内 AtJ3 与 PKS5 存在相互作用(Myc:Myc 标签;FLAG:FLAG 标签;TTG1 为对照);(C) AtJ3－GFP 在原生质体(上部)与转基因植株(下部)中的定位;(D) YFP－PKS5 在原生质体(上部)与转基因植株(下部)中定位

2.5 *atj*3－1 与 *atj*3－2 萌发期 ABA 表型

AtJ3 与 PKS5 体内存在相互作用。*pks*5－3 和 *pks*5－4 表现对 ABA 的敏感表型。为测试 AtJ3 是否也参与了植物 ABA 响应过程,使用 *atj*3－1 与 *atj*3－2 功能缺失突变体(图 5:A)对其 ABA 表型进行考察。结果表明:*atj*3－1 与 *atj*3－2 在含 0.3$\mu mol \cdot L^{-1}$ ABA 的 MS 平板上也显示对 ABA 的敏感表型(图 5:B、C)。在 0.3$\mu mol \cdot L^{-1}$ ABA 处理下,与野生型相比较,*atj*3－1 与 *atj*3－2 的萌发率降低(图 5:B)。萌发后 9 d,*atj*3－1 与 *atj*3－2 的幼苗生长矮小,黄化程度增加(图 5:C),呈现与 *pks*5－3 和 *pks*5－4 在相同浓度 ABA 下的相似表型。*atj*3－1 与 *atj*3－2 杂交 F1 子代与其亲代 ABA 表型相同,表明 AtJ3 的确参与了 ABA 响应过程。*atj*3－1 与 *atj*3－2 在不同 ABA 浓度下的萌发率曲线表明,*atj*3－1 与 *atj*3－2 的萌发率随 ABA 浓度的增加而下降(图 5:D),在 0.15－0.75$\mu mol \cdot L^{-1}$ ABA 浓度区间内,*atj*3－1 与 *atj*3－2 在相同 ABA 浓度下的萌发率无显著差异($p<0.05$)。

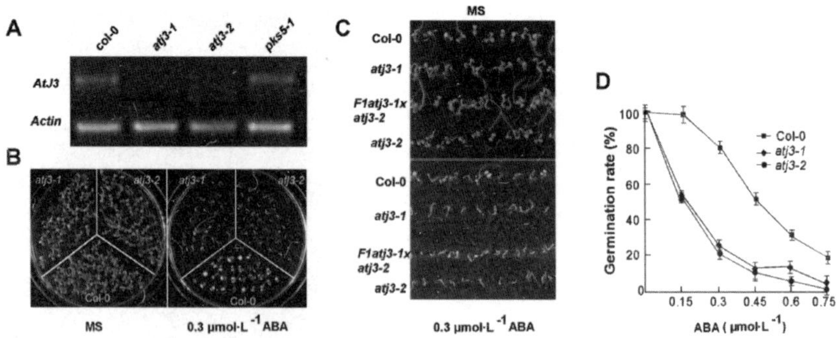

图 5　*atj3* 功能缺失突变体 ABA 表型

(A) *AtJ3* 功能缺失突变体鉴定 (B) *atj3*－1 与 *atj3*－2 功能缺失突变体在含 0.3μmol·L^{-1} ABA 的 MS 平板上生长 9 d 后表现出 ABA 敏感表型 (C) *atj3*－1 与 *atj3*－2 杂交一代生长 9 d 后表现与亲代相似的 ABA 表型 (D) 不同 ABA 浓度下 *AtJ3* 突变体萌发率曲线

2.6　*AtJ3* 与 *PKS5* 双突变体 ABA 表型

为确认 *AtJ3* 是否在遗传上与 *PKS5* 相互作用，共同参与植物 ABA 响应过程。将 *atj3*－1 与 *PKS5* 不同类型突变体进行杂交，构建 *atj3*－1*pks5*－1、*atj3*－1*pks5*－3 和 *atj3*－1*pks5*－4 双突变体，并对其 ABA 表型进行考察，表型测试结果如图 6 所示。

在含 0.3μmol·L^{-1} ABA 的 MS 平板上，*atj3*－1*pks5*－3 和 *atj3*－1*pks5*－4 双突变体对 ABA 处理的表型一致，均表现为对 ABA 的敏感表型。生长 12 d 的幼苗出现生长矮小，*atj3*－1*pks5*－3 出现严重黄化。双突变体与杂交亲本 ABA 的表型相比较，*atj3*－1*pks5*－3 和 *atj3*－1*pks5*－4 对 ABA 的反应比各自的亲本更加强烈（图 6：B、C）。而 *atj3*－1*pks5*－1 双突变体对 ABA 表型与 *atj3*－1*pks5*－3 和 *atj3*－1*pks5*－4 不同。*atj3*－1*pks5*－1 与其单突变体对 ABA 的敏感性较 *atj3*－1*pks5*－3 和 *atj3*－1*pks5*－4 对各自单突变体的 ABA 降低（图 6：A）。体外激酶活性测试表明，PKS5－3 与 PKS5－4 较野生型 PKS5 有着更高的磷酸化活性（图 2）。*pks5*－3 和 *pks5*－4 为功能获得性突变体，*pks5*－1 则为功能缺失突变体。此表明对于 *PKS5* 而言，其不同的突变形式对 ABA 的响应存在差异。不同类型 *PKS5* 突变体对 ABA 的响应与其体内激酶活性的差异相关。*atj3*－1 与不同形式 *PKS5* 双突变体 ABA 表型与相应的 *PKS5* 突变体相似，表明 *AtJ3* 在遗传上处于 *PKS5* 的上游起作用。*AtJ3* 与 *PKS5* 双突变体 ABA 表型分析也进一步说明，*AtJ3* 与 *PKS5* 存在遗传上的相互作用，*AtJ3* 处于 *PKS5* 的上游，在 ABA 同一信号途径起作用。

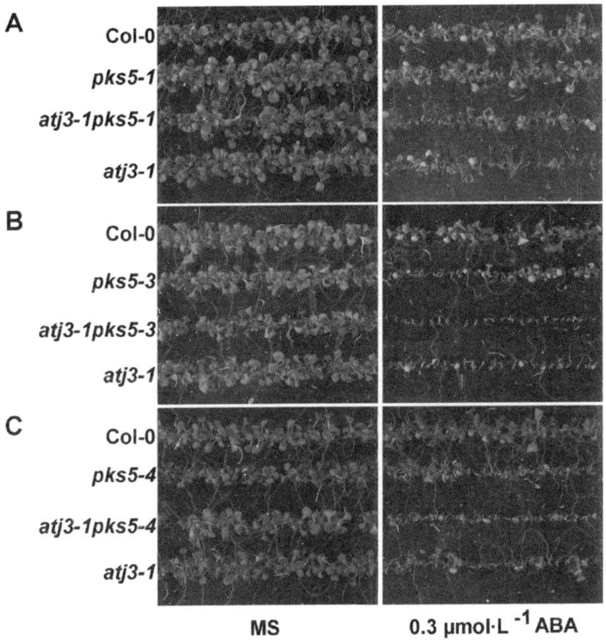

图 6　*atj*3 –1*pks*5 –1,*atj*3 –1*pks*5 –3 与 *atj*3 –1*pks*5 –4 双突变体 ABA 表型

(A)*atj*3 –1*pks*5 –1 双突变体生长 12 d 的 ABA 表型;(B)*atj*3 –1*pks*5 –3 双突变体生长 12 d 的 ABA 的表型;(C)*atj*3 –1*pks*5 –4 双突变体生长 12 d 的 ABA 表型

2.7　AtJ3 对 PKS5 激酶活性的影响

AtJ3 与 PKS5 存在相互作用,共同参与拟南芥 ABA 响应,且 *AtJ*3 在遗传上处于 *PKS*5 上游起作用。为考察 AtJ3 与 PKS5 两者间在植物 ABA 响应中的分子作用机理,对 AtJ3 是否影响 PKS5 的激酶活性进行了测试。在体外 PKS5 激酶活性测试中加入重组 AtJ3,测试 PKS5 激酶对通用底物 MBP 的磷酸化程度变化。图 7A 显示:当 PKS5 激酶活性测试体系加入 AtJ3 重组蛋白后,PKS5 对 MBP 的磷酸化程度降低,表明 AtJ3 可抑制 PKS5 的激酶活性。随着体系中 AtJ3 含量的增加,PKS5 的激酶活性下降。PKS5 – 3 与 PKS5 – 4 的激酶活性较 PKS5 高,AtJ3 对 PKS5 及其两种突变形式的激酶活性呈现相同的抑制模式,而 AtJ3 不能对与 PKS5 同源的 SOS2 激酶活性进行抑制,进一步说明 AtJ3 可对 PKS5 激酶活性以特异的抑制方式起作用。

为确认体内 AtJ3 对 PKS5 激酶活性的抑制作用,构建 *AtJ*3 及 *SCABP*1 的 *FLAG* 标签融合植物表达载体,转化拟南芥野生型(Col –0)原生质体进行 FLAG 融合蛋白表达。使用 IP 方法从原生质体中纯化植物表达 AtJ3 与 SCABP1 蛋白(图 7:B 上部),将其加入体外 PKS5 激酶活性测试体系中进行 PKS5 激酶活性测试。结果

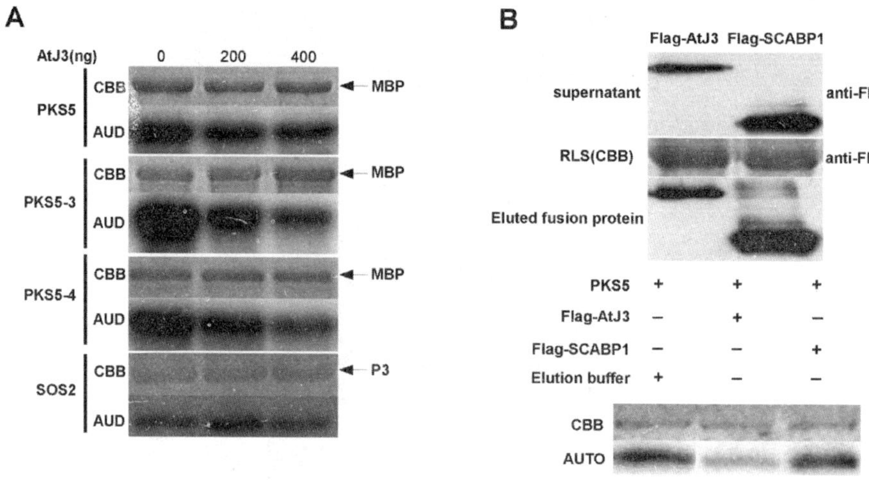

图 7　AtJ3 抑制 PKS5 激酶活性

(A)重组 AtJ3 体外抑制 PKS5、PKS5 − 3 和 PKS5 − 4 激酶活性(PKS5、PKS5 − 3 与 PKS5 − 4 活性测定使用 MBP 为激酶作用底物;SOS2 为对照,以 P3 为激酶作用底物;数字表示加入的不同量 AtJ3 蛋白;CBB:考马斯亮蓝染色;AUD:放射自显影;AUTO:PKS5 蛋白激酶自磷酸化活性);(B)瞬时表达 AtJ3 的 IP 产物抑制 PKS5 激酶自磷酸化活性(RLS:Rubisco 大亚基;supernatant:原生质体裂解上清液;Eluted fusion protein:洗脱的不同标签融合蛋白;Elution buffer:洗脱缓冲液;SCABP1:表达的 SCABP1 蛋白,作为阴性对照;CBB:考马斯亮蓝染色;AUTO:PKS5 蛋白激酶自磷酸化活性)

表明:植物表达 AtJ3 对 PKS5 激酶活性同样具有抑制作用(图 7:B 下部),与重组 AtJ3 对 PKS5 活性抑制作用相同。植物表达阴性对照 SCABP1 不能抑制 PKS5 激酶活性,说明植物表达 AtJ3 同样对 PKS5 活性抑制作用具有特异性。重组与植物表达 AtJ3 均对 PKS5 激酶活性具有抑制作用,表明在植物对 ABA 响应中,AtJ3 通过对 PKS5 激酶活性的抑制方式参与 ABA 信号途径。

3　讨论

ABA 参与植物发育过程及对外界各种环境胁迫,现已鉴定了多种参与 ABA 信号途径的元件。其中之一为蛋白激酶。PKSes 属于 CDPK − SnRK 超家族中 SnRK3 亚家族。已发现 PKS 家族 PKS3、PKS12、PKS18 及 PKS24 为植物 ABA 信号途径中的成员,参与 ABA 响应过程。猜测 PKS5 也可能参与 ABA 响应过程。

为验证 PKS5 是否也参与 ABA 的响应过程,本研究使用 *PKS5* 不同类型突变体测试其在萌发期 ABA 处理下的表型。结果显示:在 0.3μmol · L^{-1} ABA 处理下,*PKS5* 功能获得性突变体 *pks*5 − 3 与 *pks*5 − 4 在萌发期生长表现出对 ABA 的敏感表型。突变体萌发率下降,萌发后幼苗生长矮小,表现黄化现象。此说明 PKS5 在植物萌发与幼苗期参与植物的 ABA 响应过程。为获得与 PKS5 可能相互作用并参与 ABA 信号途径的元件,使用酵母双杂交技术,从拟南芥表达库中筛选到编码类似 DnaJ 热激蛋白的分子伴侣 AtJ3,并对其两个功能缺失突变体 *atj*3 - 1 与 *atj*3 - 2 在萌发与幼苗期 ABA 表型进行了测试。*atj*3 - 1 与 *atj*3 - 2 在萌发与幼苗期的 ABA 表型与 PKS5 突变体相似,也具有对 ABA 的敏感表型。对 *PKS*5 与 *AtJ*3 的杂交子代纯合双突变体的 ABA 表型测试表明:*PKS*5 与 *AtJ*3 存在遗传上的相互作用,*AtJ*3 处于 *PKS*5 上游起作用,共同参与对外源 ABA 处理的响应过程。结合体外 PKS5 与 AtJ3 相互作用证据,可确认 PKS5 与 AtJ3 在植物体内的确存在相互作用,二者共同参与 ABA 响应过程。重组与植物表达 AtJ3 均对 PKS5 的激酶活性具有抑制作用,表明在植物体内 AtJ3 以对 PKS5 激酶活性抑制的方式参与对 PKS5 的调节,响应外界 ABA 信号。本研究结果为 ABA 信号转导途径鉴定了新的参与元件,并提供了元件间的相互作用证据与分子作用机理。

在 *PKS*5 不同类型突变体对 ABA 处理的表型测试中,功能缺失突变体 *pks*5 - 1 与 *pks*5 - 1*atj*3 - 1 双突变体对 ABA 处理的敏感性低于其它类型的突变体。*pks*5 - 3、*pks*5 - 4 及 *atj*3 - 1*pks*5 - 3、*atj*3 - 1*pks*5 - 4 则表现出萌发与幼苗期的 ABA 超敏感表型,其中 *pks*5 - 3 较 *pks*5 - 4 的 ABA 表型更为明显。*pks*5 - 3 与 *pks*5 - 4 为功能获得性突变体,而 *pks*5 - 1 为功能缺失突变体,推测植物体内还存在处于 PKS5 下游的其它参与 ABA 基因调节的元件,当 PKS5 蛋白缺失时,未知对下游 ABA 元件的调节作用互补了 PKS5 的功能,只有当 PKS5 蛋白存在且其激酶活性发生变化时,PKS5 对下游 ABA 元件的调节功能主要由 PKS5 承担,参与 ABA 的响应过程。

已有研究表明:PKS3 与 SCaBP5 相互作用并在拟南芥 ABA 响应中起作用[30]。SCaBP1 具有感应细胞内钙离子水平功能,作为钙的感应器可与 PKS 家族成员相互作用,激活与招募 PKS 家族成员至质膜上,最终激活目标调节子[31-34]。本研究确认 AtJ3 与 PKS5 相互作用并参与 ABA 信号过程,推测也同样存在 SCaBP 家族成员与 PKS5 相互作用共同参与 ABA 的响应过程。此外,已有报道表明:AtJ3 与 PKS5 可对质膜质子泵进行调节,参与植物对外界高盐与 pH 响应过程[25]。由此,保持细胞内 pH 内平衡是否与 ABA 响应过程相关,AtJ3 与 PKS5 是否在不同信号途径中具有交叉作用是本研究的进一步工作。

4 结论

研究使用拟南芥 *PKS*5 不同类型突变体确认 PKS5 蛋白激酶参与植物 ABA 响

应,其为植物 ABA 信号转导途径中作用元件之一。此外,以 PKS5 为诱饵蛋白,通过酵母双杂交技术筛选到与 PKS5 相互作用的蛋白分子伴侣 AtJ3。对 2 个拟南芥 *AtJ3* 功能缺失突变体的研究表明:AtJ3 也参与植物 ABA 响应过程。对 AtJ3 与 PKS5 的遗传与生化测试表明:AtJ3 与 PKS5 在体内存在着相互作用,AtJ3 通过对 PKS5 蛋白激酶活性的抑制作用方式共同参与植物 ABA 响应过程。

参考文献

[1] Koornneef M, Leon - Kloosterziel K, Schwartz S H, et al. The genetic and molecular dissection of abscisic acid biosynthesis and signal transduction in*Arabidopsis*[J]. Plant Physiol Biochem, 1998, 36:83 - 89.

[2] Leung J, Giraudat J. Abscisic acid signal transduction[J]. Annu Rev Plant Physiol Plant Mol Biol, 1998, 49:199 - 222.

[3] McCourt P. Genetic analysis of hormone signaling. Annu Rev Plant Physiol Plant Mol Biol, 1999, 50:219 - 243. 4. Shinozaki K, Yamaguchi S K. Molecular responses to dehydration and low temperature: Differences and cross - talk between two stress signaling pathways[J]. Curr Opin Plant Biol, 2000, 3:217 - 223.

[4] Hrabak E M, Chan C W, Gribskov M, et al. The *Arabidopsis* CDPK - SnRK superfamily of protein kinases[J]. Plant Physiology, 2003132:666 - 680.

[5] Weinl S, Kudla J. The CBL - CIPK Ca^{2+} - decoding signaling network: function and perspectives[J]. New Phytologist, 2009, 184:517 - 528.

[6] Jen S, Li Z, Jang Z, et al. Sugars as signaling molecules[J]. Curr Opin Plant Biol, 1999, 2:410 - 418.

[7] Gong D M, Zhang C Q, Chen X Y, et al. Constitutive activation and transgenic evaluation of the function of an *Arabidopsis* PKS protein kinase[J]. J Bio Chem, 2002, 227:42088 - 42096.

[8] Cheong Y H, Kim K N, Pandey G K, et al. CBL1, a calcium sensor that differentially regulates salt, drought, and cold responses in Arabidopsis[J]. Plant Cell, 2003, 15:1833 - 1845.

[9] Qin YZ, Guo M, Li X, Xiong XY, He CZ, Nie XZ, Liu XM(2010). Stress responsive gene *CIPK*14 is involved in phytochrome A - mediated far - red light inhibition of greening in *Arabidopsis*. Science China Life Sciences40, 970 - 977.

[10] Qin Y Z, Li X, Guo M, et al. Regulation of salt and ABA responses by CIPK14, a calcium sensor interacting protein kinase in *Arabidopsis*[J]. Science in China Series C Life Sciences, 2008, 38:446 - 457.

[11] Fuglsang A T, Guo Y, Cuin T A, et al. Arabidopsis protein kinase PKS5 inhibits the plasma membrane H^+ - ATPase by preventing interaction with 14 - 3 - 3 protein[J]. Plant Cell, 2007, 19:1617 - 1634.

[12]Xie C G, Lin H X, Deng X W, et al. Roles of SCaBP8 in salt stress response[J]. Plant Signaling & Behavior, 2009, 4:956 – 958.

[13]Xie C, Zhou X, Deng X, Guo Y. PKS5, a SNF1 – related kinase, interacts with and phosphorylates NPR1, and modulates expression of WRKY38 and WRKY62[J]. J Genet Genomics, 2010, 37:359 – 369.

[14]Boston R S, Viitanen P V, Vierling E. Molecular chaperones and protein folding in plants[J]. Plant Mol Biol, 1996, 32:191 – 222.

[15]Waters E R, Lee G J, Vierling E. Evolution, structure and function of the small heat shock proteins in plants[J]. J Exp Bot, 1996, 47:325 – 338.

[16]Wang W, Vinocur B, Shoseyov O, et al. Role of plant heat – shock proteins and molecular chaperones in the abiotic stress response[J]. Trends Plant Sci, 2004, 9:244 – 252.

[17]Caplan A J, Cyr D M, Douglas M G. Eukaryotic homologues of Escherichia coli dnaJ: A diverse protein family that functions with hsp70 stress proteins[J]. Mol Biol Cell, 1993, 4:555 – 563.

[18]Silver P A, Way J C. Eukaryotic DnaJ homologs and thespecificity of Hsp70 activity[J]. Cell, 1993, 74:5 – 6.

[19]Qian Y Q, Patel D, Hartl F U, et al. Nuclear magnetic resonance solution structure of the human Hsp40(HDJ – 1)J – domain[J]. J Mol Biol, 1996, 260:224 – 235.

[20]Qiu X B, Shao Y M, Miao S, et al. The diversity of the DnaJ/Hsp40 family, the crucial partners for Hsp70 chaperones[J]. Cell Mol Life Sci, 2006, 63:2560 – 2570.

[21]Miernyk J A. The J – domain proteins of Arabidopsis thaliana: An unexpectedly large and diverse family of chaperones[J]. Cell StressChaperones, 2001, 6:209 – 218.

[22]Shen L, Yu H. J3 regulation of flowering time is mainly contributed by its activity in leaves[J]. Plant Signal Behav, 2011, 6:601 – 603.

[23]Shen L, Kang YG, Liu L, et al. The J – domain protein J3 mediates the integration of flowering signals in Arabidopsis[J]. Plant Cell, 2011, 23:499 – 514.

[24]Yang Y Q, Qin Y X, Xie C G, et al. The Arabidopsis chaperone J3 regulates the plasma membrane H^+ – ATPase through interaction with the PKS5 kinase[J]. Plant cell, 2010, 22:1313 – 1332.

[25]Jen S. Signal transduction in maize and Arabidopsis mesophyllprotoplasts[J]. Plant Physiol, 2001, 127:1466 – 1475.

[26]Quan R, Lin H, Mendoza I, et al. SCABP8/CBL10, a putative calcium sensor, interacts with the protein kinase SOS2 to protect Arabidopsis shoots from salt stress[J]. Plant Cell, 2007, 19:1415 – 1431.

[27]Guo Y, Halfter U, Ishitani M, et al. Molecular characterization of functional domains in the protein kinase SOS2 that is required for plant salt tolerance[J]. Plant Cell, 2001, 13:1383

-1399.

[28]Gong D, Guo Y, Schumaker K S, et al. The SOS3 family of calcium sensors and SOS2 family of protein kinases in Arabidopsis[J]. Plant Physiol, 2004, 134:919-926.

[29]Guo Y, Xiong L, Song C P, et al. A calcium sensor and its interacting protein kinase are global regulators of abscisic acid signaling in Arabidopsis[J]. Dev Cell, 2002, 3:233-244.

[30]Halfter U, Ishitani M, Zhu J K. The Arabidopsis SOS2 protein kinase physically interacts with and is activated by the calcium-binding protein SOS3[J]. Proc Natl Acad Sci USA, 2000, 97:3735-3740.

[31]Quintero F J, Ohta M, Shi H, et al. Reconstitution in yeast of the Arabidopsis SOS signaling pathway for Na^+ homeostasis[J]. Proc Natl Acad Sci USA, 2002, 99:9061-9066.

[32]Lin H, Yang Y, Quan R, et al. Phosphorylation of SOS3-LIKE CALCIUM BINDING PROTEIN8 by SOS2 protein kinase stabilizes their protein complex and regulates salt t olerance in Arabidopsis[J]. Plant Cell, 2009, 21:1607-1619.

[33]Xu J, Li H D, Chen L Q, et al. A protein kinase, interacting withtwo calcineurin B-like proteins, regulates K^+ transporter AKT1 in Arabidopsis[J]. Cell, 2006, 125:1347-1360.

注:本文曾发表在2016年36卷1期"植物研究"期刊上。

过表达拟南芥点突变乙酰羟酸合成酶基因改变植物对缬氨酸的抗性及增强缬氨酸合成

赵菲佚　焦成瑾　王太术　田春芳　谢尚强　刘亚萍*

拟南芥乙酰羟酸合成酶(acetohydroxyacid synthase,AHAS)在支链氨基酸合成中具有重要的作用。为考察AHAS不同亚基关键位点突变对植物缬氨酸抗性与缬氨酸合成的影响,对AHAS大小亚基上特定位点进行体外突变,构建AHAS点突变过表达转基因植物,研究AHAS不同亚基点突变转基因植物对缬氨酸抗性及其合成的影响。研究结果表明:AHAS小亚基G88D突变将解除终端产物对该酶的反馈抑制作用,使转基因植物缬氨酸含量提高。大亚基E305D突变增强小亚基G88D突变效应,而大亚基E482D突变对G88D突变具有相反的作用。AHAS全酶E305DG88D双突变转基因植物较E482DG88D具有更强的缬氨酸抗性表型和更高的缬氨酸含量。这些结果提示AHAS大小亚基间存在着相互作用,大小亚基不同位点突变对AHAS全酶活性具有不同的影响。

缬氨酸、亮氨酸和异亮氨酸由于具有支链碳骨架被称为支链氨基酸,在动物生长与发育过程中有着不可或缺的作用(Harris等2005;Nair和Short2005;Brosnan和Brosnan 2006)。已有研究表明:在动物体内,支链氨基酸可被如mTOR(mammalian target of rapamycin)信号蛋白感应,在促进蛋白合成与抑制蛋白降解中起着重要的作用。此外,支链氨基酸也直接或间接参与神经递质谷氨酸的合成与区室化;对胺类神经递质5-羟色胺、儿茶酚胺类多巴胺和去甲肾上腺素的合成也有影响(Hutson等2001;Fernstrom 2005;Nair和Short 2005)。然而,动物体内不能进行支链氨基酸的从头合成,只能从饮食中获取,植物是动物获取支链氨基酸的重要来源。因而,植物支链氨基酸的合成与调控研究具有重要的理论与实际意义。

---

* 作者简介:赵菲佚(1972—),男,江苏镇江人,现为天水师范学院生物工程与技术学院教授,博士,主要从事植物分子生物学研究。

植物、细菌与真菌通过一个保守的代谢途径进行支链氨基酸的合成,该合成途径中的合成酶活性不仅受到底物的严格调控,而且受终端产物的反馈抑制(Singh 和 Shaner 1995;Duggleby 和 Pang 2000;Chipman 等 2005)。乙酰羟酸合成酶(AHAS,也称为乙酰乳酸合成酶)为支链氨基酸合成途径的第 1 个合成酶。在植物中,3 种支链氨基酸对 AHAS 的活性均有抑制作用,但缬氨酸显示出较亮氨酸与异亮氨酸更为明显的抑制效应(Miflin 和 Cave 1972)。

至今,人们对不同物种的 AHAS 已进行了比较详细的研究。细菌 AHAS 为 4 聚体,由 2 个催化亚基(CSUs)和 2 个调节亚基(RSUs)组成。调节亚基上包含 1 个高度保守的 ACT 功能结构域,该结构域使调节亚基有稳定和增强催化亚基活性的功能(Vyazmensky 等 1996)。另外,调节亚基与催化亚基相互作用介导终端支链氨基酸对 AHAS 酶活性的反馈抑制作用(Lee 和 Duggleby 2001,2002;Mendel 等 2001,2003;Kaplun 等 2006)。与细菌 AHAS 不同,植物 AHAS 催化亚基含有 2 个 ACT 功能结构域,使植物 AHAS 可能有与细菌 AHAS 不同的拓扑属性,导致与细菌 AHAS 功能的差异(Hershey 等 1999)。

过表达植物 AHAS 催化亚基不能提高支链氨基酸含量(Tourneur 等 1993),表明植物 AHAS 的调控方式可能更复杂。最近发现拟南芥 RSU 的 T119L 显性突变可解除缬氨酸对 AHAS 的反馈抑制,使 *vat*1 突变体有较高的缬氨酸含量(Chen 等 2010)。但该研究仅对 AHAS 调节亚基上点突变对植物支链氨基酸的合成影响进行了探讨。对原核 AHAS 研究表明:AHAS 大小亚基同时突变导致酶蛋白产生变构效应,进而影响酶活性。细菌 AHAS 大亚基 H132、K155、E213、D217、E221、E389、E393 和 S414 为关键性活性位点,除 E213D 外,其余位点突变使细菌 AHAS 酶活性完全丧失。E213D 突变使细菌 AHAS 的酶活性只有野生酶蛋白的 25%(Rajiv 等 2005)。菠菜 AHAS 大亚基体外突变分析结果与细菌的相似,大亚基 R258、E311、D315、E319、E488、E492、S518 和 T520 组成了 AHAS 的活性中心,除 E488D 突变外,其他突变导致酶活性丧失,E488D 突变的酶活性为野生型的 48%(Renaud 等 1995)。蛋白序列比对发现,细菌 AHAS 大亚基 E213 对应拟南芥 E305,菠菜 E488 对应拟南芥 E482。细菌 AHAS 小亚基 ilvH 上 N11A、G14D、N29H、T34L、A36V 及 Q59L 突变对细菌 AHAS 的缬氨酸抗性效应与酶活性均有影响,G14D 突变酶活性为野生型的 92%,并表现出最强的缬氨酸抗性。细菌 G14D 突变对应拟南芥小亚基 G88D 突变(Mendel 等 2001)。

本研究对拟南芥 AHAS 大小亚基上相应于细菌与菠菜的关键位点进行体外突变,构建 AHAS 点突变转基因植物,观察其缬氨酸抗性表型并对转基因植物缬氨酸含量进行测定,考察 AHAS 大小亚基关键位点突变对植物支链氨基酸合成的影响。研究结果对揭示 AHAS 大小亚基相互作用方式及提高支链氨基酸含量方法具有指导意义。

1　材料与方法

1.1　引物设计

使用 Primer Premier 5.0 软件进行引物设计,研究用所有引物由上海生物工程公司进行合成。引物序列见表 1。

表 1　引物序列

Table 1　Primer sequences used in this study

引物名称	引物序列(5′→3′)	限制性酶
ALU－F	CTAGTCTAGAATGGCGGCGGCTACTTCATCCATC	*Xba*I
ALU－R	ACGCGTCGACTCAGTTGCTAGATTGACGCAAC	*Sal*I
ASU－F	CGCGGATCCATGGCGGCGACGACGACTGCTAC	*Bam*HI
ASU－R	CCCAAGCTTCTACAAAGGAAGAGAGTATCCACG	*Hin*dIII
ALUE305D－F	ACTCTTGAACAGGATTACAGGAGTGAC	
ALUE305D－R	GTCACTCCTGTAATCCTGTTCAAGAGT	
ALUE482D－F	CACTCTTACTCAGATATCATCAACGAG	
ALUE482D－R	CTCGTTGATGATATCTGAGTAAGAGTG	
ASUG88D－F	GGCGATGAGAGCGATATAATAAATAGA	
ASUG88D－R	TCTATTTATTATATCGCTCTCATCGCC	
E305D－F	GCCGCCGATGTTGCATTGGGATG	
E305D－R	AAAGATGTCACTCCAGTA	
E482D－F	CGCGTCAGGAAGTCCAGACCTG	
E482D－R	CACGGATTCAATCACACTCTCGTTGATGAA	
G88D－F	GACGGGAGGAAGATGAGGAATGC	
G88D－R	TACTCCAGCAATTCTATTTATTAGA	
35S－F	CTCAAGCAATCAAGCATTCTAC	
ASUN－F	GTCTAGCTGTTGGTCCTGCTG	
ASUN－R	GATGCTGAAACAACGTGCA	
ALUN－F	GTCCAGACCTGCTGGTGACT	
ALUN－R	ACTCTCGACTTAATACTCCG	

序列中下划线为酶作用位点或突变碱基。

1.2 AHAS 点突变植物表达载体构建

拟南芥总 RNA 提取选用 MS 平板上生长 7 d 的整株幼苗(Col－0),使用上海生物工程公司总 RNA 提取试剂盒进行(Sangon,SK1312),提取总 RNA 使用 Invitrogen 公司反转录酶进行 cDNA 第一链反转录(Invitrogen,18080－093),以 cDNA 为模板,PCR 高保真酶扩增 AHAS 大小亚基 CDS(Trans,AP221－01)。野生型 AHAS 大亚基使用 ALU－F 和 ALU－R 引物对进行扩增。AHAS 小亚基扩增正向引物为 ASU－F,反向引物为 ASU－R。AHAS 大亚基第 305 位 E305D 的正向突变引物为 ALUE305D－F,反向突变引物为 ALUE305D－R。大亚基第 482 位 E482D 正向突变引物为 ALUE482D－F,反向引物为 ALUE482D－R。AHAS 小亚基第 88 位 G88D 正向突变引物为 ASUG88D－F,反向突变引物为 ASUG88D－R。突变引物中下划线指示各突变碱基(表 1)。PCR 方式引入目标突变,以 AHAS 大小亚基基因全长正向引物与目标突变位点反向突变引物或全长反向引物与目标突变正向突变引物为引物对,以 cDNA 为模板进行 PCR 扩增,分别扩增的 PCR 产物经胶回收柱(Sangon,SK8132)回收后等量混合作为突变全长 PCR 模板,再次以基因全长两端引物进行扩增,扩增产物回收后进行双酶切,再次纯化后连接到 pCAMBIA1307 植物表达载体上。转化 DH5α 宿主菌,每个突变载体挑选 2 个不同克隆送上海生物工程公司测序验证,每个克隆重复测 2 次,以 2 个不同克隆 4 次测序结果与基因标准参考序列进行比对,如与参考序列相同并在目标突变位点引入正确突变,则突变载体构建成功,否则,重新构建突变载体,直至正确为止。

1.3 植物材料、生长条件、植物转化与鉴定

植物材料使用拟南芥(*Arabidopsis thaliana* L.)野生型(Col－0),营养钵中 AHAS 点突变转基因植物生长表型观察使用 22℃下长日照(16h 光/8h 暗)条件培养。含不同缬氨酸浓度 MS 竖直平板上转基因植物初生根长表型分析使用 22℃下连续光照培养。

植物转化方法采用农杆菌浸蘸法,拟转化农杆菌 GV3101 在含 50 $\mu g \cdot L^{-1}$的利复平和 50$\mu g \cdot L^{-1}$卡那霉素的 LB 培养基中小量过夜培养,次日接种至 250 mL 与小量培养液相同的 LB 中培养液中,当菌液 OD_{600}至 2 左右时,4000 $r \cdot min^{-1}$常温离心 10 min 收集菌体,菌体沉淀重悬浮于 150 mL 转化液(每升含 MS 盐 2.22 g、蔗糖 50 g、SilwetL－77 200 μL、6－BA 200 μL)中,拟被转化植物倒置浸入转化液中 20～30 s 后用保鲜膜包裹,水平放置于培养盘中 24 h 后转入正常培养。转基因植物 T_0代种子在 MS 固体平板(含 40$\mu g \cdot L^{-1}$潮霉素和 50$\mu g \cdot L^{-1}$的头孢氨苄)上进行筛选。

用衍生型酶切扩增多态性序列(Derived Cleaved Amplified Polymorphic Sequence,dCAPS)分子标记技术鉴定阳性候选植株。AHAS 大亚基 E305D 及 E482D,小亚基 G88D 的 dCAPS 鉴定引物由 dCAPS Finder 2.0(http://helix.wustl.edu/dcaps/dcaps.html)设计如下:AHAS 大亚基 E305D 突变鉴定引物为 E305D-F 和 E305D-R(*Sca*I 切野生型)。AHAS 大亚基 E482D 突变鉴定引物为 E482D-F 和 E482D-R(*Eco*RI 切野生型)。小亚基 G88D 突变鉴定引物使用 G88D-F 和 G88D-R(*Dpn*I 切突变体)。引物中下划线为 dCAPS Finder 2.0 引入的突变碱基。dCAPS 鉴定过程为:提取转化植株或野生型基因组 DNA 作为模板,对 AHAS 点突变转基因植物使用 CaMV35S 正向引物 35S-F 与突变基因全长反向引物进行 PCR 扩增,扩增产物作为模板,以设计的上述 AHAS 大小亚基上点突变位点 dCAPS 鉴定特异引物进行 PCR 扩增;野生型对照则以提取的基因组 DNA 为模板,直接以 dCAPS 突变鉴定引物进行扩增,扩增产物回收后以设计的内切酶进行酶切,酶切产物经 5% 琼脂糖凝胶电泳分离。

1.4 RNA 提取、分离及 Northern blot 分析

以拟南芥野生型及 AHAS 各转基因植株叶片为材料,总 RNA 提取采用上海生物工程公司试剂盒(SK8661),并按照推荐提取方法进行。所提取总 RNA 定量后,取 30μg 总 RNA 在含甲醛的 1.0% 琼脂糖凝胶中进行电泳,后转到 Hybond N^+ 尼龙膜上进行杂交,杂交过程参考 Liu 等(2000)方法。探针使用 Takara 公司试剂盒(D6045)以 $\alpha^{32}P$-dCTP,植物表达载体为模板进行标记,对 AHAS 小亚基使用引物对 ASUNF 和 ASUNR;大亚基使用引物 ALUNF 和 ALUNR。杂交在 42℃下进行,rRNA 的 EB 染色作为上样对照,杂交信号用 Phosphoimager 扫描。

1.5 AHAS 点突变转基因植物表型分析

AHAS 点突变转基因植物种子 4℃春化 3 d 后,用 20% 次氯酸钠消毒液(含 0.1% TritonX-100)在 EP 管中消毒 10 min,灭菌 ddH_2O 洗 5 次后点种于 MS 固体培养基平板上,置于 22℃连续光照下竖直培养。当幼苗根长至 1.5~2cm 时将相同大小幼苗移入含"品氏育苗"基质(Pindstrup,丹麦品氏托普公司)营养钵中,置于 22℃长日照(16h 光/8h 暗)条件下生长,2 周后进行生长表型观察。同时将相同大小幼苗转移至不同浓度缬氨酸的 MS 处理平板上,22℃连续光照培养条件竖直培养,转板后 6 d 对转基因植物拍照,使用 ImageJ(http://rsb.info.nih.gov/nih-image/)软件对初生根长度进行测量。

1.6 自由缬氨酸含量测定

自由氨基酸含量测定使用营养钵中生长 4 周的野生型(Col-0)与转基因拟南芥幼苗为材料,测定方法参考 Hacham 等(2002)的方法进行。约 200 mg 幼苗在液氮中研磨至粉末状,加入 600 μL 提取液(水:氯仿:甲醇=3:5:12,*V/V*),再

次充分研磨至匀浆样转入离心管中,600 μL 提取液漂洗研钵合并至离心管中,12000 r·min^{-1}离心 10 min,将上清转入另一离心管中,加入 300 μL 氯仿和 450 μL 水,再次 12000 r·min^{-1}离心 10 min,收集上部水与甲醇相冷冻干燥,产物重新溶于 100 μL 的 20 mmol·L^{-1} HCl 中,取出 10 μL 使用 AccQ 衍生试剂盒进行衍生化(Waters,http://www.waters.com/)。衍生化后的氨基酸使用 ACQUITY UPLC 系统进行自由氨基酸分析(Waters),用缬氨酸标准品定量。

2 实验结果

2.1 拟南芥 AHAS 大小亚基基因与蛋白结构及其突变位置

拟南芥 AHAS 由 2 个大亚基和 2 个小亚基构成,大亚基为催化亚基,小亚基为调节亚基。AHAS 大亚基(ALU)基因定位于第 3 条染色体上(At3g58610),基因全长 3658bp,包含 11 个外显子和 10 个内含子(图 1-A),CDS 共有 1776bp,编码蛋白含 591 个氨基酸(图 1-B)。突变第 6 个外显子上第 1611 位(以其转译起点计)G 为 T,蛋白第 305 位氨基酸由 E 转变为 D,突变第 10 个外显子第 2749 位(转译起点计)A 为 T,使蛋白第 482 位由 E 变为 D。AHAS 小亚基(ASU)基因定位于第 5 条染色体上(At5g16290),基因全长 3981bp,包含 13 个外显子和 12 个内含子(图 1-C),CDS 共有 1441bp,编码 477 个氨基酸(图 1-D),属于 ACT 超家族。该蛋白包含 2 个 ACT 结构域,分别位于 78~150 和 309~383 氨基酸区域。突变第 2 个外显子上第 733 位(以转译起点计)的 G 为 A,蛋白第 88 位由 G 变为 D。

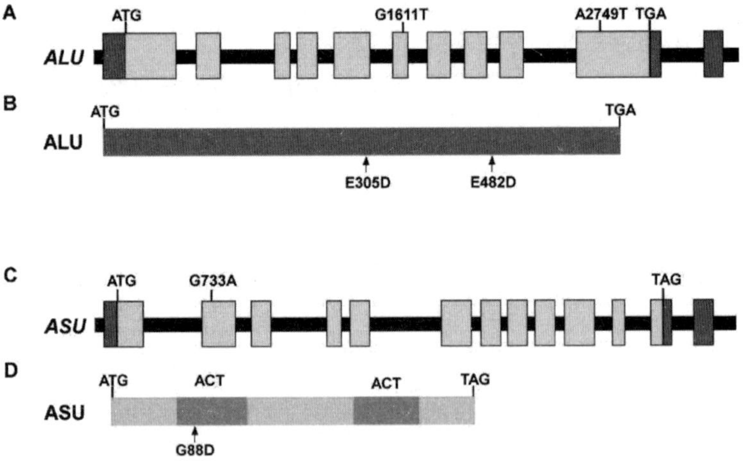

图 1 拟南芥乙酰羟酸合成酶(AHAS)基因与蛋白结构及其突变位置

A 与 C 分别示例 AHAS 大亚基(ALU)与小亚基(ASU)基因结构。图中黑色棒表示内含子,黑色方框表示外显子,红色方框表示非编码区(UTRs),数字表示基因从翻译起始点的核苷酸突变位置。B 与 D 分别示例 AHAS 大亚基(ALU)与小亚基(ASU)蛋白结构,数字表示相应于转译起始点第一个氨基酸突变位置。

2.2 AHAS 植物转化、筛选、突变鉴定与过表达分析

将 AHAS 小亚基及大小亚基点突变植物过表达载体转化拟南芥野生型(Col-0),转基因植物筛选、突变体鉴定及 AHAS 点突变基因过表达分析如图 2 所示。

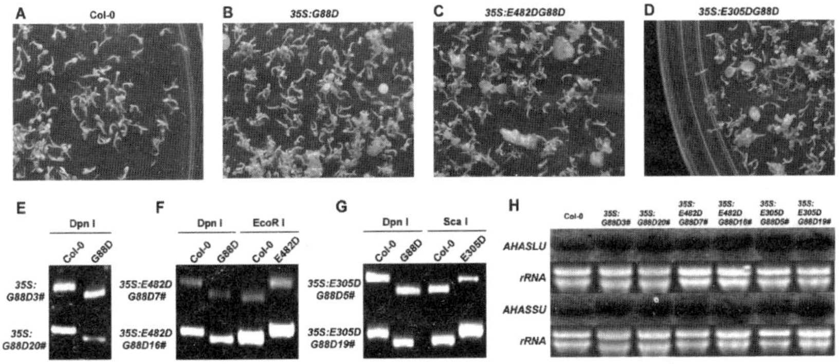

图 2 AHAS 点突变转基因植物筛选、突变鉴定与过表达分析

A ~ D:AHAS 点突变转基因植物抗性筛选,红色箭头示阳性转基因候选植株,筛选培养基为 MS + 50 mg · L^{-1} 卡那霉素;E ~ G:AHAS 点突变阳性转基因植株 dCAPS 鉴定,*Dpn*I 切 G88D 突变体,*Sca*I 与 *Eco*RI 分别切 E305D 和 E482D 野生型;H:AHAS 点突变转基因植株过表达分析。

使用含 AHAS 小亚基和大小亚基点突变植物转化载体的农杆菌对目标植物(第一次为拟南芥野生型,第二次为已转入第一种突变载体的转基因植物)分别进行两次基因转化操作,每次转化后使用卡那霉素抗性对转基因植株进行筛选,筛选过程示例如图 2 - A ~ D。从图中可看出,构建的植物过表达转化载体已将目标突变基因转入受体植物中,转化植物在抗性筛选平板上表现出对卡那霉素的抗性,说明植物至少受到一种植物转化载体的转化。从转化后代中,使用 dCAPS 鉴定 AHAS 小亚基单突变体和大小亚基双突变体,图 2 - E ~ G 表明转基因后代群体中可筛选到 AHAS 大小亚基均发生目标点突变的双突变体单株。获得 AHAS 小亚基 G88D 突变 35S:G88D3#和 35S:G88D20#两个单突变体植株、大亚基 E305D 与小亚基 G88D 双突变体 2 个单株 35S:E305DG88D5#与 35S:E305DG88D19#、大亚基 E482D 与小亚基 G88D 双突变体 2 个单株 35S:E482DG88D7#与 35S:E482DG88D16#用于后续表型分析。为检测 AHAS 大小亚基突变基因是否得到了过表达,使用 Northern 方法对转入的基因进行过表达分析,图 2 - H 表明:在不同的双突变单株中,对照野生型,AHAS 小亚基 G88D 单突变、大亚基 E305D 与 E482D 及小亚基 G88D 双突变转基因均得到了过表达。

2.3 AHAS 点突变转基因植物生长及缬氨酸处理下初生根长度表型分析

AHAS 小亚基突变及大小亚基双突变转基因植株生长表型分析如图 3 - A。小亚基单突变体及大小亚基双突变体转基因植株生长表型与野生型相同,未观察到任何异常生长表型。表明当 AHAS 小亚基发生单突变或大小亚基发生双突变后,在无外源胁迫条件时,AHAS 突变不影响植物的生长表型。

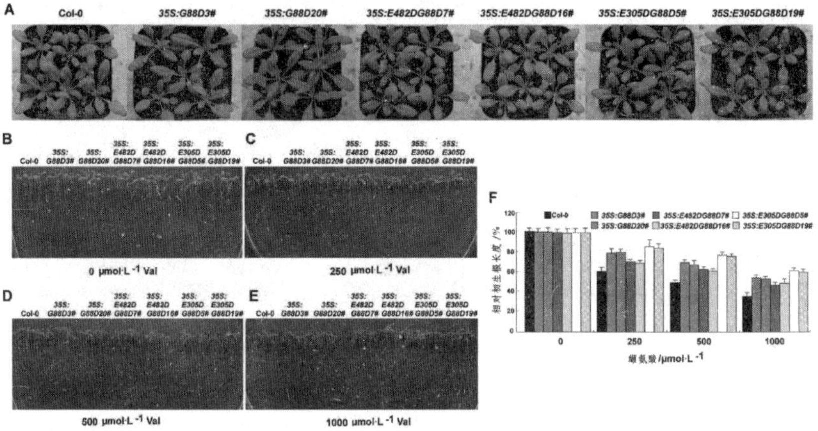

图 3　AHAS 点突变转基因植物表型分析

A:AHAS 点突变转基因植株生长 2 周后生长表型;B ~ E:AHAS 点突变转基因植株在不同浓度缬氨酸胁迫下的表型;F:图 B ~ E 中 AHAS 点突变转基因植株转板后 6 d 相对初生根长度。

AHAS 合成的 3 种支链氨基酸中,缬氨酸的反馈抑制作用最为强烈(Iris 等 2004)。为测试当 AHAS 小亚基单突变及大小亚基发生双突变后,缬氨酸对 AHAS 点突变转基因植物表型的影响,分别使用 0、250、500 和 1000 $\mu mol \cdot L^{-1}$ 外源缬氨酸对转基因植株进行处理,观察其表型变化,结果如图 3 - B ~ E。

测试结果表明,在同一缬氨酸浓度处理下,小亚基 G88D 突变的转基因植物初生根长度均大于野生型。说明 AHAS 小亚基发生 G88D 突变解除了缬氨酸对 AHAS 的反馈抑制作用(图 3 - C ~ E),使 AHAS 小亚基单突变体或大小亚基双突变体产生对缬氨酸的抗性,在初生根长度上均表现出大于野生型,此与 Chen 等(2010)的结果相同。在不同外源缬氨酸浓度处理下,与 MS 平板上各对照相比较,处理板上野生型及 AHAS 点突变转基因植物初生根生长均受到抑制,随缬氨酸处理浓度升高,抑制程度增加。不同缬氨酸处理浓度间初生根长度差异明显($P<0.05$)(图3-F)。此外,在相同浓度缬氨酸处理下,AHAS大亚基不同位置双突变体与小亚基单突变体在初生根长度上又存在差异,E482DG88D转基因植物相对初生根长度小于G88D转基因植物,而E305DG88D转基因植物相对初生根长度大于G88D转基因植物,3种转基因植物相对初生根长度差异显著($P<0.05$)(图3-F)。说明AHAS大小亚基同时发生突变后,由于AHAS大亚基上的不同突变位点对小亚基突变的影响作用存在差异,可能对AHAS酶活性产生不同影响,使E305DG88D转基因植物较E482DG88D转基因植物对缬氨酸处理更具有抗性。

2.4 AHAS点突变转基因植物缬氨酸含量测定

为考察AHAS小亚基及大小亚基点突变转基因植物中缬氨酸含量变化,对AHAS转基因植株幼苗及完全展开叶片中的缬氨酸含量进行了测定,结果如表2所示。

表2 AHAS点突变转基因植株幼苗及叶片缬氨酸含量

突变株类型	缬氨酸含量/pmol·mg^{-1}(DW)	
	幼苗	叶
Col-0	418.3±35.4[a]	330.9±7.2[A]
35S:G88D3#	2867.3±125.7[b]	1102.9±22.4[B]
35S:G88D20#	2903.6±138.4[b]	1137.5±19.1[B]
35S:E482DG88D7#	2544.4±147.6[c]	998.7±18.4[C]
35S:E482DG88D16#	2632.3±152.5[c]	995.7±16.3[C]
35S:E305DG88D5#	3079.1±201.5[d]	1320.1±36.5[D]
35S:E305DG88D19#	3124.1±207.6[d]	1285.3±29.4[D]

表中数值为3次重复测定,表示为平均值±标准差;不同字母表示差异达到显著水平($p<0.05$);幼苗中差异用小写字母表示,叶中用大写字母表示。

表2数据表明,AHAS各转基因植株幼苗缬氨酸含量升高,含量达野生型的6~7倍之多(表2)。叶片缬氨酸含量较幼苗低,但叶片缬氨酸含量变化趋势与幼苗中相同,转基因植株叶片缬氨酸含量为野生型对照的3~4倍(表2)。在AHAS大亚基不同位置突变植株幼苗与叶片中,E305DG88D与E482DG88D转基因植株的缬氨酸含量存在差异($P<0.05$)。在幼苗中,AHAS大亚基E482D转基因植株中缬氨酸含量为E305D突变的82.5%,而叶片中只有72.8%。这表明AHAS小亚基G88D突变解除了终端产物对AHAS的反馈抑制作用,使转基因植株缬氨酸

含量相对于野生型提高,而大亚基 E305D 与 E482D 不同位置突变又使转基因植株间在相应组织的缬氨酸含量存在差异。

3 讨 论

本研究发现拟南芥 AHAS 小亚基 G88D 突变可解除缬氨酸对 AHAS 的反馈抑制作用,使 AHAS 突变体内缬氨酸含量提高(表 2),并在初生根长度上表现出对外源缬氨酸抗性(图 3 - C ~ E)。此结果与在细菌及拟南芥中的研究结果一致(Mendel 等 2001;Chen 等 2010)。然而,拟南芥 AHAS 双突变 E305DG88D 较 E482DG88D 转基因植物有更强的缬氨酸抗性,体内积累了更多的缬氨酸(图 3 - C ~ E,表 2)。说明在 AHAS 小亚基上发生 G88D 突变解除终端产物反馈抑制的前提下,大亚基不同位置的突变对植物缬氨酸合成有不同的影响。对不同物种 AHAS 大亚基蛋白编码序列比对发现,ALU 含有 5 个保守结构域,E305 与 E482 分别位于第 3 与第 4 个结构域中(Rajiv 等 2005)。拟南芥 ALU 在 2 个关键结构域中突变可能造成该酶活性下降,并且不同位点突变对酶活性有不同的影响。因而,在缬氨酸抗性与缬氨酸积累上,AHAS 双突变 E305DG88D 与 E482DG88D 转基因植物存在差异是 AHAS 小亚基与大亚基上两种突变共同作用结果。

细菌 AHAS 小亚基具有稳定并调节其大亚基活性的作用,小亚基介导终端产物支链氨基酸对酶活性调节。尽管目前尚未发现小亚基结合支链氨基酸的直接证据,但在体外测试中,当存在支链氨基酸时,大亚基活性的确会发生变化(Siew 和 Ronald,2001)。推测 AHAS 活性变化与其全酶构象相关,本研究结果支持此推测的合理性。

拟南芥 AHAS 大亚基点突变中,E305D 突变较 E482D 表现出对缬氨酸抑制更具有抗性(图 3 - C ~ E)。拟南芥 AHAS 小亚基与细菌的不同,其包含 2 个 ACT 结构域(细菌中仅有 1 个)。细菌 AHAS 更倾向于形成 2 个大亚基与 2 个小亚基的 4 聚体,介导终端产物抑制的位置存在于 2 个小亚基组成的 2 聚体界面上(Pang 等 2002,2003)。拟南芥 AHAS 小亚基含有 2 个 ACT 结构域,对大亚基活性调节与细菌不同,对小亚基不同 ACT 结构域的突变分析表明,小亚基 2 个 ACT 结构域存在分子内相互作用,2 个 ACT 形成 U 型结构,且 2 个 ACT 在介导大亚基活性抑制中均起作用(Chen 等 2010)。本研究中,相同 AHAS 小亚基点突变与不同位置大亚基点突变转基因植株在缬氨酸抗性与缬氨酸积累上存在差异,提示 AHAS 大小亚基间存在相互作用。AHAS 大亚基上的不同位置突变导致不同的全酶构象,对酶活性产生不同的影响,最终反映出对外源缬氨酸的抗性与体内缬氨酸积累的不同。

AHAS 可能存在由不同位点突变产生的连续构象变化,也可能存在构象变化的阈值效应,在一定的空间变化范围内,突变效应对酶活性产生影响,超过一定的阈值后将完全消除酶活性。不能排除在 AHAS 大小亚基上存在对 AHAS 活性产生影响的其他位点。

参考文献

[1] Brosnan JT, Brosnan ME(2006). Branched – chain amino acids: enzyme and substrate regulation. J Nutr, 136(1):207 – 211.

[2] Chen H, Kristen S, Zhao F, Joyce Q, Xiong L(2010). Genetic analysis of pathway regulation for enhancing branched – chain amino acid biosynthesis in plants. Plant J, 63(4):573 – 583.

[3] Chipman DM, Duggleby RG, Tittmann K(2005). Mechanisms of acetohydroxyacid synthases. Curr Opin Chem Biol, 9(5):475 – 481.

[4] Duggleby RG, Pang SS(2000). Acetohydroxyacid synthase. J Biochem Mol Biol, 33(1): 1 – 36.

[5] Fernstrom JD(2005). Branched – chain amino acids and brain function. J Nutr, 135(6): 1539 – 1546.

[6] Hacham Y, Avraham T, Amir R(2002). The N – terminal region of *Arabidopsis* cystathionine c – synthase plays an important regulatory role in methionine metabolism. Plant Physiol, 128(2):454 – 462.

[7] Harris RA, Joshi M, Jeoung NH, Obayashi M(2005). Overview of the molecular and biochemical basis of branched – chain amino acid catabolism. J Nutr, 135(6 suppl):1527 – 1530.

[8] Hershey HP, Schwartz LJ, Gale JP, Abell LM(1999). Cloning and functional expression ofthe small subunit of acetolactate synthase from Nicotiana plumbaginifolia. Plant Mol Biol, 40(5): 795 – 806.

[9] Hutson SM, Lieth E, LaNoue KF(2001). Function of leucine in excitatory neurotransmitter metabolism in the central nervous system. J Nutr, 131(3):846 – 850.

[10] Iris P, Michael V, Maria V, Chung D, David MC, Ahmed TA, Ze'ev B(2004) Cloning and Characterization of Acetohydroxyacid Synthase from*Bacillus stearothermophilus*. J Bacterol, 186(2):570 – 574.

[11] Kaplun A, Vyazmensky M, Zherdev Y, Belenky I, Slutzker A, Mendel S, Barak Z, Chipman DM, Shaanan B(2006). Structure of the regulatory subunit of acetohydroxyacid synthase isozyme III from Escherichia coli. J Mol Biol, 357(3):951 – 963.

[12] Lee YT, Duggleby RG(2001). Identification of the regulatory subunit of*Arabidopsis thaliana* acetohydroxyacid synthase and reconstitution with its catalytic subunit. Biochemistry, 40(23): 6836 – 6844.

[13] Lee YT, Duggleby RG (2002). Regulatory interactions in *Arabidopsis thaliana* acetohydroxyacid synthase. FEBS Lett, 512(1): 180 - 184.

[14] Liu J, Ishitani M, Halfter U, Kim CS, Zhu JK (2000). The *Arabidopsis thaliana SOS2* gene encodes a protein kinasethat is required for salt tolerance. Proc Natl Acad SciUSA, 97(7): 3730 - 3734.

[15] Mendel S, Elkayam T, Sella C, Vinogradov V, Vyazmensky M, Chipman DM, Barak Z (2001). Acetohydroxyacid synthase: a proposed structure for regulatory subunits supported by evidence from mutagenesis. J Mol Biol, 307(1): 465 - 477.

[16] Mendel S, Vinogradov M, Vyazmensky M, Chipman DM, Barak Z (2003). The N - terminal domain of the regulatory subunit is sufficient for complete activation of acetohydroxyacid synthase III from Escherichia coli. J Mol Biol, 325(2): 275 - 284.

[17] Miflin BJ, Cave PR (1972). The control of leucine, isoleucine, and valine biosynthesis in a range of higher plants. J Exp Bot, 23(2): 511 - 516.

[18] Nair KS, Short KR (2005). Hormonal and signaling role of branchedchain amino acids. J Nutr, 135(6): 1547 - 1552.

[19] Pang SS, Duggleby RG, Guddat LW (2002). Crystal structure of yeast acetohydroxyacid synthase: a target for herbicidal inhibitors. J MolBiol, 317(2): 249 - 262.

[20] Pang SS, Guddat LW, Duggleby RG (2003). Molecular basis of sulfonylurea herbicide inhibition of acetohydroxyacid synthase. J Biol Chem, 278(9): 7639 - 7644.

[21] Rajiv T, Lee, YT, Luke WG, Ronald GD (2005). Probing the mechanism of the bifunctional enzyme ketol - acid reductoisomerase by site - directed mutagenesis of the active site. FEBS J, 272(2): 593 - 602.

[22] Renaud D, Marie CB, Dominique J, Roland D (1995). Evidence fortwo catalytically different magnesium - Binding sites in acetohydroxy acid isomeroreductase by site - directed mutagenesist. Biochemistry, 34(18): 6026 - 6036.

[23] Siew SP, Ronald GD (2001). Regulation of yeast acetohydroxyacid synthase by valine and ATP. Biochem J, 357(3): 749 - 757.

[24] Singh BK, Shaner DL (1995). Biosynthesis of branched chain aminoacids: from test tube to field. Plant Cell, 7(7): 935 - 944.

[25] Tourneur C, Jouanin L, Vaucheret H (1993). Over - expression of acetolactate synthase confers resistance to valine in transgenic tobacco. Plant Sci, 88(2): 159 - 168.

[26] Vyazmensky M, Sella C, Barak Z, Chipman DM (1996). Isolation and characterization of subunits of acetohydroxy acid synthase isozyme III and reconstitution of the holoenzyme. Biochemistry, 35(32): 10339 - 10346.

注：本文曾发表在 2015 年 51 卷 7 期“植物生理学报”期刊上。

平邑甜茶金属硫蛋白基因 MhMT2 的克隆和表达分析

王顺才　梁　东　马锋旺*

【目的】克隆平邑甜茶金属硫蛋白(Metallothionein,MT)基因,探讨该基因对逆境的响应机理,为苹果抗逆分子育种提供依据【方法】采用电子克隆和 RT-PCR 技术从平邑甜茶叶片中克隆 *MhMT*2 基因,利用生物信息学方法对获得的 cDNA 序列及推测氨基酸序列进行分析,通过定量 PCR 技术研究在胁迫条件和激素处理下该基因在叶片中的表达模式。【结果】平邑甜茶 *MhMT*2 基因 cDNA 全长 534 bp,具有一个 243 bp 的完整开放阅读框,编码含 80 个氨基酸的多肽,其分子量为 7.87 kDa。同源性比对显示,*MhMT*2 编码的氨基酸序列与其它植物的 MT2 蛋白有很高的相似性,其中与小金海棠的同源性最高,达到 96.4%。*MhMT*2 编码的多肽包含 14 个 Cys 残基,其 N 端富含 Cys 的结构域具有 C-C、C-X-C 和 C-X-X-C 基序,而 C 端仅有 C-X-C 基序。聚类分析显示,该基因属于植物 MT2 基因家族。定量 PCR 分析表明,*MhMT*2 基因在叶中表达量最高,其次是根,在茎中表达量最低。在 ABA、H_2O_2、机械损伤及 NaCl 处理下,该基因在叶片中的表达都上调,其中以 H_2O_2 最为明显。【结论】对 *MhMT*2 基因在 ABA、H_2O_2、机械损伤及盐胁迫下的表达模式分析,预示该基因表达水平在植物抗逆反应中起着重要作用,为今后进一步探讨其在逆境胁迫中发挥的作用提供了重要信息。

1　引言

【研究意义】金属硫蛋白(Metallothioneins,MTs)是一类富含半胱氨酸(Cys)、具有金属结合能力的低分子量(4~8 kDa)蛋白质。早在 1957 年,在马肾皮质中首次发现了 MT[1],植物 MT 是于 1977 年从大豆根中最早发现的[2],现已发现 MTs

* 作者简介:王顺才(1975—　),男,甘肃陇西人,现为天水师范学院生物工程与技术学院副教授,博士,主要从事果树逆境生理研究。

广泛分布于动物、植物、真菌及蓝细菌等生物体内[3-4]。据研究报道,MTs 与植物的生长发育、胚发育、果实成熟、衰老、抗逆反应以及基因调控等过程有关[3,5-6]。因此,克隆 *MTs* 基因并分析该基因的表达特征,对探讨平邑甜茶对逆境的响应机理及抗逆分子育种具有重要的价值。【前人研究进展】人们已从拟南芥(*AtMT*1 ~ *AtMT*4)、水稻(*OsMT*1 ~ *OsMT*4)、烟草(*Nicotiana glutinosa*)、橡树(*Quercus suber*)、棉花(*Gossypium hirsutum*)、荞麦(*Fagopyrum esculentum*)、莲花(*Nelumbo nucifera*)等植物中克隆了编码 MTs 蛋白的基因或 cDNA 序列[6-12]。然而,由于 MTs 蛋白的高巯基含量使之在空气中极不稳定,在自然状态下分离纯化该蛋白非常困难,迄今为止,除了小麦胚乳 E_c - 1 等极少数植物 MTs 蛋白外,其他绝大数的植物 MTs 蛋白尚未得到分离[5]。与动物 MTs 相比,对植物 MTs 功能的研究远远不足,而对植物 MTs 功能的研究主要集中在基因表达、酵母互补测验及基因敲除等方面[5,13-14]。近年来,对植物 *MTs* 基因的研究在组织器官特异性、表达特征、基因组结构和染色体定位等方面得到快速发展[3]。尽管已从植物中克隆到许多编码 MTs 的基因,但对苹果砧木 *MTs* 基因的研究报道很少[15]。【本研究切入点】苹果是我国北方栽培面积最大的落叶果树之一。苹果主要靠嫁接繁殖,其抗逆性主要来源于砧木。平邑甜茶(*Malushupehensis*)是苹果生产栽培的优良砧木之一。本研究通过筛选干旱诱导的楸子叶 SSH cDNA 文库,获得一段编码 MT 的基因序列,经电子克隆(in silico cloning)和 RT - PCR 扩增,从平邑甜茶叶片中克隆 *MT*2 基因全长序列,进而分析该基因的序列特征、组织表达特异性以及表达模式。【拟解决的关键问题】利用荧光定量 PCR 技术考察 ABA、H_2O_2、机械损伤和盐处理对平邑甜茶 *MT*2 基因的诱导表达变化规律,进而探讨 *MT*2 基因在植物抗逆反应中的可能机理。

2 材料与方法

2.1 材料

干旱处理的楸子叶 SSH cDNA 文库,由西北农林科技大学果树逆境生物学实验室构建保存,该文库的构建和筛选方法见参考文献[16]。

两年生的平邑甜茶盆栽苗,选自西北农林科技大学园艺场苹果资源圃。盆栽苗试验于 2010 年 7 月下旬在西北农林科技大学园艺场大棚内进行。取长势良好一致的盆栽苗各 9 株,用 50 μm ABA 或 10 mM H_2O_2 均匀喷植株顶部叶片的正背面,对照各 6 株用清水喷施。用止血钳子垂直挤压叶片中部 2 ~ 3 次造成机械损伤,对照未挤压。盐胁迫试验在园艺学院水培室进行。一年生的平邑甜茶幼苗在 Hoagland 营养液中正常培养 20 天,每隔半小时通气 3 min,5 d 换一次营养液;20

天后选高度一致的幼苗(约 10 cm)转入添加 100 mM NaCl 的营养液中,盐胁迫处理 10d,对照转置新的营养液中正常生长。取样后立即投入液氮带回实验室,置于 -70℃冰箱备用。

2.2 方法

2.2.1 总 RNA 的提取与 cDNA 合成

总 RNA 提取采用改良 CTAB - LiCl 法[17]。总 RNA 用 RNase - free DNase(Takara,China)37℃处理 30 min 后用于 cDNA 合成。以 Oligo $d(T)_{18}$为反转录引物,用 M - MLV 反转录酶(Invitrogen,USA)试剂盒合成 cDNA 第一链,用于 *MT* 基因的全长 cDNA 克隆。

2.2.2 *MT* 基因的全长 cDNA 克隆

通过 PCR 方法筛选 SSH cDNA 文库时,获得编码 MT 的 EST 序列,以该 EST 序列为种子序列,利用 NCBI 的 EST 数据库进行 EST Contig 的拼接和延伸,利用 NCBI 的 ORF finder 软件结合 BLASTX 程序,进一步筛选可能的全长 cDNA 序列。根据通过电子克隆拼接得到的 *MT* 基因序列,在 ORF 的两端设计基因特异引物(表 1)。PCR 反应程序为:94℃ 5 min,35 个循环(94℃ 30 s,57℃ 30 s,72℃ 1 min),72℃延伸 10 min。PCR 扩增产物经琼脂糖凝胶电泳分析后切下目的条带,用 PCR 纯化试剂盒(TaKaRa)回收。将回收产物与克隆载体 pMD19 - T 连接,转化感受态细胞 DH5 - α 后,涂于添加 100 μg/mL Amp 的 LB 平板过夜培养,挑取阳性克隆进行双向测序。

2.2.3 生物信息学分析

正反向测序结果去除载体序列后,利用 DNAman 和 DNAStar 软件进行序列校对,拼接成一条 cDNA 序列。利用 NCBI(http://ncbi. nlm. nih. gov/)的 ORF finder 软件预测最大开放阅读框,用 BLASTn 和 BLASTp 进行序列比对分析。采用 Compute pI/MW tool(http://www. expasy. org/tools/pi_tool. html)计算蛋白质的等电点和分子量;使用 TMpred(http://www. ch. embnet. org/software/TMPRED_form. html)进行氨基酸跨膜预测;利用 ProtScale(http://expasy. org/tools/protscale. html)与 ProtParam(http://expasy. org/tools/protparam. html)进行疏水性/亲水性预测;利用 superfamily(http://supfam. org/superfamily/hmm. html)和 Pfam(http://pfam. sanger. ac. uk/)进行基因家族预测。采用 MEGA4. 0 软件(http://www. megasoftware. net/)构建系统发生树。

表1 *MT*基因克隆及定量PCR引物

Table 1 Primers for cloning and real - time RT - PCR of *MT* gene

编号 code	引物 Primer sequences(5′ - 3′)	用途 Remark
F1	TTGTCATCAATCCAAAAATGTCGTC	全长cDNA克隆 Full - length cDNA cloning
R1	CAAGTTTTTCATTCATTTCATCACA	
F2	CAACGGGTGCGGGATGGCTCCTGAT	实时定量表达 *MT*Real - time PCR amplification
R2	GCAGGGGTTGCAGGTGCAGCTATCC	
EF - 1α - F	ATTCAAGTATGCCTGGGTGC	内参扩增 EF - 1α Real - time PCR amplification
EF - 1α - R	CAGTCAGCCTGTGATGTTCC	
Actin - F	CCAAAGGCTAATCGGGAGAA	内参扩增 *Actin* Real - time PCR amplification
Actin - R	ACCACTGGCGTAGAGGGAAAG	

2.2.4 平邑甜茶*MT*2基因的定量分析

利用定量RT - PCR方法,对平邑甜茶*MT*2(*MhMT*2)基因的组织特异性表达进行分析,进而分析平邑甜茶叶片*MhMT*2基因在ABA、机械损伤、H_2O_2及盐胁迫下的表达特征。平邑甜茶根、茎和叶的总RNA提取方法同上。RNA完整性及浓度分别用1%琼脂糖凝胶电泳和NanoDrop™分光光度计进行检测。定量分析按PrimeScript© RT reagent Kit With gDNA Eraser试剂盒方法(TaKaRa)进行,该试剂盒方法中包含RNA样品中基因组DNA的去除及cDNA第一链的合成。定量PCR反应程序为:95℃3 min,然后95℃20 s、58℃20 s、72℃20 s进行40个循环,每个循环结束时采集荧光信号。40个循环后PCR扩增产物进行溶解曲线分析(58 ~ 95℃)。利用iQ5.0实时定量PCR仪(Bio - Rad,USA)自带软件分析实时定量PCR数据。用苹果EF - 1α(DQ341381)和*Actin*(GQ339778)作为内参基因以确定*MT*2基因的相对表达量(表1)。

3 结果与分析

3.1 *MhMT*2基因的克隆与分析

从楸子SSH cDNA文库获得一段大小为318 bp的EST序列,经电子克隆拼接后得到了611 bp的cDNA序列,其最大开放阅读框为243 bp。跨ORF设计一对特异引物,通过RT - PCR在平邑甜茶叶片中扩增*MT*2基因,其大小约为500bp(图1)。对目的片段的阳性克隆双向测序后得到该基因的cDNA序列,大小为534 bp,包含一个243 bp的ORF,编码生成一条80个氨基酸的多肽(图2)。该蛋白中Gly(15个)和Cys(14个)和Ser(9个)残基含量较高,而His残基仅有1。该蛋白的相对分子质量为7.87 kDa,理论等电点为5.09,属于酸性蛋白。

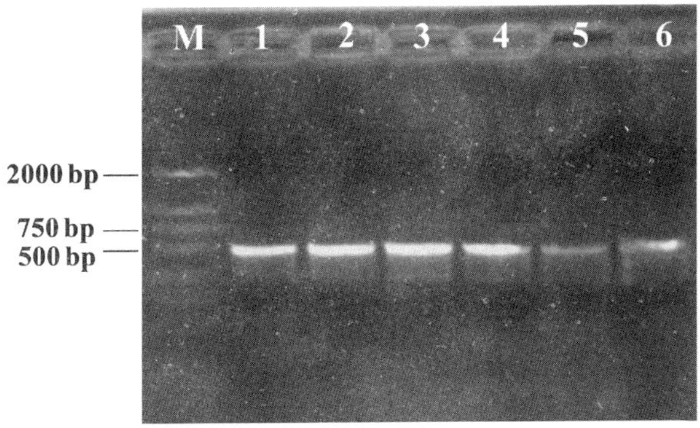

M:DNA 分子量 DL 2000;1～4:CDS 序列。

图 1　平邑甜茶 *MT*2 基因的 PCR 扩增电泳图谱

```
1     TT GTC ATC AAT CCA AAA ATG TCG TCG TGC TGC GGT GGT AAA TGT GGT TGC GGG TCC GGC TGC
                   Primer F1
                               M   S   S   C   C   G   G   K   C   G   C   G   S   G   C   15
63    GGC TGC GGC AGT GGC TGC AAC GGG TGC GGG ATG GCT CCT GAT CTG AGC TAC ATG GAG GGG TCC
                                 Primer F2
      G   C   G   S   G   C   N   G   C   G   M   A   P   D   L   S   Y   M   E   G   S   36
126   ACC ACT GAG ACC CTT GTC ATG GGA GTT GCT CCC CAG AAG TCG CAC TTT GAG GCA TCT GAG ATG
      T   T   E   T   L   V   M   G   V   A   P   Q   K   S   H   F   E   A   S   E   M   57
189   GGA GTT GCA GCT GAG AAC GGA TGC AAG TGC GGG GAT AGC TGC ACC TGC AAC CCC TGC AAG TGC
                                                       Primer R2
      G   V   A   A   E   N   G   C   K   C   G   D   S   C   T   C   N   P   C   K   C   78
252   ACC AAG TGA GTG ACA GAT GCC ACC TTC TGA CGA AGC AAG AGA ATC GTA GGG CTT CTA TGT TTA
      T   K   *                                                                         80
315   GCA GTA GTA ATC TAT GTG TGT GTC TGG TAA TTA TGA TCC TAA TGT ACC AAG GTC GTG TTG TTA
378   GAA TAA AGT AGC CAT GGC TTG CCC TCT GAC CTG CGG ATC CAC TAG GCT AGG ATC TGC TGA GGA
441   TGA TCC TAG GTT ATG GCT GCT TGT GTC TGT GTT TCT ATC ATG TGT TGT GTT GAG TTG GAT CAT
504   ACA GTG TGT GAT GAA ATG AAT GAA AAA CTT G
                   Primer R1
```

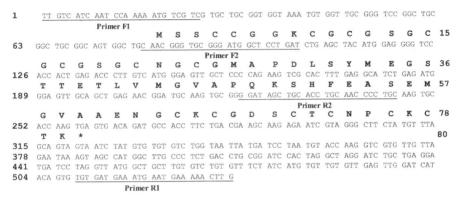

图 2　*MhMT*2 基因 cDNA 全长序列及其推测的氨基酸序列

由图 3 可知,平邑甜茶 ORF 编码蛋白与栽培苹果 MT2(AEJ37038)、小金海棠 MT2(ACX49138)及拟南芥 AtMT2A(AEE74760)与 AtMT2B(AED90465)具有相同的保守序列,其中,平邑甜茶 MT2 与小金海棠的同源性最高,达到 96.4%;从 Cys 残基的排列方式可知,它们都具有典型的 CC、CXC 和 CXXC 排列方式(C 代表 Cys,X 代表其他氨基酸)。

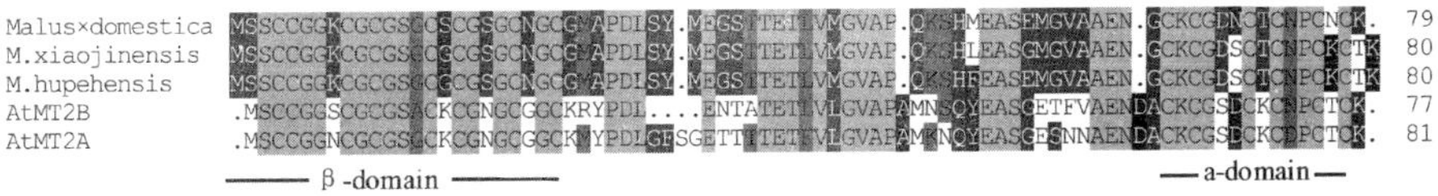

图 3　MhMT2 与其他植物 MT2 的氨基酸序列比对

信号肽预测表明，该蛋白无信号肽结构，为非分泌性蛋白。疏水性/亲水性预测表明，该多肽的总平均亲水性值（GRAVY）为负值，为亲水蛋白。由 Superfamily 预测表明，该基因的 ORF 编码蛋白属于 MT 蛋白家族的可能性较高；经 Pfam 预测显示，该多肽与 Metallothio_2 相似性最高，据 Pfam 注释，该蛋白属于 II 类 MTs。

3.2 *MhMT2* 基因的聚类分析

对平邑甜茶 *MhMT2* 基因编码蛋白序列与来自 33 个植物物种的 51 条 MTs 蛋白序列进行聚类分析（图 4）。聚类分析结果显示，所有植物 MTs 分成 4 个分支，每一分支中来自同一属或物种 MTs 的亲缘关系大于不同属物种的 MTs。在 MT2 分支中，苹果属植物形成独立的一簇，其中平邑甜茶 *MhMT2* 与小金海棠（ACX49138）的亲缘关系最近，而与栽培苹果（AEJ37038）的较近。这可能与栽培苹果的起源及其与苹果属植物不同物种之间的进化关系有关。

研究表明，MTs 在生物进化上较为保守。植物 MT1、MT2 和 MT3 多肽的 N 端和 C 端都具有富含 Cys 的两个结构域，其中的 Cys 按 CC、CXC 或 CXXC 的方式排列。MT1 和 MT3 中的 Cys 一般按 CXC 方式排列，而 MT2 中 Cys 为 CC、CXC 和 CXXC 排列方式；与 MT1 和 MT2 相比，MT3 多肽链中间区的氨基酸含量较少，故序列较短。植物 MT4（Pec 蛋白）中通常具有三个富含 Cys 的结构域和高度保守的两个 His 残基，靠近 C 端的富含 Cys 的结构域高度保守，严格按 CXCXXXCXCXXCXC 方式排列，而其他结构域的 Cys 一般按 CXC 方式排列，散布在整个肽链中[3,13]。由图 4 可以看出，MT1 分支与 MT2、MT3 分支之间的亲缘关系要比 MT4 分支的近些。

以上结果表明，该聚类树是以物种及其 MTs 一级结构为依据而构建形成的，*MhMT2* 基因属于植物 MT2 基因家族。这与上述研究结果相一致。

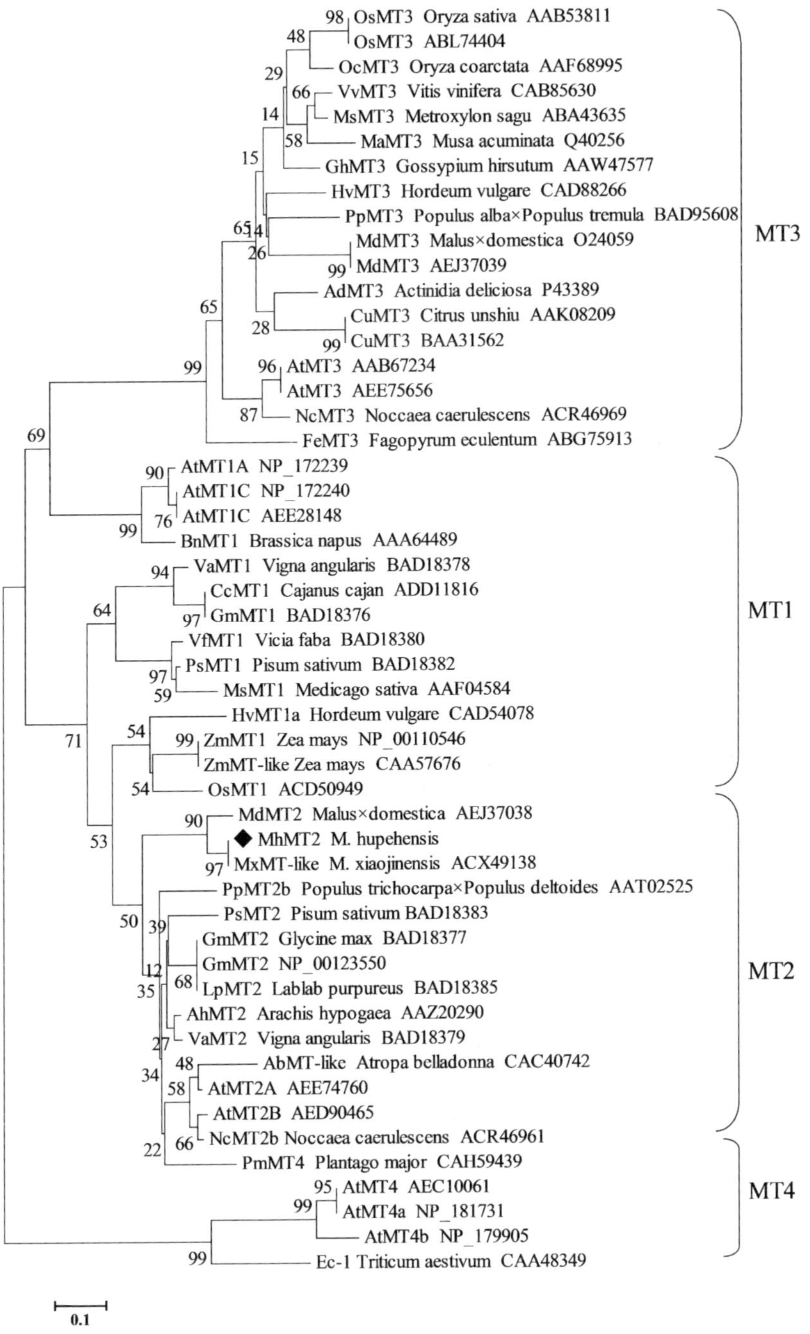

图 4　植物 MTs 的聚类分析

3.3 *MhMT2* 基因在不同组织中的表达分析

结果表明，*MhMT2* 基因在叶中表达最强，其次是根，茎中最弱。其中，叶的表达量是茎的5.2倍，而根为茎的1.9倍（图5）。不同组织中 *MhMT2* 的表达水平不同，这可能与植物不同组织的生长发育调控有关。

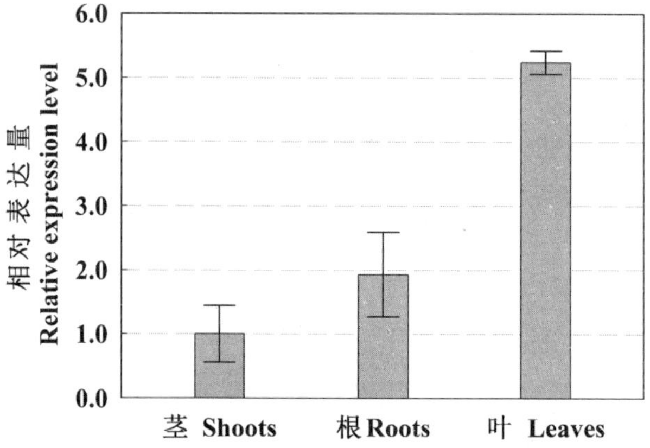

图5 *MhMT2* 基因在不同组织中的表达分析

Fig.5 Expression analysis of*MhMT2* gene in various tissues of *M. hupehensis* seedling

3.4 *MhMT2* 基因在ABA、机械损伤和 H_2O_2 处理下的表达分析

由图6可知，ABA处理3h时，平邑甜茶叶片 *MhMT2* 基因的表达量达到峰值，比对照增加了5.2倍，随后逐渐降低，24 h时仍高于对照表达水平。机械损伤3h时，*MhMT2* 基因的表达水平略有增加，6 h时，其表达水平达到最大值，比对照增加了5.5倍，然后迅速降低并低于原来的表达水平。H_2O_2 处理6 h时，*MhMT2* 基因的表达量达到峰值，比对照增加了13.0倍，随后逐渐降低且高于对照。这表明，*MhMT2* 基因受到ABA、机械损伤及 H_2O_2 处理后均上调表达，但该基因的表达时序及其表达丰度有所不同。

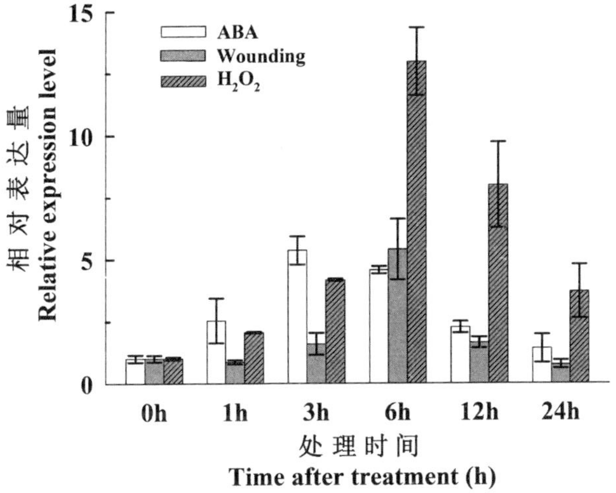

图6 ABA、机械损伤与 H_2O_2 处理下 *MhMT2* 在叶片中的表达分析

3.5 *MhMT2* 基因在盐胁迫下的表达分析

结果表明,盐处理 8d 时,幼苗叶片 *MhMT2* 基因的表达量达到峰值,比对照增加了 3.3 倍,随后逐渐降低,10 d 时其表达水平仍高于对照(图 7)。

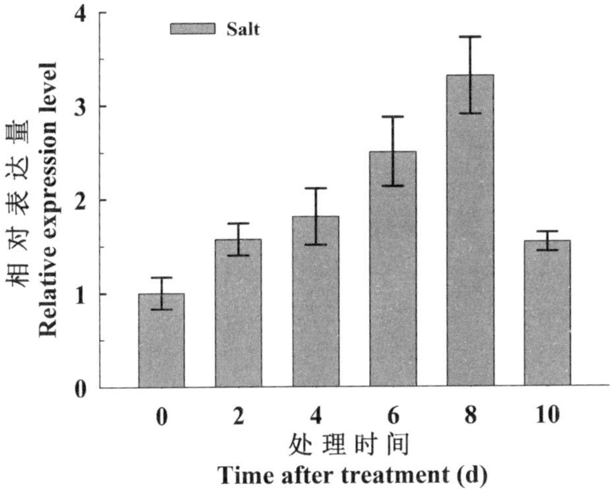

图7 盐处理下 *MhMT2* 在叶片中的表达分析

4 讨论

基于对 MTs 多肽的结构分析,许多学者提出了不同的命名与分类方法。Fowler 等(1987)根据 MTs 中 Cys 残基的排列方式,将所有生物的 MTs 分为三类

（Ⅰ、Ⅱ和Ⅲ类）[18]。而 Binz 和 Kägi（1999）将所有的 MTs 定义为一个超家族（superfamily），并基于其序列相似性及进化关系，将其细分为脊椎动物、软体动物、甲壳动物、棘皮动物、双翅目、线虫、纤毛虫、真菌（Ⅰ～Ⅵ）、原核生物及植物共 15 个家族（family），其中植物 MTs 被分在第 15 家族[19]。最初，Binz 和 Kägi 将植物 MTs 家族分为 p1、p2、p2v、p3、pec 和 p21 六个亚家族（subfamily），后来，人们又将植物 MTs 分为四类：MT1（p1）、MT2（p2）、MT3（p3）与 MT4（pec/E_c蛋白）[3,7,13,20]。近些年，随着多种植物 *MTs* 基因的克隆和分离，人们发现植物 MTs 与其他生物的 MTs 具有一定的相似性，但也有明显的差异[5,21]。植物 MTs 基因家族成员的多肽长度为 45～87 个氨基酸残基，Cys 残基含量在 10～17 个之间，而 His 残基含量很低[3]。在本试验中，平邑甜茶 *MhMT*2 基因编码 80 个氨基酸的多肽，其中含有 14 个 Cys 残基，占总氨基酸数的 17.5%；与小金海棠相比，平邑甜茶 *MhMT*2 多肽在第 52 位 Phe 和第 56 位 Glu 氨基酸残基发生变异，即 Leu_{52}（L）→Phe_{52}（F）、Gly_{56}（G）→Glu_{56}（E），这可能与苹果属植物种间的进化关系有关（图 3）。

研究报道，植物 MT2 多肽 N 端通常具有高度的保守序列 MSCCGGNCGCGS，Cys 残基有典型的 CC、CXC 和 CXXC 排列方式，而且 C 端包含三个按 CXC 方式排列的 Cys 基序（Motif）[20]。本试验中，*MhMT*2 基因编码蛋白含有与之相似的保守序列 $M(S)_2$CCGGKCGCGS，N 端包含 1 个 CC、1 个 CXXC 及 2 个 CXC 基序，在 C 端也有三个 CXC 基序（图 3）。这与水稻 OsMT2（AJ277599）的研究结果相一致[8]。序列比对发现，MhMT2 蛋白与已鉴定的苹果（AEJ37038，94.3%）、大豆（NP_001235506，74.1%）、花生（AAZ20290，70.9%）及拟南芥（AEE74760，86.1%）等植物 MT2 蛋白具有很高的同源性。聚类分析表明，*MhMT*2 与同为苹果属植物的小金海棠、栽培苹果组成一簇，并与豆类、花生和拟南芥等植物 MT2 形成一大分支（图 4）。鉴于以上阐述，笔者认为 *MhMT*2 基因为植物 MT2 基因家族成员。

现已表明，植物 *MTs* 基因的器官表达特异性差别很大。拟南芥 *AtMT*1 在根中表达丰富，在叶中的表达量低；而 *AtMT*2 在叶中的表达量比根中的高；*AtMT*3 主要在果实和叶中表达，而 E_c 主要在种子中表达[3,5,22]。水稻幼苗 *OsMT*1*a* 基因主要在根中表达，在叶和花中的表达量较低，而在茎中的表达量最低[23]。Northern 杂交结果显示，菊芋（*Helianthus tuberosus* L.）*htMT*2 在叶、叶柄、茎及块茎中均有表达，在茎中的表达水平较高，而在根中未检测到杂交信号[24]。但也相反的报道，RNA 原位杂交显示，橡树 *QsMT*2 在种子胚根及幼苗根和茎中均有表达[8]；而苹果 *MT*2 在叶片中几乎检测不到表达，而在花序和果实发育早期表达丰富；在幼果中其表达量随着果实的发育而下降，在成熟期降至最低[5]。最近报道，荞麦 *FeMT*3 在种子胚乳发育过程中均有表达，在叶片保卫细胞和根系维管组织中的杂交信号非常强烈,而在叶片表皮细胞中检测不到杂交信号[12]。在本试验中,*MhMT2* 在根、茎、叶中均有表达,但其表达量明显不同(图 5)。这表明,*MhMT2* 基因的表达具有明显的组织特异性。植物 *MT* 基因表达特性各异的原因可能与不同物种组织的生长阶段、不同的基因类型及发育调控机理有关[22,25]。

据报道,*MTs* 基因表达受金属离子、环境胁迫、激素、损伤、病毒侵染等因素的调节[3,6,25],其中研究最多的是金属离子对 *MTs* 基因表达的影响[13,23,27]。在本试验中,机械损伤使平邑甜茶叶片 *MhMT2* 上调表达(图 6)。这与 Choi 等对烟草 *MT2* 的研究结果相一致[9]。RNA 原位杂交显示,用 H_2O_2处理橡树未成熟胚诱发氧化胁迫,*MT2* 在胚根、胚芽分生组织的杂交信号较强烈[8]。荞麦叶片经 10 mM H_2O_2处理 4.5 小时后,*FeMT3* 的 mRNA 水平增加了 2.4 倍[12]。这与本试验对 H_2O_2处理的研究结果相一致(图 6)。研究发现,植物 MT2 能直接清除细胞体内 O_2^-和 -OH 等活性氧自由基(ROS),在植物抗逆反应中具有重要的作用[3,5,8,28]。在清除 ROS 的过程中,MT 中的 Cys 残基与 ROS 结合而释放出金属离子,金属离子可能参与到植物细胞信号级联反应[3]。在本试验中,平邑甜茶叶面喷施 50μm ABA 诱导 *MhMT2* 上调表达(图 6)。最近报道在拟南芥种子发育过程中,ABA 在 *AtMT4a* 和 *AtMT4b* 的表达调控和功能中起着重要的作用[22]。

最近研究发现,莲花种子 *NnMT2a*、*NnMT2b* 和 *NnMT3* 在盐胁迫处理下,基因表达水平明显上调;转 *NnMT2a* 和 *NnMT3* 基因的拟南芥植株增强了抗衰老的能力[25]。转柽柳(*Tamarix*sp.)*MT1* 基因的烟草植株增强了耐盐胁迫的能力[29]。在本试验中,盐胁迫处理使 *MhMT2* 上调表达(图 7)。但也有相反的报道,水稻 *OsMT2b* 在根中的表达水平经 NaCl 或细胞分裂素处理后,其表达水平明显下降,*OsMT2b* 基因可能参与细胞分裂素的信号传导途径[6]。

综上所述,当植物遭受逆境胁迫时致使 *MT2* 基因表达水平发生改变,从而减轻或抵抗逆境对植物的不良影响。迄今为止,人们对生物 MTs 蛋白的研究时间已超过半个世纪,但其确切的功能仍不清楚[30]。近年来,人们对植物 *MTs* 基因及其蛋白在非生物胁迫中的功能有了初步的认识[14,25-27],但植物 *MTs* 基因是如何调控植物参与逆境反应的分子机制尚不清楚[22,26]。因此,对植物 MTs 的研究仍将是一个重要的方向[5]。

5 结论

克隆并分析了平邑甜茶 *MhMT2* 基因序列。生物信息学分析结果表明,该基因属于植物 MT2 基因家族。组织表达特异性分析表明,*MhMT2* 基因表达水平为叶 > 根 > 茎。定量 PCR 分析表明,在盐胁迫、机械损伤、ABA 和 H_2O_2处理下,*Mh-*

MT2 基因在叶片中均上调表达,其中以 H_2O_2 处理尤为明显,植物可能通过提高该基因表达水平来抵抗或适应逆境的影响。

REFERENCES

[1]Margoshes M,Vallee B L. A cadmium protein from equine kidney cortex. *Journal of the American Chemical Society*,1957,79(17):4813 - 4814.

[2]Casterline J L,Barnett N M. Isolation and characterization of cadmium binding components in soybean plants. *Plant Physiology*,1977,59:S - 124.

[3]Hassinen V H,Tervahauta A I,Schat H,Karenlampi S O. Plant metallothioneins - - metal chelators with ROS scavenging activity? *Plant Biology*(Stuttg),2011,13(2):225 - 232.

[4]Vasak M,Meloni G. Chemistry and biology of mammalian metallothioneins. *Journal of Biological Inorganic Chemistry*,2011,16(7):1067 - 1078.

[5]Freisinger E. Plant MTs - long neglected members of the metallothionein superfamily. *Dalton Transactions*,2008,47:6663 - 6675.

[6]Yuan J,Chen D,Ren Y,Zhang X,Zhao J. Characteristic and expression analysis of a metallothionein gene, OsMT2b, down - regulated by cytokinin suggests functions in root development and seed embryo germination of rice. *Plant Physiology*,2008,146(4):1637 - 1650.

[7]Xue T,Li X,Zhu W,Wu C,Yang G,Zheng C. Cotton metallothionein GhMT3a,a reactive oxygen species scavenger,increased tolerance against abiotic stress in transgenic tobacco and yeast. *Journal of Experimental Botany*,2009,60(1):339 - 349.

[8]Mir G,Domenech J,Huguet G,Guo W J,Goldsbrough P,Atrian S,Molinas M. A plant type 2 metallothionein(MT)from cork tissue responds to oxidative stress. *Journal of Experimental Botany*,2004,55(408):2483 - 2493.

[9]Choi D,Kim H M,Yun H K,Park J A,Kim W T,Bok S H. Molecular cloning of a metallothionein - like gene from Nicotiana glutinosa L. and its induction by wounding and tobacco mosaic virus infection. *Plant Physiology*,1996,112(1):353 - 359.

[10]Zhou G,Xu Y,Li J,Yang L,Liu J Y. Molecular analyses of the metallothionein gene family in rice (*Oryza sativa* L.). *Journal of biochemistry and molecular biology*, 2006, 39 (5): 595 - 606.

[11]Guo W - J,Bundithya W,Goldsbrough P B. Characterization of the *Arabidopsis* metallothionein gene family:tissue - specific expression and induction during senescence and in response to copper. *New Phytologist*,2003,159:369 - 381.

[12]Samardzic J T,Nikolic D B,Timotijevic G S,Jovanovic Z S,Milisavljevic M D,Maksimovic V R. Tissue expression analysis of FeMT3,a drought and oxidative stress related metallothionein gene from buckwheat(*Fagopyrum esculentum*). *Journal of Plant Biology*,2010,167(16):1407

-1411.

[13]Freisinger E. The metal - thiolate clusters of plant metallothioneins. *Chimia* (Aarau), 2010,64(4):217 -224.

[14]Yang J,Wang Y,Liu G,Yang C,Li C. Tamarix hispida metallothionein - like ThMT3,a reactive oxygen species scavenger,increases tolerance against Cd^{2+},Zn^{2+},Cu^{2+},and NaCl in transgenic yeast. *Molecular Biology Reports*,2011,38(3):1567 - 1574.

[15]张玉刚,韩振海.小金海棠金属硫蛋白基因 MxMT2 克隆与生物信息学分析[J].华北农学报,2010,25(2):60 -63.

ZHANG Y G,HAN Z H. Cloning and bioinformatic analysis of metallothionein gene(*MxMT2*) in *Malus xiaojinensis*. *Acta Agriculturae Boreali - Sinica*,2010,25(2):60 -63(in chinese)

[16]Wang S,Liang D,Shi S,Wang R,Ma F,ShuH. Isolation and characterization of a novel drought responsive gene encoding a glycine - rich RNA - binding protein in *Malus prunifolia* (Willd.)Borkh. *Plant Molecular Biology Reporter*,2011,29:125 - 134.

[17]Chang S,Puryear J,Cairney J. Simple and efficient method for isolating RNA from pine trees. *Plant Molecular Biology Reporter*,1993,11(2):113 - 116.

[18]Fowler B A,Hildebrand C E,Kojima Y,Webb M. Nomenclature of metallothionein. *Experientia Suppl*,1987,52:19 -22.

[19]Binz P A,Kägi J H R. Metallothionein:molecular evolution and classification. *In*:*Klaassen C.* (*Ed.*),*Metalloth - ionein* Ⅳ,*Birkhäuser Verlag Basel*,Switzerland,1999:pp. 7 - 13.

[20]Cobbett C,Goldsbrough P. Phytochelatins and metallothioneins:roles in heavy metal detoxification and homeostasis. *Annual Review of Plant Biology*,2002,53:159 - 182.

[21]Wan X,Freisinger E. The plant metallothionein 2 from Cicer arietinum forms a single metal - thiolate cluster. *Metallomics*,2009,1(6):489 -500.

[22]Ren Y,Liu Y,Chen H,Li G,Zhang X,Zhao J. Type 4 metallothionein genes are involved in regulating Zn ion accumulation in late embryo and in controlling early seedling growth in *Arabidopsis*. *Plant*,*Cell and Environment*,2011,35(4):770 -789.

[23]Yang Z,Wu Y,Li Y,Ling H Q,Chu C. OsMT1a,a type 1 metallothionein,plays the pivotal role in zinc homeostasis and drought tolerance in rice. *Plant Molecular Biology*,2009,70(1 - 2):219 -229.

[24]Chang T - J,Chen L,Lu Z - X,Chen W - X,Liu X,Zhu Z. Cloning and Expression Patterns of a Metallothionein - like Gene htMT2 of *Helianthus tuberosus*. *Acta Botanica Sinica*,2002,44(10):118 - 1193.

[25]Zhou Y,Chu P,Chen H,Li Y,Liu J,Ding Y,Tsang E W,Jiang L,Wu K,Huang S. Overexpression of *Nelumbo nucifera* metallothioneins 2a and 3 enhances seed germination vigor in *Arabidopsis*. *Planta*,2011,235(3):523 -537.

[26]Singh R K,Anandhan S,Singh S,Patade V Y,Ahmed Z,Pande V. Metallothionein - like

gene from Cicer microphyllum is regulated by multiple abiotic stresses. *Protoplasma*,2011,248(4):839 - 847.

[27] Osobova M, Urban V, Jedelsky P L, Borovicka J, Gryndler M, Ruml T, Kotrba P. Three metallothionein isoforms and sequestration of intracellular silver in the hyperaccumulator *Amanita strobiliformis*. *NewPhytologist*,2011,190(4):916 - 926.

[28] Zhu W, Zhao DX, Miao Q, Xue TT, Li XZ, Zheng CC. *Arabidopsis thaliana* Metallothionein, AtMT2a, Mediates ROS Balance during Oxidative Stress. *Journal of Plant Biology*, 2009, 52(6):585 - 592.

[29] 周博如,王雷,吴丽丽,姜廷波. 转金属硫蛋白基因(MT1)烟草耐 NaCl 胁迫能力[J]. 生态学报,2010,30(15):4103 - 4108.

ZHOU B R, WANG L, WU L L, JIANG T B. Characterization of transgenic tobacco overexpressing metallothionein gene(MT1) onNaCl stress. *Acta Ecologica Sinica*, 2010, 30(15):4103 - 4108(in chinese)

[30] Sutherland D E, Stillman M J. The "magic numbers" of metallothionein. *Metallomics*, 2011, 3(5):444 - 463.

本论文已发表在《中国农业科学》,2012 年 45 卷 14 期,pp:2904 - 2912.

干旱胁迫对3种苹果属植物叶片解剖结构、微形态特征及叶绿体超微结构的影响

王顺才　邹养军　马锋旺*

采用树脂包埋块半薄/超薄切片技术，通过光镜(LM)、扫描电镜(SEM)和透射电镜(TEM)方法，研究了轻度、中度和严重干旱胁迫对楸子(*Malus prunifolia*)、新疆野苹果(*M. sieversii*)和平邑甜茶(*M. hupehensis*)叶片组织解剖结构、表皮微形态特征(气孔密度、大小及角质层厚度)及叶绿体超微结构的影响。光镜观察结果表明，与对照相比，干旱胁迫条件下3种苹果属植物叶片厚度、栅栏组织厚度及叶肉组织结构紧密度(CTR)都显著减小($P<0.05$)，而海绵组织厚度与叶肉组织结构疏松度(SR)均显著增加($P<0.05$)。扫描电镜观察结果显示，3种苹果属植物幼叶气孔密度在干旱胁迫下显著增大($P<0.05$)，而气孔宽度、开张比及其开张度明显下降。透射电镜观察结果表明，在干旱胁迫下，楸子和新疆野苹果上下角质层厚度逐渐增加，而平邑甜茶的随干旱胁迫程度增加呈先增后减的变化；在轻度和中度水分胁迫下，叶绿体膨胀变形，淀粉粒变小消失，基粒片层排列松散减少，类囊体腔扩大；在严重胁迫条件下，叶绿体膨胀近圆形，叶绿体膜破裂，类囊体严重泡化开始解体。与平邑甜茶相比，严重水分胁迫下楸子和新疆野苹果叶绿体超微结构损伤较小，能较好地保持细胞结构的完整性。

苹果是我国第一大果品产业，也是世界栽培面积最广的树种之一。苹果优势产区多为干旱和半干旱地区，水分胁迫成为制约这些地区苹果生长发育和产量的重要非生物因子。我国是苹果属植物起源中心之一，拥有丰富的苹果抗性种质资

---

* 作者简介：王顺才(1975—　)，男，甘肃陇西人，现为天水师范学院生物工程与技术学院副教授，博士，主要从事果树逆境生理研究。

源，进行苹果属植物抗逆性研究对苹果生产及推广具有重大的意义。苹果（*Malus domesticaBorkh.*）主要靠嫁接繁殖，其砧木是树体生长的基础，对于控制树体大小、增强果树抗逆性及抗病能力都有重要的影响[1-2]。干旱对苹果属植物的影响既有直观的形态解剖学变化，又有生理生化反应，还有基因表达的差异[3-6]。目前，尽管苹果属植物抗旱性研究已有较多报道，但多数集中在生理生化方面的抗性比较[7-8]，而干旱胁迫条件下苹果属植物叶片显微及超微结构，特别是抗旱材料与不抗旱材料间叶片微形态特征、解剖结构及超微结构差异的研究依然缺乏[9]。楸子（*M. prunifolia*（Willd.）Borkh.）、新疆野苹果（*M. sieversii* Ledeb.）和平邑甜茶（*M. hupehensis* Rehd.）是苹果生产中应用较广的砧木材料。平邑甜茶抗旱性较差，而楸子和新疆野苹果抗旱性较强[4-5,8,10]。本研究以3种苹果属植物为研究对象，应用树脂包埋块半薄/超薄切片技术，通过光镜、扫描电镜及透射电镜方法，观察在轻度、中度和重度干旱胁迫下植物叶片解剖结构、角质层厚度、气孔特征及叶绿体超微结构的变化及其差异，探明3种植物抗旱生物学结构基础，旨在揭示苹果属植物对干旱胁迫的响应和适应性机制，并为抗旱苹果品种的选育提供理论依据。

1 材料与方法

1.1 试验材料

楸子（陕西富平，34°75′N，109°15′E）、新疆野苹果（新疆巩留，43°15′N，82°51′E）和平邑甜茶（山东平邑，35°07′N，117°25′E）种子均从原产地采调，经4℃层积处理发芽后，选生长整齐一致的幼苗进行盆栽，移栽于塑料盆（30 cm×26.5 cm×22 cm）中，每盆1株，培养基质由田园土、细砂和腐熟羊粪按5:1:1（V:V:V）配制。在西北农林科技大学园艺场日光温室中进行常规管理1年。

1.2 试验处理

2009年6月底进行干旱处理。每树种分别选取长势一致（株高约1米）的2年生盆栽苗各90盆（株），分成两组，一组正常供水（对照30株）；另一组控水处理（60株），直至植株叶片出现永久性凋萎，共12 d。试验期间，观察叶片变化及受害程度，测定土壤容积含水量和叶片相对含水量，隔天分别采样。为了防止自然雨水的影响，干旱处理在温室内进行。

1.3 测试指标

1.3.1 土壤容积含水量和叶片相对含水量测定

利用HH_2 Moisture Meter便携式土壤水分测定仪（Delta - T Devices，Cambridge，UK）分别测定3种苹果植株幼苗的土壤容积含水量，每隔1 d测定1次，每次每种植物测6~8株(盆),每株重复3次。同期植株叶片相对含水量参照Barrs和Weatherley的方法(1962)测定[11],每次每种植物选6~8株,于每株顶端切取3片完全展开叶片,分别计算平均值。

1.3.2 透射电镜样品的制备与观察

取对照及干旱处理植株新梢顶端完全展开的第4片幼叶(形态学顶端自上而下),在叶片主叶脉两侧用刀片切取大小为1 mm × 1 mm的组织块,立即投入用0.1 mol·L^{-1}磷酸缓冲液(PBS,pH6.8,下同)配制的1%戊二醛中抽气5 min,组织块下沉后,用PBS配制的4%戊二醛于4℃下固定3 h,PBS冲洗4~6次、30 min/次;再用PBS配制的1%锇酸在4℃下后固定2 h,用PBS冲洗4~6次、30 min/次;丙酮梯度脱水,依次用30%、50%、70%、80%、90%和95%的丙酮浸泡,每梯度脱水30 min,再用100%丙酮脱水3次、30 min/次;脱水后进行渗透与包埋,100%丙酮:包埋剂=3:1渗透3 h,1:1渗透5 h,1:3渗透12 h,然后换成纯包埋剂包埋48 h,每24 h更换1次;最后在30℃下聚合12 h,60℃聚合48 h。包埋剂由10 ml环氧树脂Epon812 + 7 ml软化剂MNA +4 ml硬化剂DDSA +0.3 ml催化剂DMP-30配制而成。

用LeicaultracutUCT超薄切片机(Leica Microsystems GmbH, Wetzlar, Germany)切出厚度约为70 nm的超薄切片,用醋酸双氧铀和柠檬酸铅双重染色后,用JEM-1230型透射电镜(JEOL Ltd., Tokyo, Japan)对叶片上下角质层及叶绿体等亚细胞超微结构进行观测。超薄切片每种植物每个处理观察20个视野并照相,用ImageJ软件(http://rsbweb.nih.gov/ij/)进行测量。

1.3.3 扫描电镜样品的制备与观察

与透射电镜相同取样,叶片切取大小为2mm×2mm,快速投入用0.1mol·L^{-1} PBS配制的1%戊二醛中,抽气使组织块下沉后,用PBS配制的4%戊二醛于4℃下固定3 h,用PBS冲洗残留的固定液4~6次、30 min/次。然后依次用30%、50%、70%、80%、90%和95%丙酮梯度脱水,每一梯度脱水30 min,再用100%丙酮梯度脱水3次、30 min/次,然后用醋酸异戊酯置换2次、30 min/次,再用Hitachi HCP-2型临界点干燥器进行CO_2临界点干燥,喷金后用JSM-6360LV型扫描电镜(JEOL Ltd., Tokyo, Japan)对叶片表面进行观测照相,每种植物每个处理观察20个视野。用ImageJ软件对气孔数目、大小(长度×宽度)及开张度(用气孔开口横径表示)进行测量和计算。

1.3.4 显微结构样品的制备与观察

选取上述透射电镜制样块,切出厚度约为1100 nm的半薄横切切片,用1%甲苯胺蓝染色20~25 s,用蒸馏水冲洗后置于60℃烘干,在光学显微镜CX31-

12C02(OlympusCo.,Ltd.)下观察叶片上下表皮厚度、栅栏组织厚度、海绵组织厚度和叶片厚度并拍照,用 ImageJ 软件进行测量。并计算:叶片组织结构紧密度 CTR% = 栅栏组织厚度/叶片总厚度 ×100%,叶片组织结构疏松度 SR% = 海绵组织厚度/叶片总厚度 ×100%。

1.4 数据处理与统计分析

试验数据采用 SPSS16.0 软件进行单因素方差分析(One - way ANOVA)及 Tukey's 多重比较。

2 结果与分析

2.1 干旱胁迫下土壤容积含水量和叶片相对含水量的变化

本试验通过控水方式使盆栽苗逐渐处于缺水状态,结合土壤湿度和叶片组织的水分状况反映植株水分亏缺程度。正常供水条件下,平邑甜茶、楸子和新疆野苹果盆栽苗的土壤容积含水量(SVWC)相差不大,变化范围在 33.4 ± 0.8% 之间。干旱条件下,随着控水时间的延长,其 SVWC 值逐渐下降;控水 4、8 和 12 d 时,与对照相比,SVWC 值降幅种间差异不显著,降幅平均值为 6.9%、13.5% 和 21.0%(数据未显示)。Hsiao(1973)将中生植物按水分饱和以及亏缺 10% 以下、10% ~15% 和 15% 以上三个标准划分出轻度、中度和严重三个水分胁迫级别[12]。本试验认为,控水 4、8 和 12 d 时,3 种苹果属植物分别处于轻度、中度和重度水分胁迫阶段。这与我们前期试验结果相一致[4]。

控水 4、8 和 12 d 时,与对照相比,平邑甜茶叶片相对含水量(LRWC)下降了 12.1%、27.4% 和 34.5%,楸子 LRWC 值下降了 5.3%、18.6% 和 27.5%,而新疆野苹果 LRWC 值下降了 7.2%、19.9% 和 30.7%(数据未显示)。结果表明,平邑甜茶叶片对土壤水分亏缺的反应要比楸子和新疆野苹果的敏感。

2.2 干旱胁迫下叶片解剖结构的变化

与对照相比,在轻度、中度和重度干旱胁迫下,楸子叶厚下降了 2.6%、8.5% 和 10.9%($P<0.01$),新疆野苹果叶厚下降了 1.7%、4.4% 和 6.4%($P<0.05$),而平邑甜茶叶厚下降了 0.5%、2.2% 和 11.3%($P<0.01$);楸子栅栏组织厚度的降幅为 11.0%、25.1% 和 28.1%($P<0.01$),新疆野苹果的降幅为 5.5%、14.2% 和 21.5%($P<0.01$),而平邑甜茶的降幅为 7.6%、22.8% 和 36.0%($P<0.01$);楸子海绵组织增加了 1.0%、2.2% 和 9.8%,新疆野苹果的增加了 1.5%、3.5% 和 9.5%($P<0.05$),而平邑甜茶的增加了 6.6%、10.4%($P<0.01$)和 2.4%(表 1)。

表 1　干旱胁迫下 3 种苹果属植物叶片解剖结构的变化

种 Species	处理 Treatment	栅栏组织厚度 Thickness of palisade(μm)	海绵组织厚度 Thickness of spongy(μm)	上表皮 Upper epidermis (μm)	下表皮 Lowerepidermis (μm)	叶片厚度 Thickness of leaf(μm)	栅栏/叶厚 Cell tightness rate (CTR%)	海绵/叶厚 Scattered rate (SR%)
楸子 *M. prunifolia*	CK	93.4 ±3.5a	91.9 ±2.5b	14.8 ±1.0b	12.1 ±0.8b	212.2 ±6.1a	44.0% a	43.3% c
	LD	83.1 ±8.4b*	92.8 ±6.3ab	16.9 ±1.5a**	13.6 ±1.3a**	206.7 ±4.2a	40.2% b**	44.9% c
	MD	69.9 ±8.0c**	93.9 ±5.5ab	16.6 ±1.3a**	13.5 ±0.9a**	194.2 ±8.6b**	36.0% c**	48.4% b**
	SD	67.1 ±6.3c**	101.0 ±9.4a	14.5 ±1.0b	11.2 ±1.4b	189.1 ±6.7b**	34.4% c**	52.3% a**
新疆野苹果 *M. sieversii*	CK	110.1 ±4.6a	113.9 ±3.5b	18.3 ±3.1a	12.8 ±1.5a	255.5 ±5.7a	43.1% a	44.7% c
	LD	104.1 ±5.4a	115.6 ±6.4b	18.0 ±2.7a	13.2 ±1.5a	251.0 ±10.8ab	41.5% ab	45.8% c
	MD	94.5 ±3.2b**	117.8 ±3.2ab	15.2 ±1.3a	12.8 ±0.9a	244.3 ±8.6ab	39.3 % b**	49.0% b**
	SD	86.4 ±5.5c**	124.7 ±6.7a*	15.0 ±2.8a	12.8 ±1.7a	239.2 ±8.9b*	36.1% c**	52.3% a**
平邑甜茶 *M. hupehensis*	CK	98.4 ±7.1a	136.9 ±2.9b	15.7 ±0.9a	12.1 ±0.7b	263.0 ±6.9a	37.4% a	52.1% c
	LD	90.9 ±4.3b	143.7 ±7.2ab	16.0 ±1.5a	12.3 ±0.8ab	261.8 ±5.9a	34.3% b**	55.3% b**
	MD	76.0 ±3.7c**	151.1 ±4.1a**	16.4 ±1.9a	13.3 ±1.3ab	257.3 ±6.4a	29.5% c**	58.7% a**
	SD	63.0 ±2.9d**	140.1 ±4.9b	16.3 ±1.2a	13.8 ±1.2a*	233.3 ±6.8b**	27.0% d**	60.1% a**

注：CK、LD、MD 和 SD 分别代表对照、轻度、中度和重度水分胁迫。数据以平均数 ± 标准差(SD)表示，同列数据后不同的字母表示显著水平 $P<0.05$。* 表示与对照相比，在 $P<0.05$ 水平上显著，** 表示在 $P<0.01$ 水平上极显著，下同。

干旱胁迫下，楸子上下表皮厚度呈先增后减的变化，在轻度和中度干旱胁迫下，楸子上下表皮厚度变化与对照达极显著水平($P<0.01$)；新疆野苹果上表皮厚度呈下降趋势，下表皮厚度略有增加，与对照均未达显著水平；与对照相比，平邑甜茶上下表皮厚度略有增加，仅在重度胁迫下其下表皮厚度变化达显著水平($P<0.05$)(表1)。

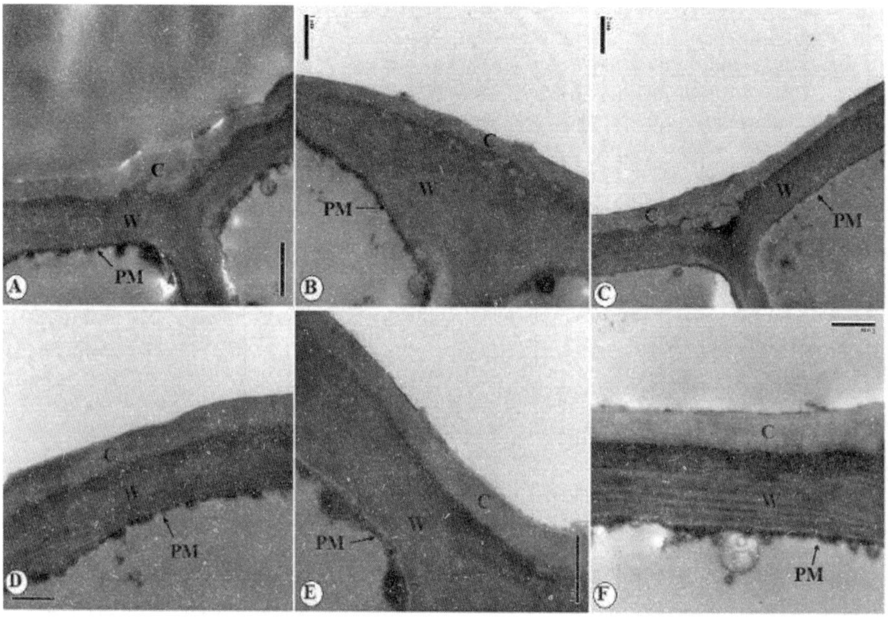

图1　干旱胁迫下3种苹果属植物叶片角质层厚度的变化

图版说明：A－C为楸子、平邑甜茶与新疆野苹果的对照上角质层；D－F为楸子、平邑甜茶与新疆野苹果在重度干旱胁迫下的上角质层。C－角质层，W－细胞壁，PM－质膜。Bar A－C：2 μm，D－F：1 μm。

2.3　干旱胁迫下叶片角质层厚度的变化

正常条件下，3种苹果属植物叶片的上下角质层厚度分别在502～529 nm和305～403 nm之间(表2)。在干旱胁迫下，楸子和新疆野苹果上下角质层逐渐增厚，而平邑甜茶上下角质层厚度呈先增后减的变化。在重度胁迫下，楸子和新疆野苹果上角质层厚度比对照分别增加了47.2%和35.9%($P<0.01$)，而下角质层厚度分别增加了57.8%和55.6%($P<0.01$)；平邑甜茶下角质层厚度比对照增加了13.4%($P<0.01$)，但上角质层厚度比对照减少了20.4%($P<0.01$)。

表2　干旱胁迫下3种苹果属植物叶片角质层厚度的变化

种 Species	组织 Tissue	处理 Treatment			
		CK	LD	MD	SD
楸子 *M. prunifolia*	上角质层 UC(nm)	502 ±41c	533 ±31bc	584 ±25b**	739 ±34a**
	下角质层 LC(nm)	403 ±32c	465 ±18b*	501 ±33b**	636 ±55a**
新疆野苹果 *M. sieversii*	上角质层 UC(nm)	512 ±20c	602 ±17b**	612 ±29b**	696 ±48a**
	下角质层 LC(nm)	340 ±10c	438 ±33b**	562 ±34a**	529 ±35a**
平邑甜茶 *M. hupehensis*	上角质层 UC(nm)	529 ±29a	582 ±23a	432 ±19b**	421 ±25b**
	下角质层 LC(nm)	305 ±34d	399 ±29b**	450 ±14a**	346 ±33c**

注:UC和LC分别代表上、下角质层厚度。数据以平均数±标准差表示,同行数据后不同的字母表示显著水平 $P < 0.05$。

2.4　干旱胁迫下叶片气孔形态特征的变化

扫描电镜观察发现,3种苹果属植物叶片气孔仅分布于下表皮。正常条件下,楸子和新疆野苹果下表皮蜡质纹饰较少(图版2A、2C),而平邑甜茶下表皮表面平坦光滑,纹饰不明显(图版2B);干旱胁迫下,楸子和新疆野苹果气孔周围条索状纹饰明显增多变粗,其中楸子尤为明显(图版2D、2F),而平邑甜茶下表皮表面变的凹凸不平,气孔明显下陷闭合(图版2E)。中度干旱胁迫下,楸子和新疆野苹果上表皮细胞呈不规则多边形,表面无纹饰(图版2G、2I),而平邑甜茶上表皮布满条状纹饰(图版2H)。

正常条件下,3种苹果属植物气孔密度在313.6~344.9个/mm^{2}之间,气孔大小(长度×宽度)依次为新疆野苹果(26.2×18.2)>楸子(19.5×16.4)>平邑甜茶(18.9×15.6)(表3)。在轻度、中度和重度水分胁迫下,楸子气孔密度的增幅为54.1%、94.2%和98.0%,新疆野苹果的增幅为29.6%、44.5%和61.8%,而平邑甜茶的增幅为5.9%、12.5%和33.8%。与对照相比,在中度和重度干旱胁迫下,楸子和新疆野苹果气孔长度略有增加但变化不显著,而气孔宽度显著($P<0.05$)或极显著降低($P<0.01$);干旱胁迫条件下平邑甜茶气孔长度和宽度呈下降变化趋势,在重度胁迫下达到极显著水平($P<0.01$)。在重度胁迫下,气孔宽度降

幅依次为楸子(16.3%)>平邑甜茶(12.3%)>新疆野苹果(7.18%)。

在中度和重度干旱胁迫下,楸子和平邑甜茶的气孔开张比和开张度显著降低($P<0.05$),而新疆野苹果的降幅变化不显著。重度胁迫下,楸子和平邑甜茶气孔大多数几乎完全关闭(图版2D、2E),而新疆野苹果尚有部分气孔处于半闭合或半开张状态(图版2F)。

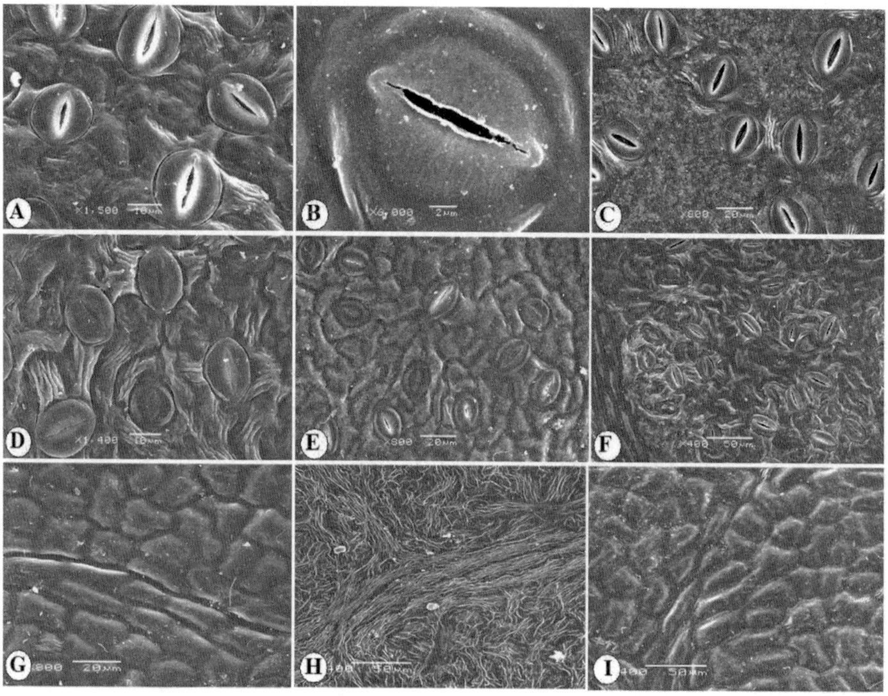

图版2 干旱胁迫下3种苹果属植物气孔及叶表皮形态特征的变化

图版说明:A-C为楸子、平邑甜茶与新疆野苹果的对照;D-F为楸子、平邑甜茶与新疆野苹果在重度干旱下的下表皮面;G-I为楸子、平邑甜茶与新疆野苹果在中度干旱下的上表皮面。

表 3　干旱胁迫对 3 种苹果属植物叶片气孔特征、叶绿体及淀粉粒大小的影响

种 Species	处理 Treatment	气孔密度 Density of stomata (n/mm^2)	气孔开张比 Stomatal opening rate (%)	气孔长 Stomata length (μm)	气孔宽 Stomata width (μm)	气孔开张度 Stomatal aperture (μm)	叶绿体大小 Chloroplast size		淀粉粒大小 Starch grainsize	
							长 Length (μm)	宽 Width (μm)	长 Length (μm)	宽 Width (μm)
楸子 *M. prunifolia*	CK	313.6 ± 30.3c	93.9 ± 4.6a	19.5 ± 1.6a	16.4 ± 0.7a	2.6 ± 0.2a	5.0a	1.6c	2.5	0.8
	LD	483.4 ± 22.4b**	92.8 ± 5.3a	19.7 ± 1.3a	15.1 ± 1.0ab	2.2 ± 0.3b*	4.9a	1.7c	1.2**	0.3**
	MD	610.9 ± 38.2a**	66.2 ± 5.2b**	19.5 ± 0.2a	14.7 ± 1.0b*	1.9 ± 0.2b**	4.7a	1.9b**	ND	ND
	SD	622.8 ± 25.6a**	26.2 ± 2.4c**	20.4 ± 1.0a	13.7 ± 1.3b**	0.9 ± 0.1c**	3.5b*	2.3a**	ND	ND
新疆野苹果 *M. sieversii*	CK	319.9 ± 29.5c	96.2 ± 2.8a	26.2 ± 0.8b	18.2 ± 0.3a	3.1 ± 0.4a	5.4a	1.9c	2.6	0.8
	LD	414.7 ± 37.0b**	95.6 ± 2.7ab	26.7 ± 0.9ab	17.7 ± 0.8ab	2.7 ± 0.2b*	5.2a	2.1b*	1.9**	0.4**
	MD	462.1 ± 52.3b**	90.0 ± 3.0bc	27.0 ± 1.0ab	17.1 ± 0.7b*	2.3 ± 0.2bc**	4.2b**	2.4a**	ND	ND
	SD	517.7 ± 41.8a**	85.2 ± 3.3c**	27.6 ± 0.7a	16.9 ± 0.8b*	2.2 ± 0.2c**	3.4c**	1.4d**	ND	ND
平邑甜茶 *M. hupehensis*	CK	344.9 ± 13.5c	94.7 ± 2.8a	18.9 ± 0.9a	15.6 ± 0.8a	2.2 ± 0.2a	5.9a	1.5b	2.0	0.6
	LD	365.2 ± 14.4bc	90.6 ± 2.1a	18.2 ± 0.7a	14.6 ± 0.6b*	2.1 ± 0.3a	5.6a	1.5b	1.1**	0.3**
	MD	388.1 ± 33.1b*	81.2 ± 3.8b**	18.1 ± 0.8a	13.9 ± 0.8bc**	1.8 ± 0.1b**	4.3b**	1.7a**	ND	ND
	SD	461.4 ± 49.7a**	22.9 ± 1.9c**	16.2 ± 1.0b**	13.6 ± 0.6c**	1.1 ± 0.2c**	2.2c**	1.0c**	ND	ND

注：数据用平均数 ± 标准差表示，同列数据后不同的字母表示显著水平 $P < 0.05$。ND 表示无数据。

2.5 干旱胁迫下3种苹果属植物叶绿体超微结构的变化

正常条件下,3种苹果属植物叶肉细胞内,叶绿体紧贴细胞壁分布,每个叶绿体上有1~2个淀粉粒清晰可见(图版3A、3E和3I);干旱胁迫下,叶绿体超微结构的变化表现出许多类似的共同特征:在轻度胁迫下,叶绿体轻微膨胀,淀粉粒大小明显变小、数量减少(表3),基质片层变稀薄模糊(图版3B、3F和3J);在中度胁迫下,叶绿体基质片层排列松散,形成弯曲状,淀粉粒消失,基粒类囊体腔扩大,基质片层出现囊泡,平邑甜茶叶肉细胞出现明显的质壁分离现象(图版3G),而楸子和新疆野苹果的不明显(图版3C、3K);在重度胁迫下,叶绿体膨胀成近圆形,相互聚集且排列混乱,叶绿体基质片层严重囊泡化,基粒类囊体解体呈现空泡化,垛叠片层数量减少甚至消失(图版3D、3L),其中平邑甜茶基质片层泡化较严重,叶绿体膜部分破裂(图版3H)。

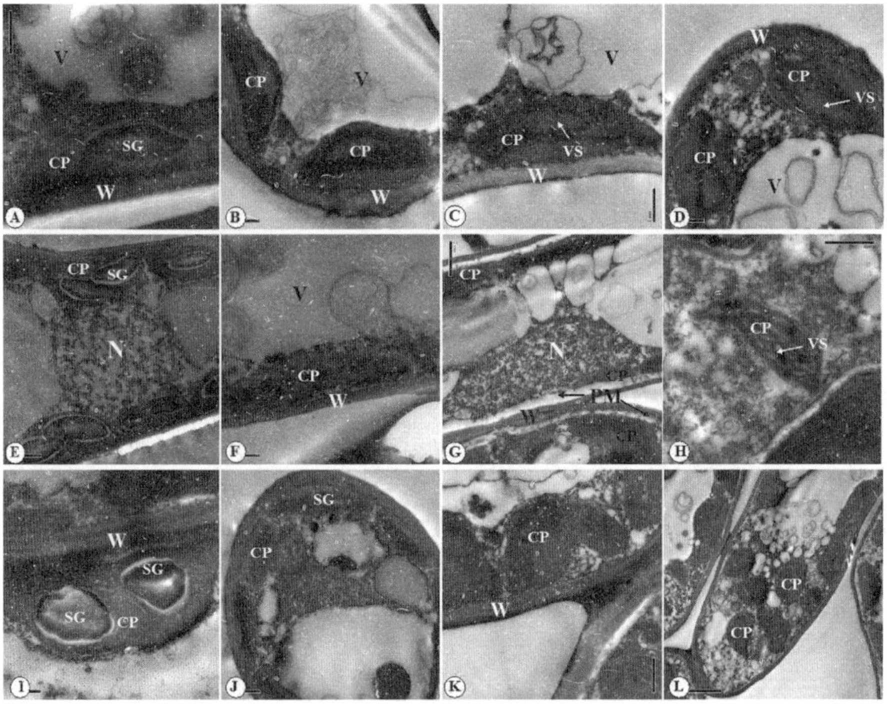

图版3 干旱胁迫对3种苹果属植物叶绿体超微结构的影响

图版说明:A-D为楸子的对照、轻度、中度和重度干旱胁迫;E-H和I-L分别代表平邑甜茶和新疆野苹果的对照、轻度、中度和重度干旱胁迫。CP-叶绿体,SG-淀粉粒,VS-基质片层囊泡化,N-细胞核,V-液泡,W-细胞壁,PM-质膜。

3 讨论

3.1 干旱胁迫对3种苹果属植物叶片解剖结构的影响

水分胁迫下,植物细胞结构及代谢活动都会受到不同程度的损害,如果这种细胞伤害是不可逆的,就会导致植物死亡。光镜观察结果发现,正常条件下,楸子和新疆野苹果栅栏组织排列紧凑较整齐,而平邑甜茶的排列疏松。干旱胁迫下,3种苹果属植物叶片栅栏组织排列趋紧变短,海绵组织变薄、排列疏松;栅栏组织和海绵组织细胞明显变小,细胞紧密程度减小。总体表现为,叶片及栅栏组织厚度逐渐减小,海绵组织厚度逐渐增加(重度胁迫下平邑甜茶除外),CTR 值逐渐减小,而 SR 值逐渐增加。这与前人对桃和苹果品种的研究结果相一致[13-14]。

研究表明,发达的栅栏组织在水分适宜时增加植物的蒸腾效率,在干旱时可阻止组织水分蒸发;叶片 CTR 值大,叶片细胞变小可减少因干旱导致细胞收缩产生的机械损伤[14]。本试验结果表明,与对照相比,中度和重度干旱胁迫下,3种苹果属植物叶片栅栏组织厚度的降幅极显著($P<0.01$),是造成其叶片厚度显著降低的主要原因($P<0.05$);与楸子和新疆野苹果不同的是,在重度水分胁迫下,平邑甜茶海绵组织厚度由增变减,进而导致其叶厚极显著下降($P<0.01$)。与平邑甜茶相比,新疆野苹果和楸子具有栅栏组织较发达、叶肉组织结构较紧密以及 CTR 值高而 SR 值低的特点,更具抗旱的细胞结构特征,故在干旱环境胁迫中显示出较强的适应性和抗旱能力。这与 Bacelar 等(2004)在不同油橄榄品种(*Olea europaea* L.)上的研究结果相一致[15]。

3.2 干旱胁迫对3种苹果属植物叶表皮微形态特征的影响

3.2.1 干旱胁迫对3种苹果属植物叶片上下角质层厚度的影响

植物角质层(也称角质膜)是在植物表面覆盖的、由脂肪酸及其衍生物构成的疏水性物质。植物角质层主要由蜡质和角质组成。角质层蜡质嵌入或沉积在角质基质中,具有限制植物表面水分非气孔性散失,保护植物内部组织免受紫外辐射及病菌侵染[16]。干旱环境条件下,植物叶片的气孔导度降低,植物通过气孔散失的水分减少,进而通过表皮角质层调节水分散失[17]。透射电镜观察结果表明,3种苹果属植物叶片上下表皮细胞均被有角质层(图版1),且上角质层厚度明显大于下角质层厚度。这可能与上表皮层一直对内部组织提供保护作用有关[15]。干旱条件下,楸子和新疆野苹果上下角质层随着胁迫强度的增加而逐渐加厚,而平邑甜茶上、下角质厚度分别在中度和重度干旱胁迫下开始降低(表2)。这表明,不同苹果属植物角质层对水分亏缺的反应能力与干旱胁迫程度有关,而且种间差异明显。由此我们推测,角质层发育程度的差异可能是构成苹果属植物抗旱能力强弱的原因之一。有研究表明,角质层的组分及其渗透性在很大程度上可决定植物自我保护抵御外界胁迫的能力,而植物角质层蜡质的化学组分及生理特性因物种、组织和器官发育阶段不同而存在差异[18],且受内部调节机制及外部环境因素(如ABA、干旱、光照和湿度)的影响[19]。研究报道,在水分胁迫条件下,植物过量表达蜡质相关转录因子或诱导表达蜡质合成相关基因,蜡质含量增加,增强了植株的抗旱性[20-21]。目前,一些角质和蜡质合成、调控及转运途径相关的基因在拟南芥、水稻等模式植物中被不断地克隆和鉴定[22-24],有关角质层蜡质生物合成及转运途径的部分具体环节已被阐明[25-26],但是有关植物角质层生物合成与调控的具体作用机理尚不清楚[16,27]。

3.2.2 干旱胁迫对3种苹果属植物气孔形态特征的影响

气孔是叶片和外界环境进行气体和水分交换的重要通道,气孔开度与气孔密度对植物的光合作用和蒸腾作用具有重要的影响[28]。通过气孔调节而提高水分利用效率是植物在干旱胁迫下忍耐饥饿能力的一种适应方式。气孔开口变小、密度变大是植物对干旱气候环境响应的典型特征[29],而植物气孔数量对干旱胁迫的响应不仅与物种有关,而且与遭受的干旱胁迫程度有关[30]。对于许多植物而言,既可短期调节气孔开度又可长期控制气孔密度的发育和形态来适应环境变化[31]。扫描电镜观察发现,在干旱胁迫下,3种苹果属植物叶片气孔器整体下陷,气孔数量明显增加,气孔开度及开张比降低。这表明,干旱胁迫下,苹果属植物气孔密度和气孔开度(或大小)呈负相关变化。这与人们对杨树[32]和拟南芥[33]的研究结果相一致。

研究表明,光照和CO_2浓度在控制气孔发育中起着重要的作用[34],植物气孔开度与气孔密度的负相关性主要与叶片的CO_2导度有关,具有较多小气孔的植物叶片更能适应CO_2受到限制的环境[35]。然而,长时间气孔关闭导致CO_2吸收速率降低,进而影响光合同化速率和植物的发育[36]。本试验研究发现,新疆野苹果的气孔较大,因水分亏缺而引起的气孔关闭反应不如楸子与平邑甜茶的敏感。这在一定程度上反映了苹果属植物适应气候生态环境的差异,即来自潮湿或较湿生境的平邑甜茶与楸子植物,其气孔调节能力高于来自干燥气候环境的新疆野苹果植物。新疆野苹果和楸子气孔对干旱环境变化的反应与抗旱性并不完全一致,这可能与它们水分利用效率的差异有关[37]。

3.3 干旱胁迫对3种苹果属植物叶绿体超微结构的影响

透射电镜观察发现,正常水分条件下,3种苹果属植物叶绿体紧贴细胞壁分布,这种排列方式有利于CO_2从大气中向叶绿体中扩散,提高光合作用[38]。水分胁迫下,细胞液泡收缩,核染色质凝聚,细胞发生质壁分离;随着水分亏缺程度加强,淀粉粒消失,叶绿体弯曲膨胀呈无规则排列,基粒类囊体空泡化,叶绿体膜结构遭到破坏。与平邑甜茶相比,楸子和新疆野苹果叶绿体超微结构受损程度相对较低(图版3)。研究表明,叶绿体的结构和功能对干旱反应较敏感,由于活性氧(ROS)大量积累而导致氧化胁迫。叶绿体 PSI 通过酶促或非酶促的 O_2- 歧化反应形成 H_2O_2,叶绿素分子也可通过能态转变释放能量产生$^1O_2$[39]。叶绿体类囊体膜不饱和脂肪酸(PUFA)含量较高,易与$^1O_2$发生质膜过氧化反应,导致叶绿体膜结构的破坏;高浓度 O_2^- 导致基粒类囊体的松散或崩裂[40]。

综上所述,干旱胁迫对 3 种苹果属植物叶片细胞结构的影响都有一个从适应到伤害的过程。在轻度和中度水分胁迫下,3 种苹果属植物通过改变自身的细胞结构来适应干旱环境,叶片显微及超微结构都发生了适应性变化;在重度胁迫下,抗旱性较强的楸子和新疆野苹果叶细胞结构受损程度较轻,而抗旱性较弱的平邑甜茶遭受不可逆损伤的程度较重。不同苹果属植物叶片结构对干旱胁迫的响应与适应性变化不仅与所遭受的水分胁迫程度相关,而且与其抗旱内在生理机制有关。这需要进一步深入探讨。

参考文献

[1]Jensen P J,Halbrendt N,Fazio G,Makalowska I,Altman N,Praul C,Maximova S N,Ngugi H K,Crassweller R M,Travis J W,McNellis T W. Rootstock - regulated gene expression patterns associated with fire blight resistance in apple[J]. BMC Genomics,2012,13:9.

[2]王忆,许雪峰,孔瑾,李天忠,韩振海. 苹果属植物的应用分型及其应用[J]. 果树学报,2007,24(4):502 - 505.

[3]Jensen P J,Makalowska I,Altman N,Fazio G,Praul C,Naximova S N,Crassweller R M,Travis J W,McNellis T W. Rootstock - regulated gene expression patterns in apple tree scions[J]. Tree Genetics & Genomes,2010,6(1):57 - 72.

[4]Wang S,Liang D,Li C,Hao Y,Ma F,Shu H. Influence of drought stress on the cellular ultrastructure and antioxidant system in leaves of drought - tolerant and drought - sensitive apple rootstocks[J]. Plant Physiology and Biochemistry,2012,51:81 - 89.

[5]Peng L X,Gu L K,Zheng C C,Li D Q,Shu H R. Expressionof *MaMAPK* gene in seedlings of *Malus* L. under water stress[J]. Acta Biochimica et Biophysica Sinica,2006,38(4):281 - 286.

[6]谭冬梅. 干旱胁迫对新疆野苹果及平邑甜茶生理生化特性的影响[J]. 中国农业科学,2007,40(5):980 - 986.

[7]马春花,李明军,李翠英,邵建辉,马锋旺. 不同抗性苹果砧木叶片抗坏血酸代谢对干旱胁迫的响应[J]. 西北植物学报,2011,31(8):1596 - 1602.

[8]Liu B,Li M,Cheng L,Liang D,Zou Y,Ma F. Influence of rootstock on antioxidant system

in leaves and roots of young apple trees in response to drought stress[J]. Plant Growth Regulation, 2012, 67(3): 247 - 256.

[9] Bauerle T L, Centinari M, Bauerle W L. Shifts in xylem vessel diameter and embolisms in grafted apple trees of differing rootstock growth potential in response to drought[J]. Planta, 2011, 234(5): 1045 - 1054.

[10] Yan G, Long H, Song W, Chen R. Genetic polymorphism of *Malus sieversii* populations in Xinjiang, China[J]. Genetic Resources and Crop Evolution, 2008, 55(1): 171 - 181.

[11] Barrs H D, Weatherley P E. A re - examination of the relative turgidity technique for estimating water deficits in leaves[J]. Australian Journal of Biological Sciences, 1962, 15: 413 - 428.

[12] Hsiao TC. Plant responses to water stress[J]. Annual Review of Plant Physiology, 1973, 24(1): 519 - 570.

[13] 姚允聪,高遐虹,程继鸿. 苹果种质资源抗旱性鉴定研究Ⅶ. 干旱条件下苹果幼树生长与叶片形态特征变化[J]. 北京农学院学报, 2001, 16(2): 16 - 21.

[14] 孟庆杰,王光全,董绍锋,张丽,龚正道. 桃叶片组织解剖结构特征与其抗旱性关系的研究[J]. 干旱地区农业研究, 2004, 22(3): 123 - 126.

[15] Bacelar E A, Correia C M, Moutinho - Pereira J M, GonçalvesBC, LopesJI, Torres - PereiraJM. Sclerophylly and leaf anatomical traits of five field - grown olive cultivars growing under drought conditions[J]. TreePhysiology, 2004, 24(2): 233 - 239.

[16] Lee S B, SuhM C. Recent advances in cuticular wax biosynthesis and its regulation in *Arabidopsis*[J]. Molecular Plant, 2013, 6(2): 246 - 249.

[17] Cavender - Bares J, Sack L, Savage J. Atmospheric and soil drought reduce nocturnal conductance in live oaks[J]. Tree Physiology, 2007, 27(4): 611 - 620.

[18] Buschhaus C, Jetter R. Composition and physiological function of the wax layers coating *Arabidopsis* leaves: β - amyrin negatively affects the intracuticular water barrier[J]. Plant Physiology, 2012, 160(2): 1120 - 1129.

[19] MackováJ, VaškováM, MacekP, HronkováM, SchreiberL, Šantr ? ček J. Plant response to drought stress simulated by ABA application: Changes in chemical composition of cuticular waxes [J]. Environmental and Experimental Botany, 2013, 86: 70 - 75.

[20] Seo P J, Park C M. Cuticular wax biosynthesis as a way of inducing drought resistance [J]. Plant Signaling &Behavior, 2011, 6(7): 1043 - 1045.

[21] Zhang J Y, Broeckling C D, Blancaflor E B, Sledge M K, Sumner L W, Wang Z Y. Overexpression of *WXP1*, a putative *Medicago truncatula* AP2 domain - containing transcription factor gene, increases cuticular wax accumulation and enhances drought tolerance in transgenic alfalfa (*Medicago sativa*)[J]. The Plant Journal, 2005, 42(5): 689 - 707.

[22] Wang Y, Wan L, Zhang L, Zhang Z, Zhang H, Quan R, Zhou S, Huang R. An ethylene response factor OsWR1 responsive to drought stress transcriptionally activates wax synthesis related

genes and increases wax production in rice[J]. Plant Molecular Biology,2012,78(3):275 - 288.

[23]YangJ, Isabel OrdizM, JaworskiJG, BeachyRN. Induced accumulation of cuticular waxes enhances drought tolerance in *Arabidopsis* by changes in development of stomata[J]. Plant Physiology and Biochemistry,2011,49(12):1448 - 1455.

[24]Lam P,Zhao L,McFarlane H E,Aiga M,Lam V,Hooker T S,Kunst L. RDR1 and SGS3, components of RNA - mediated gene silencing,are required for the regulation of cuticular wax biosynthesis in developing inflorescence stems of *Arabidopsis*[J]. Plant Physiology, 2012, 159(4): 1385 - 1395.

[25]Seo P J,Lee S B,Suh M C,Park M J,Go Y S,Park C M. The MYB96 transcription factor regulates cuticular wax biosynthesis under drought conditions in *Arabidopsis*[J]. Plant Cell,2011, 23(3):1138 - 1152.

[26]KimH,Lee S B,Kim H J,MinM K,HwangI,Suh MC. Characterization of glycosylphosphatidylinositol - anchored lipid transfer protein 2(LTPG2) and overlapping function between LTPG/LTPG1 and LTPG2 in cuticular wax export or accumulation in *Arabidopsis thaliana*[J]. Plant and Cell Physiology,2012,53(8):1391 - 1403.

[27]Bernard A,Joubès J. *Arabidopsis* cuticular waxes:Advances in synthesis,export and regulation[J]. Progress in lipid research,2013,52(1):110 - 129.

[28]Dong J,Bergmann D C. Stomatal patterning and development[J]. Current Topics in Developmental Biology,2010,91:267 - 297.

[29]Galmés J,Medrano H,Flexas J. Photosynthetic limitations in response to water stress and recovery in Mediterranean plants with different growth forms[J]. New Phytologist,2007,175(1): 81 - 93.

[30]Hamanishi E T,Thomas B R,Campbell M M. Drought induces alterations in the stomatal development program in *Populus*[J]. Journal of Experimental Botany,2012,63(13):4959 - 4971.

[31]Haworth M,Elliott - Kingston C,McElwain J C. Stomatal control as a driver of plantevolution[J]. Journal of Experimental Botany,2011,62(8):2419 - 2423.

[32]Pearce D W,Millard S,Bray D F,Rood S B. Stomatal characteristics of riparian poplar species in a semi - arid environment[J]. Tree Physiology,2006,26(2):211 - 218.

[33]Doheny - Adams T, Hunt L, Franks P J, BeerlingDJ, GrayJE. Genetic manipulation of stomatal density influences stomatalsize,plant growth and toleranceto restricted water supply across a growth carbon dioxide gradient[J]. Philosophical Transactions of the Royal Society B:Biological Sciences,2012,367(1588):547 - 555.

[34]Casson S A,Hetherington A M. Environmental regulation of stomatal development[J]. Current Opinion in Plant Biology,2010,13:90 - 95.

[35]Lammertsma E I,de Boer H J,Dekker S C,Dilcher D L,Lotter A F,Wagner - Cremer F. Global CO_2 rise leads to reduced maximum stomatal conductance in Florida vegetation[J]. Proceed-

ings of the National Academy of Sciences,2011,108(10):4035-4040.

[36]Robichaux R H,Grace J,Rundel P W,EhleringerJR. Plant water balance[J]. BioScience,1987,37(1):30-37.

[37]Ma X,Ma F,Li C,MiY,BaiT,ShuH. Biomass accumulation,allocation,and water-use efficiency in 10 *Malus* rootstocks under two watering regimes[J]. Agroforestry Systems,2010,80(2):283-294

[38]VonCaemmererS,EvansJR. EnhancingC_3Photosynthesis[J]. Plant Physiology,2010,154:589-592.

[39]Smirnoff N. The role of active oxygen in the response of plants to water deficit and desiccation[J]. New Phytologist,1993:27-58.

[40]Gill S S,Tuteja N. Reactive oxygen species and antioxidant machinery in abiotic stress tolerance in crop plants[J]. Plant Physiology and Biochemistry,2010,48(12):909-930.

本论文已发表在《干旱地区农业研究》,2014年32卷3期,pp:15-23.

越冬低温及栽期对偏低海拔区当归生长的影响

贾　贞　狄胜强　张娟娟　赵菲轶　李三相*

【目的】探讨偏低海拔区当归苗的越冬温度和成药生长光温对早薹及成药根产量的影响。【方法】采用甘肃漳县当归苗，越冬期按根直径分类后分别进行0和-5℃处理，次年与当地同期、晚15 d和晚30 d 3个时期进行分期栽植，以观察早薹及成药根农艺性状对越冬温度和分期栽植的反应。【结果】根直径大于0.45cm的当归苗具有明显的早薹现象，根直径≥0.55～<0.65cm的当归苗早薹高于≥0.45～<0.55cm的苗，根直径小于0.65cm的同类大小苗经0和-5℃冬储处理及分期栽植后，相同大小的各类苗间早薹差异不明显。当归成药根主根直径、鲜质量和干质量与苗大小成正比，苗大小不同、冬储温度不同形成的侧根数也不同。当归成药根的产量随栽期的延后而减小。【结论】根直径0.65cm以下的当归苗可在0与-5℃温区越冬，偏低海拔区适宜栽植根直径偏小的苗，栽前萌动、栽期严重影响当归的产量，随栽期延后成药根产量减小。

当归(*Angelica sinensis* (Oliv.) Diels)属伞形科植物，其干燥的根为常用中药材，始载于《神农本草经》[1]。我国当归主产于甘肃省岷县、漳县、宕昌、渭源等海拔1500～3000m的高寒阴湿山区，以岷县当归最为传统地道且产量最大[2]，栽培历史长达1700多年。当归是多年生草本药用植物，生产上分育苗、成药和种子繁育3个阶段[3]：第1阶段，夏至前后(6月中下旬)播种育苗，寒露(10月中下旬)起苗贮藏；第2阶段，次年清明前后(4月上旬)移栽，至霜降(10月下旬)收挖药用根；第3阶段，选择生长健壮的成药期植株繁种，根不收挖，次年抽薹开花、结子。成药期部分植株往往抽薹开花而得不到具有药用价值的根，此即当归的早薹现象，严重时可减产50%～90%[4-5]，是影响当归产量的关键因子。当归产区极其狭窄，

* 作者简介：贾贞(1972—　)，男，甘肃秦安人，现为天水师范学院生物工程与技术学院副教授、博士，主要从事作物成花发育研究。

随着当归药、食保健品等市场需求的增大,提高原产区产量、扩大栽培范围是提高当归产量的必由之路,但随着引种和栽培范围的扩大早薹率随之提高[6]。有关当归春化早薹研究始于20世纪七八十年代,一般认为早薹是当归苗低温春化所致。王文杰等[7-9]研究表明,春化是当归早薹的原因之一,比较适宜的春化温度为0~5℃,-10℃可抑制当归的早薹,但当归早薹的温度处理范围比较宽泛,不能用于指导实践。光温反应是大宗作物大豆[10]等作物栽培适应研究的重要方面,但对当归光温反应的栽培适应研究极少。本研究采用甘肃省漳县一农户的当归苗,于越冬期分别经0和-5℃处理,次年分期移栽至该农户大田,创设栽期不同而引起的苗期温度和光照差异,通过分析冬储低温、苗期光温对成药期当归早薹、农艺生长性状的影响,揭示当归冬储及成药生长的光温反应特性,最后结合不同海拔区有经验药农培育当归苗大小的调查结果,以期探讨当归栽培的适应机制及栽培策略。

1 材料与方法

1.1 试验设计与处理

试验用当归苗由甘肃漳县光明村一药农提供,试验地设在该农户大田。10月底起苗,起苗后带至实验室分类储藏:用游标卡尺测量根头下1cm处的直径,按根直径≥0.55~<0.65cm、≥0.45~<0.55cm、≥0.35~<0.45 cm及<0.35cm分为4类(直径≥0.65cm的苗数极少,未列入试验范围)。分类后选取无病害、无机械损伤的健康苗均分为2大组,分别进行0和-5℃冬储,每组又分为3个小组用于分期栽植,每小组设3个重复,每重复30株。用无污染的田间湿土将苗包埋于大小适宜的塑料盒内,置于变温冰箱(澳柯玛BCD-267MDG)的特定温区储藏,各处理于2012-11-22开始冬储,共冬储约4个月。

1.2 试验方法

试验用地前茬为蚕豆,经深耕平整,施用优质腐熟的农家肥作底肥,每处理3次重复,随机区组设计,行距40cm,四周留60cm保护行,采用1 m宽白色塑料膜覆盖栽培。2013-03-15开始将不同低温处理当归苗栽植于试验地,此后每隔15d即于03-30和04-15定期栽植,分3批栽植完毕。根直径≥0.45~<0.65cm的苗1株/穴、≥0.35~<0.45cm的苗2株/穴,<0.35cm的苗3株/穴,穴距30cm。栽植方法和田间管理与当地药农相同。成苗后开始观察记载抽薹现象,收挖后(2013-10-26),将试验当归根带至实验室,测定主根长、根干质量、根鲜质量、主根直径、侧根数等指标。

1.3 数据处理

所有试验数据用Excel和DPS数据处理系统进行分析,用Origin 9.0绘图[11]。

2 结果与分析

2.1 冬储温度及栽期对当归早薹的影响

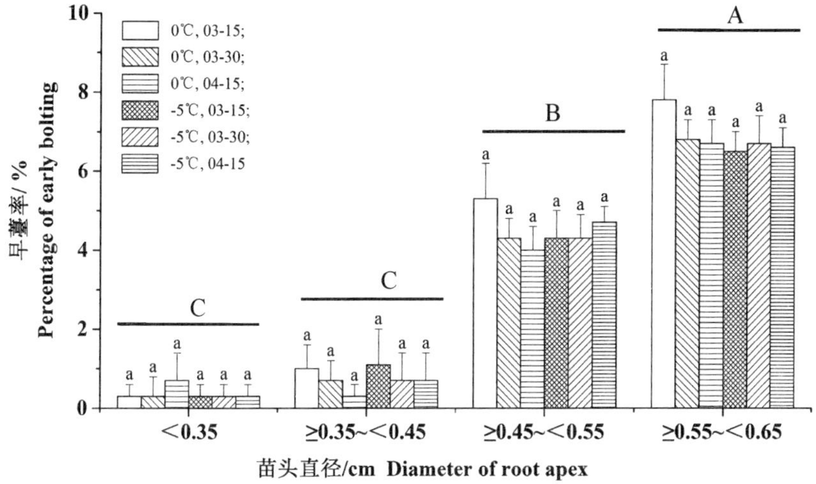

图1 冬储温度及栽期对当归早薹率的影响

图注上标不同大写字母表示不同大小苗处理间差异显著,小写字母表示同类大小苗不同处理间差异显著,下图同。

由图1可见,经历不同温度冬储和栽期处理后,根直径0.45cm以下当归苗的早薹率极低,根直径≥0.45~<0.65cm当归苗的早薹率显著高于前者,且该根直径范围内以0℃冬储同步栽植的早薹率较高,可能与其生长时间长、植株大有关。根直径0.55cm以下的当归苗,0和-5℃冬储处理对其早薹率影响不明显,观察发现其与该药农同块地当归的抽薹率接近。根直径≥0.55~<0.65cm当归苗的早薹率高于≥0.45~<0.55cm当归苗,该范围内的当归苗经0℃冬储且栽植早的早薹率最高,-5℃冬储同期栽植苗早薹率较低,且冬储处理苗各栽期间差异不显著。可见,同类大小不同处理当归苗早薹率差异不大,说明本研究所用根直径小于0.65cm的当归苗对该海拔条件的光温反应不十分敏感。

2.2 冬储温度和栽期对当归农艺性状的影响

2.2.1 主根长由图2可见,根直径<0.35cm的当归苗经0或-5℃冬储期栽植(03-15)后,成药主根显著短于其他各处理,且晚栽者产生的主根较短,说明晚栽不利于促进该类当归苗的主根伸长,这可能与该温度下冬储主根未能充分萌动有关。根直径≥0.35~<0.45cm当归苗的成药根除-5℃冬储、晚栽15d植株的主根较短($P<0.05$)外,其他各处理均无显著差异。对根直径≥0.45~<0.55cm

的当归苗而言,各冬储及栽期处理成药根主根长无显著差异。根直径≥0.55~<0.65cm的当归苗0℃冬储后,栽期越早形成的主根越长,而-5℃冬储苗,晚栽15d植株的主根最短,晚栽30d的主根最长。总体来看,当归苗越大0℃冬储处理有利于形成长的主根,-5℃冬储则相反,这可能与栽植前该温度处理下当归苗主根未能充分萌动有关。由此可知,收获当归成株的主根长易受当归苗冬储温度的影响,有利于主根萌动的偏高温更有助于根的伸长生长。

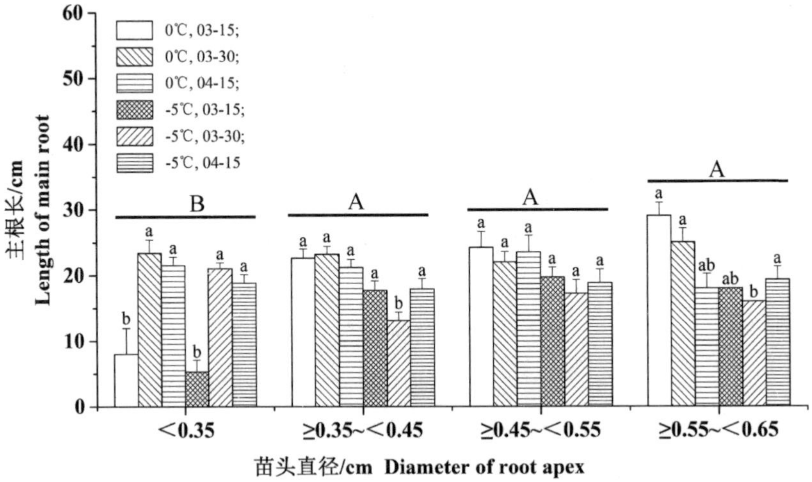

图2 冬储温度与栽期对当归主根长的影响

2.2.2 主根直径由图3可见,根直径在0.35cm以下当归苗的成药根直径除-5℃同期栽植较小外,其余各处理间差异不显著。根直径≥0.35~<0.45cm的苗经-5℃冬储晚栽后,成药根直径小于按期栽植和0℃冬储苗。根直径≥0.45~<0.55cm的当归苗成药根主根直径随栽期延后而变小,0℃冬储晚30d栽植(04-15)和-5℃冬储晚15及30d栽植处理的主根直径显著变小。根直径≥0.55~<0.65cm的当归苗经0℃冬储后成药根主根直径随栽期延后而变小,其中晚30d栽植(04-15)的主根直径显著变小,其他各处理间差异不显著。说明当归苗的栽期不同,各类苗成药后的主根直径基本表现为根苗越大成药根主根直径越粗。基本表现为苗大即根直径较大者成药根主根较粗。

2.2.3 侧根数侧根是成药当归的另一个外观指标,侧根少则主根相对较大。由图4可见,根直径<0.35cm的当归苗,成药根不同处理间侧根数差异不显著,但以0℃冬储后与当地同期栽植(03-15)处理的侧根数最多。根直径≥0.35~<0.45cm的当归苗也以0℃冬储后与当地同期栽植苗形成的侧根数最多,显著多于该类苗的其他各处理,其余各处理间差异不显著。根直径≥0.45~<0.55cm

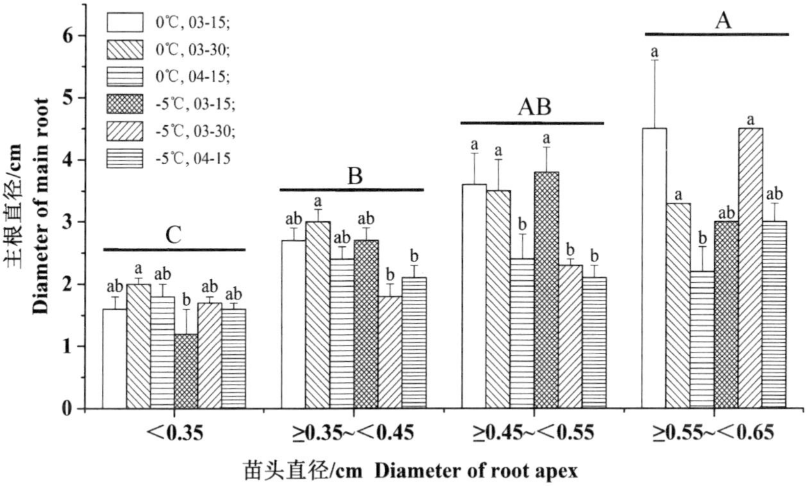

图 3　冬储温度与栽期对成药当归主根直径的影响

当归苗成药根的平均侧根数较多,但各处理间侧根数差异不大。与根直径<0.45cm 的当归苗相同,根直径≥0.55 ~ <0.65cm 的当归苗仍以 0℃冬储与当地同期栽植(03 - 15)处理成药根的侧根数最多,且与其他各处理差异显著,其他各处理侧根数较少,且处理间差异较大,另外各处理的标准误也较大,说明同一处理内不同植株间侧根数变化较大,是当归根最不稳定的性状,也是当归的一个普遍现象。由此可见,不同类当归苗经不同温度的冬储和栽期处理后,成药侧根数的形成能力并不相同。

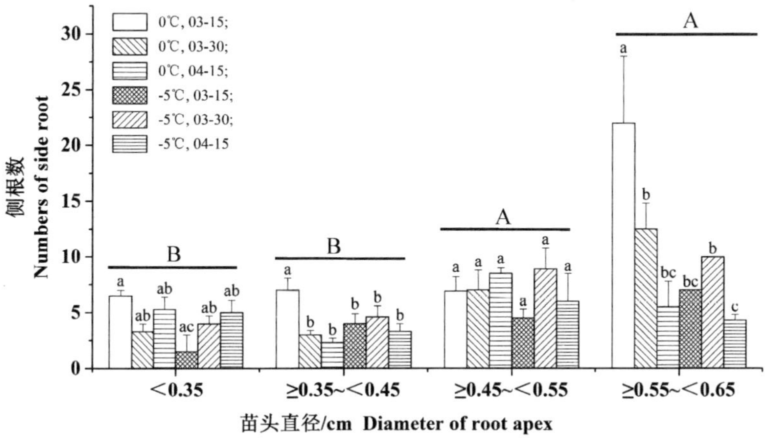

图 4　冬储温度与栽期对成药当归侧根数的影响

2.2.4　根鲜质量鲜质量是当归产量最直接的衡量指标之一。由图5可见，根直径<0.35cm的当归苗，各处理成药根鲜质量差异不显著。根直径≥0.35~<0.45cm及≥0.45~<0.55cm的当归苗，0℃冬储与当地同期(03－15)和晚15 d(03－30)栽植及－5℃冬储与当地同期(03－15)栽植当归苗成药根的鲜质量较高($P<0.05$)，在同类大小的苗内与其他各处理当归苗的成药根鲜质量间差异显著。根直径≥0.55~<0.65cm当归苗的成药根，以0℃冬储后与当地同期栽植(03－15)成药根的鲜质量最高，且0℃冬储苗成药根的鲜质量随移栽时期的延后而变小，－5℃冬储苗以晚15 d(03－30)移栽当归苗成药根的鲜质量最高，但与其余处理间差异并不显著。总体上来看，当归成药根鲜质量随移栽时期的延后而变小。

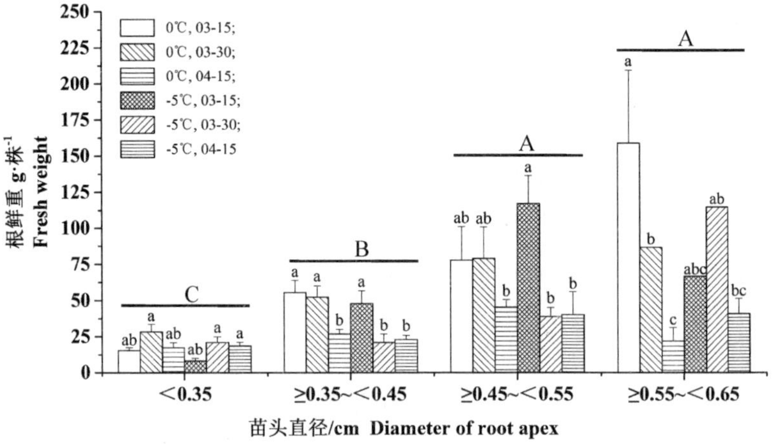

图5　冬储温度与栽期对当归根鲜质量的影响

2.2.5　根干质量由图6可见，根直径<0.35cm的当归苗成药后根干质量较轻，且各处理间差异不显著。根直径≥0.35~<0.55cm的当归苗，总体上栽期越晚成药根的干质量越小，其中0℃冬储后晚30d(04－15)栽植苗成药根的干质量显著小于前2个时期的栽植苗；－5℃冬储处理后，以按期栽植苗成药根的干质量最大，显著高于晚栽苗成药根的干质量。对根直径≥0.55~<0.65cm的当归苗而言，各冬储温度处理后的成药根干质量均随栽期的延后而减小，但同一冬储温度处理各栽期间差异并不显着。

3　讨论

3.1　当归的春化早薹

当归道地产区将当归苗于湿土中埋储越冬，使其冬季大部分时间处于低温冰冻状态，但在10月底和次年4月初，当地白天气温在0℃以上、夜间在0℃以下振

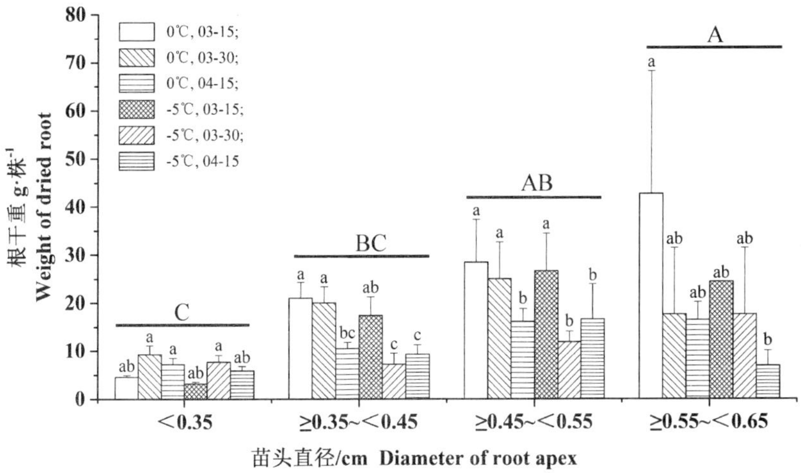

图6 冬储温度与栽期对当归根干质量的影响

荡,在振荡中逐渐升高。研究表明,越冬期1~5℃储藏可促进当归早薹[8],张恩和等[12]也分析了当归苗对4℃越冬处理的早薹反应。本课题组研究数据显示,在道地产区岷县栽培条件下,越冬期0℃冬储对当归春化早薹最有效,故本研究选择0℃作为促进春化处理的最有效温度。试验所用漳县有经验的药农培育的当归苗,根直径大于0.65cm以上的苗极少,未列入试验范围。

当归的光强度反应研究揭示,低光照强度会显著抑制当归早薹[13]。本研究结果显示,根直径0.45cm以上当归苗的早薹率显著高于根直径0.45cm以下的当归苗;与产区较为严重的当归早薹现象相比,根直径≥0.45~<0.65cm的当归苗经-5和0℃冬储后早薹率较低,其次,根直径≥0.55~<0.65cm当归苗的早薹率明显高于≥0.45~<0.55cm的当归苗,表明大苗更易于早薹。有研究显示当归的早薹与苗大小成正相关[4,14],经-5和0℃冬储处理后,同类大小苗各处理间早薹率差异不大。张恩和等[12]将根直径0.45cm以上的苗确定为具有感受低温春化早薹的苗类范围,而当归苗越小,感受低温春化早薹的能力越低。

植物种类不同开花诱导的光温反应不同,蓝针花科植物蓝针花(*Brunoniaaustralis*)和红娘花属植物马齿苋(*Portulacaoleracea* L.)从小就可感受春化刺激,但对光周期刺激的反应不同,显示出了青春期长短的不同[15]。当归的早薹极其敏感,除低温春化外,苗的播期、大小、冬储、成药期栽期、成药期生长海拔高度、土壤肥力、相对湿度、干旱和气温等都会影响当归的早薹[4,16],这可能是根直径0.45cm以下个别当归苗出现早薹的原因。当归可诱导的春化敏感状态与苗大小呈正相关[17],青春期后进入春化敏感状态的当归苗,经低温春化后次年春季在适宜的光温条件下可以早薹开花。可见,当归苗无明显的青春期阶段,当归的栽培和育种应以延长其青春期为目标。低温冬储可抑制当归早薹[8],本研究中所用苗类偏小,同类苗 0 和 -5℃冬储处理间早薹差异不显著,按期栽植和晚 30d 栽植苗间差异显著,可能因植株大小及栽期引起的苗期光温差异所致。

3.2 当归成药根的农艺性状

本研究结果表明,经低温冬储和栽期处理后,各类当归苗普遍以 0℃冬储后的成药根直径、鲜质量和干质量高于同类苗 -5℃冬储处理,其次,0℃越冬和按期栽植当归苗的生长性状好于 -5℃冬储的晚栽苗,说明当归苗的移栽宜早不宜迟,除个别情况下因规避倒春寒恶劣天气的需要而适当延迟外,当归苗的移栽应尽量做到适时。尽管冷冻低温冬储可抑制当归早薹[8],但本研究表明,当归苗 -5℃冬储将影响成药当归产量,这可能是由于 0℃冬储后移栽时苗芽已充分萌动,栽植后便能很快进入生长状态,而 -5℃冬储苗始终处于冰冻状态,栽植后才开始萌动,因此苗期生长进度慢,成药后鲜质量偏低。王文杰等[9]认为,控制当归早薹要注意做到储苗不萌动、不霉烂。本研究证明,苗萌动与否直接影响当归栽植后的生长及产量。因此,在当归苗的人工冬储中,宜在栽植前将苗转至适当温度下先进行萌动处理。

当归苗大小不同,经不同低温冬储和栽期处理后形成侧根数的能力不同,0℃冬储和按期栽植可促进当归成药根的侧根形成,其中以根直径≥0.55 ~ <0.65cm 当归苗各栽植处理的侧根数最多。-5℃冬储不利于侧根的形成,可见苗大小差异、冬储温度的不同形成侧根数的能力也不同。同一温度冬储,成药根除主根长差异不明显之外,其他各农艺性状与苗大小有一定的相关关系,即苗根较粗者移栽后主根直径、侧根数及根鲜质量、干质量相应较高,这与王兴政等[17]的结果基本一致。在本研究条件下,以根直径≥0.45 ~ <0.65cm 当归苗移栽后收获的根产量及根农艺性状最佳。

3.3 当归的育苗与用苗策略

为进一步探明不同海拔与当归苗大小间的关系,于 2013 -2015 年,选取岷县寺沟(海拔 2600m)、漳县盐井(海拔 1900m)和漳县殪虎桥(海拔 2000 m)3 个相距 100km 以内的不同海拔当归产地所育的当归苗进行分析。结果显示,高海拔地区(海拔 2600m)根直径 0.65cm 以上大苗所占比例较大,而偏低海拔地区(1900 ~ 2000m)根直径 0.65cm 以上大苗所占比例极小,即偏低海拔地区当归用苗偏小,不同海拔地区有经验药农所育当归苗的大小与当地海拔相适应。

甘肃岷县、漳县、宕昌、渭源等地海拔为 1500 ~ 3000m,各栽培地区地理条件千差万别[5]。大苗栽培时干旱、高温等气候容易引起当归早薹,而小苗的早薹反应相对较弱[8]。此外,当归早薹率与海拔高度呈负相关[18],漳县东北部海拔偏低[19],气温偏高,早薹率也较高,选用较小的当归苗栽培有利于防止早薹。但当归成药根的大小、产量与苗的大小成正相关,小苗栽培意味着较低的产量[17]。与根用糖源作物甜菜(*Beta vulgaris* L.)不同,前者的早薹受遗传[20]和特定光温影响[21],而当归的早薹主要取决于与其生长状态密切相关的苗大小特性,达到一定发育状态的当归苗在光温诱导下通过类似于拟南芥光温开花的分子调控机制实现开花[22]。因此,当归育种和栽培的目标是培育和选用早薹率低的大苗。有经验的药农会依自身栽培地的海拔、阴阳坡等条件控制当归的早薹反应,如培育选用大小适宜的苗,既可将早薹控制在一定范围内又可获得尽可能高的产量。当归产区以岷县海拔最高,是当归最传统、最地道、最适宜的产区,面积产量也最大,质量也最好。岷县周边漳县、渭源的海拔呈西南高、东北低(海拔1640m),这些偏低海拔区是当归栽培的边界地带,培育或选用较小的当归苗是减少当归早薹、保证顺利栽培的有效手段[23]。因此,在偏低海拔高温区,当归成功栽培的前提是使用小苗栽培,并将当归早薹率控制在一定的范围内,虽然小苗栽培产量较低,但仍具有一定的推广潜力。

参考文献

[1]国家药典委员会. 中华人民共和国药典一部[M]. 北京:中国医药科技出版社,2010:224.

National Pharmacopoeia Committee. Pharmacopoeia of People's Republic of China(Part Ⅰ)[M]. Beijing:Chinese Medical Science and Technology Press,2010:224.

[2]赵杨景,陈四保,高光耀,等. 道地与非道地当归栽培土壤的理化性质[J]. 中国中药杂志,200227,(1):19-22.

Zhao Y J,Chen S B,Gao G Y,et al. Study on the physicochemical properties of cultivated soil of genuine crude and no-enuine crude Chinese angelica [J]. China Journal of ChineseMateria-Medica,200227,(1):19-22.

[3]孙红梅,张本刚. 甘肃地区当归生长动态调查[J]. 中国农学通报,2010,26(17):386-389.

Sun H M,Zhang B G. Investigation on the growth activities of *Angelica sinensis*(Oliv.)Diels. in Gansu region [J]. Chinese Agricultural Bulletin,2010,26(17):386-389.

[4]李广骥. 当归的生物学特性及防止早期抽苔的研究[J]. 中药材科技,1979(2):1-10.

Li G J. Study on biological characteristics of*Angelica sinensis* and the prevention of early bolting [J]. Chinese Herbal Medicine Science and Technology,1979(2):1-10.

[5]陈瑛,明图林.当归提早抽薹问题的调查[J].植物生理学通讯,1966,13(1):9-11.

Chen Y,Ming TL. Investigation of problems in *Angelica sinensis* early bolting[J]. Plant Physiology Communication,1966,13(1):9-11.

[6]武延安,陈垣,蔺海明.当归早期抽薹研究进展[J].甘肃农业科技,2007(3):20-23.

Wu YA,Chen Y,Lin HM,et al. Study progress of *Angelica sinensis* early bolting[J]. Gansu Agricultural Science and Technology,2007(3):20-23.

[7]王文杰.对当归早期抽苔特性的分析和控制[J].西北大学学报(自然科学版),1977,7(2):32-39.

Wang WJ. Analysis and control of early bolting of *Angelica sinensis*[J]. Journal of Northwestern University(Natural Science Edition),1977,7(2):32-39.

[8]王文杰."立秋直播"当归的栽培技术和原理:控制当归早期抽苔的途径之一[J].西北大学学报(自然科学版),1977,7(2):40-44.

Wang W J. Cultivation techniquesof *Angelica* once sowing at autumn and its principles(one way to control *Angelica sinensis* early bolting)[J]. Northwest University Journal(Natural Science Edition),1977,7(2):40-44.

[9]王文杰,张正民.当归的抽薹特性和控制途径 [J].西北植物研究,1982,2(2):95-104,149.

Wang W J,Zhang Z M. Characteristics of *Angelica* bolting and its control [J]. Northwestern Plant Research,1982,2(2):95-104,149.

[10]费志宏,吴存祥,孙洪波,等.以光周期处理与分期播种试验综合鉴定大豆品种的光温反应[J].作物学报,2009,35(8):1525-1531.

Fei ZH,Wu CX,Sun HB,et al. Identification of photothermal responses in soybean by integrating photoperiod treatments with planting-date experiments processing and sowing experiment[J]. Acta Agronomica Sinica,2009,35(8):1525-1531.

[11]唐启义,冯明光.DPS数据处理系统:实验设计、统计分析及数据挖掘 [M].北京:科学技术出版社,2007.

Tang QY,Feng MG. DPS data processing system:experimental design,statistical analysis and data mining [M]. Beijing:Science Press,2007.

[12]张恩和,黄鹏.春化处理对当归苗生理活性的影响 [J].甘肃农业大学学报,1998,33(3):240-243.

Zhang EH,Huang P. Effects of vernalization treatment on physiological character of *Angelica sinensis*seedlings[J]. Journal of Gansu Agricultural University,1998,33(3):240-243.

[13]徐继振,刘效瑞,荆彦民,等.甘肃当归提前抽苔的防治研究[J].中国中药杂志,1999,24(11):660-663.

Xu JZ,Liu XR,Jing YM,et al. Study on the prevention and treatment of early bolting of *Angelica sinensis* in Gansu [J]. Journal of Traditional Chinese Medicine,1999,24(11):660-663.

[14]李明世. 防止当归早期抽苔的研究[J]. 西北植物研究,1983,3(1):70-76.

Li MS. On the control bolting in the early stage of *Angelica sinensis*(Oliv.) Diels. [J]. Acta Botanica Boreali-Occidentalia Sinica,1983,3(1):70-76.

[15]Cave R L, Birch C J, Hammer G L, et al. Juvenility and flowering of*Brunoniaaustralis*(Goodeniaceae) and *Calandrinia*sp. (Portulacaceae) in relation to vernalization and daylength[J]. Annals of Botany,2011,108(1):215-220.

[16]王春明. 当归提早抽薹综合防治技术[J]. 甘肃农业科技,2008(8):61-62.

Wang CM. comprehensive prevention and control technology of*Angelica* early bolting [J]. Gansu Agricultural Science and Technology,2008(8):61-62.

[17]王兴政,蔺海明,刘学周. 种苗大小对当归综合农艺性状及抽薹率的影响[J]. 甘肃农业大学学报,2007,42(5):59-63.

Wang XZ,Lin HM,Liu X Z. Effect of seedling size on comprehensive agricultural characteristics and bolting rate *Angelica sinensis* [J]. Journal of Gansu Agricultural University,2007,42(5):59-63.

[18]邱黛玉,蔺海明,陈垣. 种苗大小对当归成药期早期抽薹和生理变化的影响[J]. 草地学报,2010,18(6):838-843.

Qiu DY,Lin HM,Chen Y. Effects of latitude, longitude and altitude on *Angelica* growth and early bolting in medicine formation period[J]. Acta Agrestia Sinica,2010,18(6):838-843.

[19]朱国庆. 甘肃中部当归生态气候分析及适生种植区划[J]. 甘肃气象,2001,19(1):36-38.

Zhu GQ. Ecological climate analysis of *Angelica sinensis* in central Gansu and its suitable planting regionalization [J]. Gansu Meteorology,2001,19(1):36-38.

[20]Pfeiffer N,Tränkner C,Lemnian I,et al. Genetic analysis of bolting after winter in sugar beet(*Beta vulgaris* L.)[J]. Theoretical and Applied Genetics,2014,127(11):2479-2489.

[21]Al-Jbawi E,Sabsabi W,Gharibo G,et al. Effect of sowing date and plant density on bolting of four sugar beet(*Beta vulgaris* L.) varieties[J]. International Journal of Environment,2015,4(2):256-270.

[22]Lee J H,Park C M. Integration of photoperiod and cold temperature signals into flowering genetic pathways in *Arabidopsis*[J]. Plant Signaling & Behavior,2015,10(11):e1089373.

[23]林彰斌. 育好当归苗控制早抽薹 [J]. 农业科技通讯,1985(6):37.

Lin ZB. Nursing good seedling to control early bolting[J]. Agricultural Science and Technology Communication,1985(6):37.

注:该文章发表于2017年《西北农林科技大学学报》09期

野生药用植物红茂草挥发油提取及抗氧化活性研究

赵　强　王廷璞*

采用 CO_2 超临界法萃取红茂草中挥发油，对其进行 GC－MS 分析，以提取的挥发油为研究对象，确定红茂草挥发油的抗氧化能力大小。研究表明，红茂草挥发油成分复杂，存在一定的抗氧化能力。

红茂草［*Dicranostigma Leptodum*（*maxim.*）*Fedde*（DLF）］又名秃疮花、秃子花、勒马回（陕西），为罂粟科秃疮花属二年生或多年生草本植物，生于海拔 1300 ～ 2400m 的高原、山坡、丘陵、路旁等处，在我国青海、甘肃、陕西等省有广泛分布[1]。其味苦，性寒，有小毒，有清热解毒、消肿止痛、杀虫等[2－4]。挥发油又称植物精油，为植物的次生代谢物质，具有促进细胞的新陈代谢，调节内分泌器官等作用。本实验采用 CO_2 超临界萃取红茂草中挥发油，对其进行 GC－MS 分析，并以萃取的红茂草挥发油、VC、异紫堇碱、芦丁和红茂草提取液，通过对自身还原能力、·OH 自由基和·O_2^- 自由基清除能力进行测定，以比较红茂草中挥发油体外抗氧化活性能力的大小，为该植物资源的深层次开发和利用提供一定的实验依据。

1　材料与方法

1.1　材料

1.1.1　植物

天水师范学院生物园规范化种植的红茂草，经自然干燥、粉碎。

1.1.2　仪器与试剂

CO_2 超临界萃取装置（美国 Helix）、HP6890/ 5973 型气相色谱－质谱联用仪

* 作者简介：赵强（1982—　），男，甘肃天水人，现为天水师范学院生物工程与技术学院正高级工程师，博士，主要从事天然野生药用植物资源开发与利用研究。

(美国 HEWLETT - PACKARK 公司)、UV - 751GD 紫外/可见分光光度计(上海精密科学仪器有限公司)等仪器。

芦丁对照品(国药集团,批号:F20051222)、Vc 对照品(上海信然生物技术有限公司,批号:GS - E20525)、异紫堇碱标准品(ABCR Gmbh & Co. KG im Schlehert 公司,批号:166401)、无水乙醇(AR)等。

1.1.3 GC - MS 分析条件

石英毛细管柱 HP - 5MS、30 m × 0.25 mm、膜厚 0.25μm;初始温度:40℃,保留 1min,以 10℃/min 上升至 250℃,保留 5min;进样口温度 250℃、EI 电源、电离电压 70eV、离子源温度 250℃、扫描范围 33 ~ 550 amu、进样量 0.4μL、分溜比 1:71、载气 He、流速 1.0mL/min;传输线温度:280℃[3]。

1.2 方法

1.2.1 红茂草挥发油的制备

采用 CO_2 超临界萃取红茂草中挥发油。准确称取红茂草粉末 500g,在萃取温度为 40℃,萃取釜压力 17.6MPa,分离釜压力 10.8MPa 的条件下静态萃取 1.5 h,调节 CO_2 流量为 5.1 ~ 5.6 L/ h,动态萃取 1.5 h 后,收集萃取物。处理后得淡黄色油状液体,即为红茂草挥发油样品。

1.2.2 红茂草水提液的制备

准确称取红茂草干草粉末 10.0g,用 150mL 65% 的乙醇浸泡 8h,在温度为 60℃、超声频率为 350W 的条件下超声 35min 后,将处理后的红茂草草粉加入索氏提取器,回流 2.5h,得提取液。将提取液在 35℃下,用旋转蒸发仪进行减压浓缩,浓缩至无醇味;将浓缩液装入 100mL 锥形瓶,活性炭脱色,封口静置 1 天。减压浓缩抽滤,转入 50mL 容量瓶定容,即得质量浓度为 0.2g/mL 的红茂草提取液,静置备用。

1.2.3 抗氧化活性测定

1.2.3.1 还原能力测定

用差量法精确称取红茂草挥发油,用乙酸乙酯溶解并配制红茂草挥发油的梯度溶液进行实验,同时配制相应浓度梯度的红茂草提取液、芦丁、VC 和异紫堇碱溶液作为对照实验。采用普鲁士兰法[5]进行还原能力的测定,红茂草挥发油溶液质量浓度梯度为:0.05、0.10、0.15、0.20、0.25、0.30、0.35、0.40、0.45、0.50、0.60、0.70μg/mL。同法测定红茂草提取液、V_C、异紫堇碱、芦丁的还原能力,作为对照。

1.2.3.2 ·OH 自由基抗氧化活性测定

精确配制红茂草挥发油 0.05、0.10、0.15、0.20、0.25、0.30、0.35、0.40、0.45、0.50、0.60、0.70μg/mL 的梯度待测液,同时配制相应浓度的红茂草提取液、芦丁、

异紫堇碱、V_C作为对照液。采用邻二氮菲比色法[5]测定红茂草提取液、V_C、异紫堇碱、芦丁对·OH 自由基的清除能力,作为对照。

表1 试剂加样情况

吸光度(A)	样品
$A_{空白}$	PBS +邻二氮菲+底液(混匀)+ $FeSO_4$ + H_2O_2(混匀,37℃恒温 0.5h)
$A_{对照}$	PBS +邻二氮菲+底液(混匀)+ $FeSO_4$ +水(混匀,37℃恒温 0.5h)
$A_{样品}$	PBS +邻二氮菲+待测液(混匀)+ $FeSO_4$ + H_2O_2(混匀,37℃恒温 0.5h)
空白	蒸馏水+PBS+邻二氮菲+待测液(混匀)+$FeSO_4$ +H_2O_2(混匀,37℃恒温 0.5h)

注:邻二氮菲溶液浓度为 5mmol/L;磷酸盐缓冲液(PBS 液)浓度为 0.5mol/L;硫酸亚铁液浓度为 7.5mmol/L;

过氧化氢溶液浓度为 0.1%。

对·OH 自由基清除率的计算公式为:

$$清除率(\%)=\frac{A_{样品}-A_{空白}}{A_{对照}-A_{空白}}\times 100\%$$

1.2.3.3 ·O_2^-自由基抗氧化活性测定

准确配制红茂草挥发油质量浓度为 0.05、0.10、0.15、0.20、0.25、0.30、0.35、0.40、0.45、0.50、0.60、0.70μg/mL 梯度溶液进行实验,同时配制相应浓度的红茂草提取液、芦丁、异紫堇碱、V_C溶液作为对照。采用邻苯三酚自氧化法[5]测定红茂草提取液、V_C、异紫堇碱、芦丁对·O_2^-自由基的清除能力,作为对照。对·O_2^-自由基清除率的计算公式为:

$$清除率(\%)=\frac{A_{对照}-A_{样品}}{A_{对照}}\times 100\%$$

2 结果与分析

2.1 红茂草挥发油成分分析

CO_2超临界法提取红茂草挥发油的 3.16g,出油率 0.63%。对其进行气相色谱-质谱联用分析,共检出 201 个色谱峰,鉴定出相对含量在 0.20% 以上的化合物占挥发油总量的 96.59%。

表 2　红茂草挥发油成分的种类及含量

No.	Compound name		CAS Coding	Molecular Formula	Molecular weight	Retain Time (min)	Relative Content (%)
	Chinese name	English name					
1	乙基苯	Ethylbenzene	100 – 41 – 4	C_8H_{10}	106	3.08	0.31
2	对二甲苯	p – Xylene	106 – 42 – 3	C_8H_{10}	106	3.33	2.16*
3	1,3,5 – 环庚三烯	1,3,5 – Cycloheptatriene	544 – 25 – 2	C_7H_8	92	3.52	0.25
4	邻二甲苯	o – Xylene	95 – 47 – 6	C_8H_{10}	106	4.00	0.50
5	3 – 乙基甲苯	1 – ethyl – 3 – methyl – Benzene	620 – 14 – 4	C_9H_{12}	120	5.76	0.94
6	异丙基苯	1 – methylethyl – Benzene	98 – 82 – 8	C_9H_{12}	120	6.16	0.45
7	1,2,3 – 三甲基苯	1,2,3 – trimethyl – Benzene	526 – 73 – 8	C_9H_{12}	120	6.46	2.41*
8	1,2,4 – 三甲基苯	1,2,4 – trimethyl – Benzene	95 – 63 – 6	C_9H_{12}	120	7.06	1.70*
9	二氢化茚	Indane	496 – 11 – 7	C_9H_{10}	118	7.34	0.47
10	1,3,5 – 三甲基苯	1,3,5 – trimethyl – Benzene	108 – 67 – 8	C_9H_{12}	120	7.56	0.56
11	间异丙基甲苯	1 – methyl – 3 – (1 – methylethyl) – Benzene	535 – 77 – 3	$C_{10}H_{14}$	134	7.76	1.10*
12	4 – 正丙基甲苯	1 – methyl – 4 – propyl – Benzene	1074 – 55 – 1	$C_{10}H_{14}$	134	7.92	0.26
13	4 – 异丙基甲苯	1 – methyl – 4 – (1 – methylethyl) – Benzene	99 – 87 – 6	$C_{10}H_{14}$	134	8.14	0.57
14	邻异丙基甲苯	1 – methyl – 2 – (1 – methylethyl) – Benzene	527 – 84 – 4	$C_{10}H_{14}$	134	8.28	0.72
15	2,6,10 – 三甲基十二烷	2,6,10 – trimethyl – Dodecane	3891 – 98 – 3	$C_{15}H_{23}$	212	8.44	0.77

续表

No.	Compound name		CAS Coding	Molecular Formula	Molecular weight	Retain Time (min)	Relative Content (%)
	Chinese name	English name					
16	叔丁基苯	tert – butyl – Benzene	98 – 06 – 6	$C_{10}H_{14}$	134	8.66	0.78
17	3,5 – 二乙基甲苯	1,3 – diethyl – 5 – methyl – Benzene	2050 – 24 – 0	$C_{11}H_{16}$	148	9.28	1.39*
18	2,4 – 二甲基苯乙烯	2,4 – Dimethylstyrene	2234 – 20 – 0	$C_{10}H_{12}$	132	9.43	1.80*
19	(R,R) – 1 – 甲基 – 1 – 硝基 – 苯丙醇	(R,R) – 1 – methyl – 1 – nitro – Benzenepropanol	–	$C_{10}H_{13}NO_3$	195	9.64	1.03*
20	甘菊蓝	Azulene	275 – 51 – 4	$C_{10}H_8$	128	10.03	7.37**
21	4,11 – 二炔 – 氧杂环十四碳	4,11 – diyne – Oxacyclotetradeca	6568 – 32 – 7	$C_{13}H_{18}O$	190	10.58	0.81
22	6,6,11 – 三甲基 – 1,9 乙酰氧基 – 7 酮甾烷	6,6,11a – trimethyl – 1,9a – bis(acetyloxy) – 7 – one – Gonane	–	$C_{25}H_{38}O_5$	418	10.96	0.38
23	4 – 烯 – 3,20 – 二酮,11,16,22 – 三乙酰氧基雄甾	4 – ene – 3,20 – dione,11,16,22 – triacetoxy – Androst	–	$C_{27}H_{36}O_8$	488	11.17	0.78
24	1 – 亚乙基 – 苯并环丙烯	1 – ethylidene – 1H – Indene	2471 – 83 – 2	$C_{11}H_{10}$	142	11.73	1.48*
25	L – 抗坏血酸 – 2,6 – 二棕榈酸酯	1 – (+) – Ascorbic acid 2,6 – dihexadecanoate	4218 – 81 – 9	$C_{38}H_{68}O_8$	652	19.44	5.73**

续表

No.	Compound name		CAS Coding	Molecular Formula	Molecular weight	Retain Time (min)	Relative Content (%)
	Chinese name	English name					
26	硬脂酸	Octadecanoic acid	57 – 11 – 4	$C_{18}H_{36}O_2$	284	19.64	1.52*
27	熊去氧胆酸	Ursodeoxycholic acid	128 – 13 – 2	$C_{24}H_{40}O_4$	392	20.27	1.37*
28	3 – 乙基 – 5 – (2 – 乙基正丁)正十八烷	3 – ethyl – 5 – (2 – ethylbutyl) – Octadecane	55282 – 12 – 7	$C_{26}H_{54}$	366	20.61	0.39
29	(E,E)8,11 – 十八碳二烯酸甲酯	8,11 – Octadecadienoicacid, methyl ester	56599 – 58 – 7	$C_{19}H_{34}O_2$	294	21.13	37.98**
30	二十碳饱和脂肪酸	Eicosanoic acid	506 – 30 – 9	$C_{20}H_{40}O_2$	312	21.50	0.38
31	(Z,Z) – 9 – 十六碳烯酸,9 – 十八碳烯基酯	(Z,Z) – 9 – Hexadecenoicacid, 9 – octadecenyl ester	22393 – 98 – 2	$C_{34}H_{64}O_2$	504	21.82	0.49
32	20 – 酮,5,6 – 环氧 – 3,17 – 二羟基 – 16 – 甲基 – 孕甾烷	20 – one, 5, 6 – epoxy – 3, 17 – dihydroxy – 16 – methyl – Pregnan	55630 – 87 – 6	$C_{22}H_{34}O_4$	362	22.05	0.90
33	9 – 十二烷基十四碳氢菲	9 – dodecyltetradecahydro – Phenanthrene	55334 – 01 – 5	$C_{26}H_{48}$	360	22.79	1.02*
34	13 – 正十八烯醇	13(18) – ene – Olean	629 – 98 – 1	$C_{30}H_{50}$	410	23.14	0.93

续表

No.	Compound name		CAS Coding	Molecular Formula	Molecular weight	Retain Time (min)	Relative Content (%)
	Chinese name	English name					
35	[20.8.0.0(7,16)]蜂花烷-1(22),7(16)-二环氧-三环	[20.8.0.0(7,16)]triacontane,1(22),7(16)-diepoxy-tricyclo	-	$C_{30}H_{52}O_2$	444	23.38	0.74
36	E-乙酰基-10-十八碳烯	E-1-ol acetate-10-Octadecen	86390-77-4	$C_{20}H_{38}O_2$	310	23.85	4.99*
37	二十四烷	Tetracosane	646-31-1	$C_{24}H_{50}$	338	24.17	0.74
38	四十四烷	Tetratetracontane	7098-22-8	$C_{44}H_{90}$	618	24.67	3.20*
39	6-乙基辛-3-含氧基-2-乙基己基邻苯二甲酸	6-ethyloct-3-yl-2-ethylhexyl ester-Phthalic acid	-	$C_{26}H_{42}O_4$	418	25.61	1.01*

注：* 表示相对含量 >1.00%；* * 表示相对含量 >5.00%。
Note: * Relative content >1.00%; * * Relative content >5.00%.

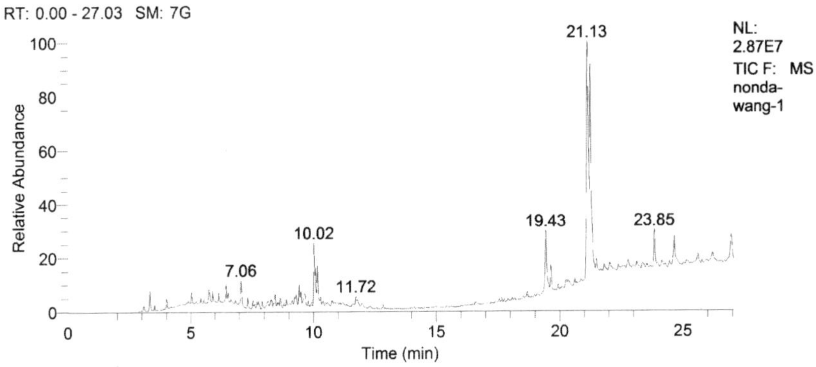

图1 红茂草挥发油总离子流图

2.1 还原能力测定

采用普鲁士兰法,选用 Vc、芦丁、异紫堇碱、红茂草提取液,对比测定红茂草挥发油的还原能力。

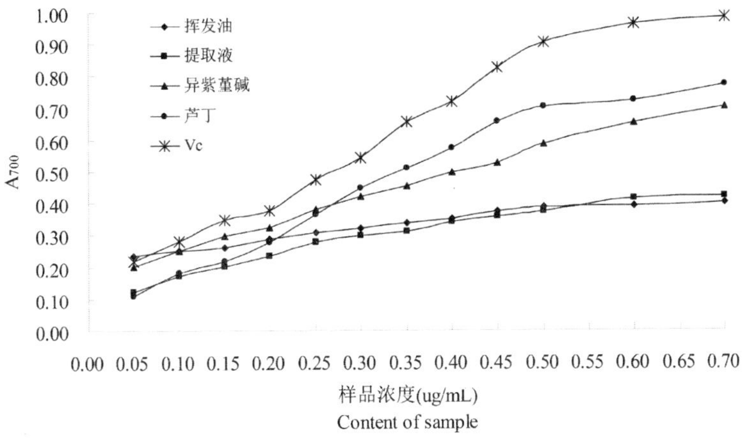

图2 不同浓度的 Vc、芦丁、异紫堇碱、红茂草提取液和挥发油的还原能力

各样品的还原能力与自身浓度成一定的量效关系,并且随着样品浓度的增加,还原能力也增加,这表明各样品具有不同程度的还原能力。当各样品溶液的浓度低于 0.45μg/mL 时,还原能力的大小为:Vc > 芦丁 > 异紫堇碱 > 挥发油 > 红茂草提取液;且当各样品溶液的浓度大于 0.60μg/mL 时,红茂草提取液还原能力大于挥发油,这可能与反应时间过久、温度变化等引起挥发油成分被氧化或分解挥发有关,其余各样品在次浓度下均保持原有趋势。

2.2 对·OH 自由基清除能力的测定

采用邻二氮菲法,选用 Vc、芦丁、异紫堇碱、红茂草提取液,对比测定红茂草挥发油清除·OH自由基的能力。

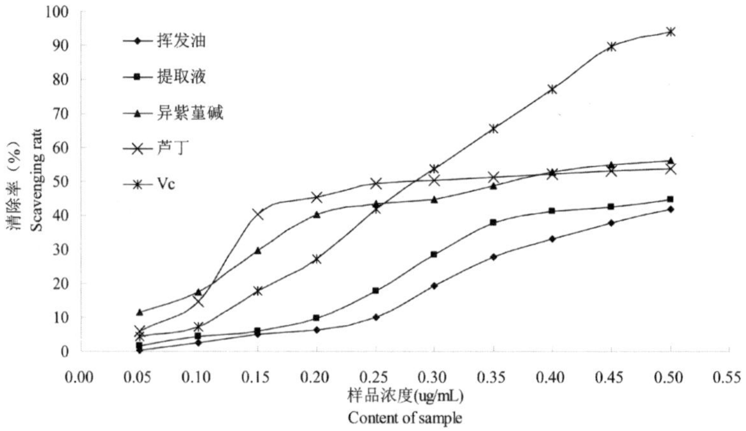

图3　不同浓度的Vc、芦丁、异紫堇碱、红茂草提取液和挥发油对·OH的清除作用

各样品清除·OH的能力与自身浓度成一定的量效关系，并且随着样品浓度的增加，·OH清除率也增加，这表明各样品对·OH具有清除作用。当各样品溶液的浓度低于0.20μg/mL时，清除·OH的能力大小为：芦丁>异紫堇碱>Vc>红茂草提取液>挥发油；而当各样品溶液的浓度为0.45μg/mL时，其清除·OH的能力大小为：Vc>异紫堇碱>芦丁>红茂草提取液>挥发油。随着各种样品浓度的增加，其对·OH的清除作用也愈渐显著，Vc浓度在0.50μg/mL时其清除率可达90%，其余样品的清除率也在40%左右。

2.3　对·O_2^-自由基清除能力的测定

采用邻苯三酚自氧化法，选用Vc、芦丁、异紫堇碱、红茂草提取液，对比测定红茂草挥发油对·O_2^-的抑制能力。

各样品的抑制作用与自身浓度成显著的量效关系，并且随着样品浓度的增加，各样品对·O_2^-的抑制作用呈明显的增加趋势，表明各样品对·O_2^-具有一定的抑制作用。当各溶液的浓度达到0.50μg/mL时，各样品清除·O_2^-的能力大小为：Vc>芦丁>挥发油>异紫堇碱>红茂草提取液，Vc在该浓度下其清除率可达75%，其余样品的清除率在65~45%之间。

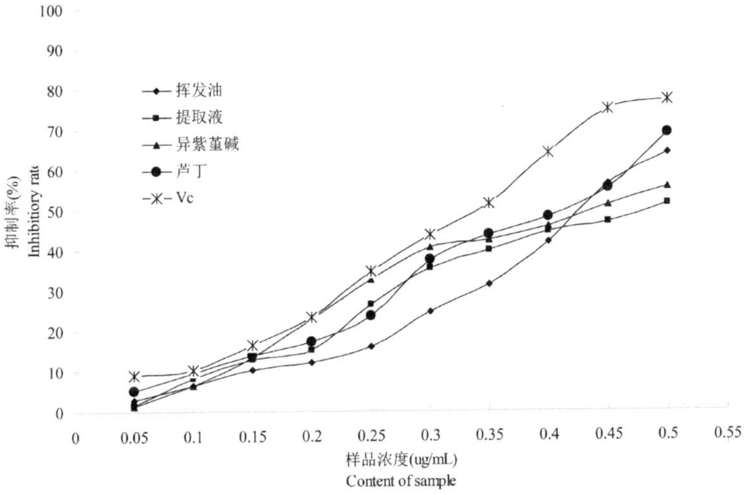

图4 不同浓度的Vc、芦丁、异紫堇碱、红茂草提取液和挥发油对·O_2^-的抑制作用

3 讨论

采用CO_2超临界法萃取红茂草挥发油的主要成分为(E,E)8,11-十八碳二烯酸甲酯(37.98%)、甘菊蓝(7.37%)、L-抗坏血酸-2,6-二棕榈酸酯(5.73%)、E-乙酰基-10-十八碳烯(4.99%)、四十四烷(3.20%),其它成分大多为二甲苯、三甲苯、苯乙烯以及不饱和烷、烯、酮、脂肪酸、脂肪酸酯和甾类等。其具有多种生物活性,有抗肿瘤、抗菌、抗病毒、抗氧化、提高机体免疫调节等功效。

以Vc、芦丁、异紫堇碱、红茂草提取液和红茂草挥发油为对照进行体外抗氧化实验,与Vc、芦丁等相比,红茂草挥发油抗氧化活性相对较低。这是由于Vc、芦丁等均为纯品,且是理想的抗氧化剂,具有较强的抗氧化能力。红茂草挥发油和提取液在对·O_2^-自由基和·OH自由基的清除作用上表现出一定差异,这与它所含的化学成分不同有关,表明抗氧化能力可能是多种成分的协同作用,而不是单一成分的作用结果[4-6]。通过分析红茂草挥发油的化学成分及抗氧化活性研究,为甘肃省陇东南地区红茂草特色药材资源的进一步深层次开发和利用提供了理论依据。

参考文献

[1]畅行若,王宏新,畅行若等.秃疮花化学成分的研究[J].药学通报.1981(2):52.

[2]赵强,王廷璞,余四九等.红茂草生物碱抑菌活性测定[J].中国兽医科学.2008,38(12):1098-1101.

[3]Cheng S S, Lin H Y, Chang S T. Chemical composition and antifungal activity of essentialOil from different tissues of Japanese cedar(Cryptomeria japonica)[J]. *J Agric Food Chem*. 2005:614 - 619.

[4]赵强,余四九,王廷璞等. 响应面法优化秃疮花中生物碱提取工艺及抑菌活性研究[J]. 草业学报,2012,21(4):206 - 214.

[5]赵强,王廷璞,余四九等. 红茂草生物碱正交提取工艺模式优化及清除自由基作用的研究[J]. 草地学报,2012,20:71 - 78.

[6]赵强. 甘肃省陇东南地区特色药用植物红茂草资源的研究与利用[D]. 甘肃农业大学. 博士学位论文. 2013,06:59 - 65.

注:该文章发表于发表于《草地学报》2016 年 02 期

Establishment of Murine Embryonic Stem Cell Line Carrying Enhanced Green Fluorescence Protein and its Differentiation into Cardiomyocyte-like Cells in vitro

JIANG Zu-Yun　YUAN Yi-Jun　CHEN Liang-Biao
LUYong-Liang　YAO Xing　DAI Li-Cheng　ZHANGMing *

摘　要:带有 GFP 基因的 ES D3 细胞系是一个良好的可以用于研究体内和体外细胞分化和组织产生的模型。用磷酸钙共沉淀法将质粒 pEGFP－N2 导入小鼠胚胎干细胞 D3 细胞系中,在荧光显微镜下以 488nm 激发光检查阳性克隆,并进行初步扩增。经 G418 筛选后,机械挑取 EGFP 强阳性表达的克隆,并在丝裂霉素 C 处理的小鼠胚胎成纤维细胞的饲养层上,在无选择性压力的条件下,进一步扩大培养,获得纯化的转染细胞系。20 代以后,转染细胞仍然表达绿色荧光蛋白。PCR 检测表明 8 代和 18 代转染细胞均携带有 GFP 标志基因。对稳定表达 EGFP 的干细胞系进行碱性磷酸酶染色、拟胚体和畸胎瘤形成的检测,证明这些细胞具有干细胞的特征。经拟胚体,可进一步分化成具有搏动能力的心肌细胞,分化百分率为 30% ~40% ,较未转染细胞 60% ~70% 的分化率低,造成低分化率机制还不清楚。这些细胞在激光共聚焦显微镜下呈绿色荧光,免疫组化染色显示具心肌细胞特异的 cTnT 分子标志。该 EGFP 标记的干细胞系带有可进行原位、实时检测的绿色荧光,可应用于细胞移植和体内分化的研究。

The availability of EGFP ES cell D3 lines provided a tractable model to study cell differentiation and tissue generation *in vivo* and *in vitro*. Plasmid pEGFP N2 was intro-

* 作者简介:袁毅君(1966—　),女,甘肃天水人,现为天水师范学院生物工程与技术学院教授,博士,主要从事中药活性成分提取分离与药理活性方面的研究。

duced into the murine embryonic stem cell D3 by standard calcium phosphate precipitation. Transfected clones were screened out under the floureseence microscope at the 488 nm emission light in the presence of G418. Strong fluorescent EGFP clones were singly picked out and further proliferated on a feeder layer of mitomycin-C treated mouse embryonic fibroblasts. One line of EGFP ES D3 cells subcultured twenty passages and still carried the EGFP DNA without the selecting pressure. It indicated that the gene might integrate into the ES genome or still dissociated in the cytoplasm. PCR analysis for EGFP DNA showed that undifferentiated EGFP ES cells at passage 8 and 18 carried the EGFP gene. Alkaline phosphatase staining embryoid body and teratoma formation were performed to analyze the differentiation status and potential of the EGFP ES D3 cells. The cells derived from embryoid body were able to differentiate into beating cardiomyocytes with green fluorescence clearly observable under the confocal laser scanning microscopy. 30% ~40% of cells from embryoid bodies were capable to differentiate into cardiomyocyte-like cells, and it appeared lower than the non-transfected ES D3 cells which could be 60% ~70% under the same conditions. The mechanism was currently unknown. Immunocytochemistry staining indicated that the contracting cells were cardiomyocytes based on the presence of cardiac specific molecular marker cTnT. Results showed that the stable EGFP positive ES cell line retained the typical characteristics of ES cells and possessed the pluripotential to differentiate into beating myocytes *in vitro*. The EGFP transfected cells stably yielding bright green fluorescence in real time and *in situ* rendered it was a powerful tool in cell transplantation and tissue engineering.

Embryonic stem cells are pluripotent cells derived from the inner cell mass of preimplantation mouse embryos. They have self-renewal ability and represent embryonic precursor cells that can differentiate into three embryonic germ layers *in vivo* and *in vitro*[1,2]. It is attractable cellular system to investigate cellular and genetic programming of early development. In terms of clinical benefit, stem cells are generating many hopes for future regenerative medicine.

Mouse embryonic stem cells are widely used for gene mutation analysis or for incorporation of marker genes. Molecular markers are useful in studying gene expression profile during differentiation and in tracing and selecting particular subsets of cells[3,4]. The MES allows for studying genes of interest in vitro, but more importantly, by creating chimeras it is an extremely useful tool in analyzing genes during the embryo develop-

ment in vivo. Eiges et al (2001) established a DNA transfection protocol for human ES cells. The transfected cells by ExGen 500 transfection system showed high levels of GFP expression limited to the undifferentiated cells, and the fluorescent cells could be separated from the differentiated cells by using a fluorescence-activated cell sorter[5]. In addition, some other GFP transfected ES cell lines are established by electroporation[6,7,8]. we describe the work that labeled the mouse ES cell lines with the green fluorescence protein through the standard calcium phosphate precipitation method and produced the stable EGFP ES D3 line. The EGFP ES D3 cell differentiated into cardiomyocytes-like cells in vitro under proper culture and treatment. The EGFP labeled clones of the ES cells will provide a useful tool for monitoring the differentiation status of the cells in vitro and in vivo.

1 Materials and Methods

1.1 Cell lines and animals

The ES cell line D3, purchased from the Institute of Biochemistry and Cell Biology, Shanghai Institutes for Biological Sciences, Chinese Academy of Sciences, was used for all studies. Feeder cells are from ICR mouse embryonic fibroblasts (MEF), prepared from 13.5 dpc mouse embryos[9]. SNL cell line, is gifted by the College of Life Science Peking University, which is derived from the mouse STO cell line and resistant to G418. Balbc/nu mice were the recipients of the EGFP transfected ES cells to form teratoma to test the *in vivo* differentiating capacity of the EGFP ES D3 cells.

ES cells were cultured on a feeder layer of mitomycin-C treated MEF in Dulbecco's modified Eagle's medium (GIBCO-BRL), supplemented with 15% newborn bovine serum (Evergreen, Hangzhou Sijiqing Biological Engineering Materials Co., Ltd.), 1 mmol/L glutamine (GIBCO-BRL), 0.1mmolL β-mercaptoethanol (Sigma), 1% nonessential amino acids stock (GIBCO-BRL), Penicillin (100 μmL), Streptomycin (100mg/mL) and leukemia inhibitory factor (LIF: 1000μmL, Chemicon International Inc., Temecula, CA, USA).

1.2 Transfection of EGFP reporter gene and establishment of transgenic cell lines

The pEGFP N2 vector, a construct expressing the enhanced green fluorescent protein (EGFP) under the control of the human cymegalovirus (hCMV) promoter (purchased from Clontech) was used for transfection. The plasmid DNA was prepared according to the maximum extraction method[10]. The construct contained an SV40-driven neo selectable marker. The use of SV40 promoterin the system was sufficient to confer G418 resistance by driving the neo gene, although it was somew hat inefficient in mouse ES cells.

Sufficiently expanded and undifferentiated murine ES cells underwent stable transfection with pEGFP N2 plasmid DNA by the standard calcium phosphate precipitation method[10]. In following 24 ~ 48h, GFP expression was monitored under the fluorescence microscope (Diaphot, Nikon) equipped with mercury burner (HBO100W2, Stromart:-DC, Osram). G418 (800 μg/mL) was then administered to the medium, allowing the selective propagation of transfected cells in culture, neo resistant fluorescent-labeled colonies were identified by a fluorescent microscope after one week (up to 10 colonies per well). Single transgenic colonies emitting strong green fluorescence were picked out by a micropipette, dissociated into small clumps, and transferred into a 96-well culture dish on a fresh MEF feeder. The cells continuously formed a large number of expanding undifferentiated colonies on a feeder layer of MEF. The different passages of the EGFP ES cells were examined under the fluorescence microscope.

1.3 DNA preparation and PCR analysis

Genomic DNA was extracted from undifferentiated EGFP ES cells at passage 8 and 18 respectively using standard method[10]. Genomic DNA from untransfected D3 cells was also extract as negative control in PCR. The primers for PCR were: 5′ CTG-GTCGAGCTGGACGGCGACG3′, and 5′ CACGAACTCCAGCAGGACCATG3′ (Clontech). PCR reaction was performed in a final volume of 50μL containing 50ng template, 1μmo/L of sense and antisense primers, 200nmo/L of dNTPs, 2μ of Taq DNA polymerase, 1 × PCR buffer and 2.5 mmo/L MgCl2 (Shanghai Sangon Co Ltd). Amplification was carried out by denaturing at 95℃ for 4min, followed by 30 cycles of 1min at 95℃, 1min at 58℃, 2min at 72℃, and finally 10 min at 72℃. PCR products were separated in 1.5% agarose gels containing ethidium bromide, and bands were visualized by ultraviolet transillumination.

1.4 Analysis of EGFP labeled ES cells

1.4.1 Alkaline phosphatase assay: The Sigma Diagnostics Alkaline Phosphatase kit was used in the assay following the manufacturer's instruction. The activity of alkaline phosphatase in the ES cells was evaluated microscopically.

1.4.2 Embryoid body formation: The EGFP positive ES cells were grown in the presence of LIF to avoid the use of MEF as feeder layer. The undifferentiated cells were digested into small clumps and were induced to differentiate *in vitro* into embryoid bodies (EBs) by omitting LIF from the growth media, allowing aggregation in petri dishes. Following the formation of simple EBs by a 5-day cell aggregation procedure.

1.4.3 Differentiation *in vivo*: About 2×10^7 cells/mL EGFP positive ES cells were trypsinized and dissociated into single cells and resuspended in 0.3 mL PBS. ES cells suspension administered subcutaneously into the oxter of the male Balbc/nu mice. Teratoma was detectable after 2 weeks. Teratoma was prominent and mature after 30 ~ 50 days. The teratomas were surgically removed from the mice and fixed in 10% neutral formalin. Routine paraffin section were performed by the standard method and stained with hematoxylin and eosin. Sections were examined and pictured under microscope.

1.5 Differentiation into cardiomyocyte-like cells

1.5.1 Cardiomyocyte-like cells derived from EGFP transfected ES D3 cell line: After about 5 days of development in suspension culture, 5 EBs with similar sizes were picked out and each replated in a well of a 0.1% gelatin-coated 24-well plate in the absence of LIF. EBs were further spread and underwent differentiation into a monolayer cells in following days. After 7 to 8 days, cardiomyocyte-like beating activity was clearly visible, the active areas were examined and photographed under the confocal laser scanning microscopy (ZEISS LSM 510).

1.5.2 Cardiomyocytes primary culture: Cardiomyocytes were isolated from 1 ~ 5 – day new born mice as previous reports[11,12]. Briefly, the hearts of newborn mice were dissected and the basal part of the ventricle was collected. They were rinsed three times with PBS and mechanically dissected in the sterile dish. 1 mL 0.08% trypsin digestion solution was then added and the minced tissue was incubated at 37℃ for 10 ~ 15 min. The tissue was gently triturated for several seconds until the tissue was dissociated into single cell or into very small clumps. After wards, cells and clusters were seeded into a culture flask containing coverslips, the flask was incubated at 37℃ overnight and used on the following day.

1.5.3 Immunocytochemistry: We used the ABC method to characterize the cardiomyocytes derived from the EGFP labeled ES cells. The differentiation cells and primary culture myocytes were washed with PBS twice and underwent fixation in 10 % neutral formalin for 10 min. A primary monoclonal antibody against Troponin T (Santa Cruz Biotechnology, Inc, cat sc-20025) was applied in a concentration of 1 : 200 in

10% NBS in PBS at RT for 1h to the fixed cells. The cells were then washed 3 times with PBS. Biotin-conjugated anti-mouse IgG(secondary antibodies) with 1 : 200 dilution in 10% NBS were added to the cells, and the mixture was incubated for 30 min at RT. The cells were washed 3 times with PBS. Reagent A and B of the ABC test kit (Sino-AmericanBiotechnologyCompany) were diluted 1 : 100 in 10% NBS in PBS. ABC compound was mixed in the same proportion of the reagent A and B and was then added to the cells and incubated at 37℃ for 30 min. AEC of the AEC kit (Sino-American Biotechnology Company) was used as a chromogenic substrate. After 8 ~ 10 min of staining, reaction was ended by washing the cells with bidistilled water.

2 Results

2.1 Transfection and EGFP positive colony screening

The plasmid pEGFP N2 were successfully introduced into murine D3 cells by standard calcium phosphate precipitation method. Expression of EGFP in the ES cells was detectable in as early as 24h after transfection. About 1% of the cells showed EGFP expression after 8-10 days culture in the presence of G418. These results indicated that calcium phosphate precipitation method is feasible for transfecting ES cells, and eytomegalovirus(CMV) promoter can drive the expression of EGFP gene in ES cells.

2.2 PCR for GFP DNA

The PCR results showed that three samples each contained EGFP DNA sequence (Fig. 1). The EGFP ES cells of passage 8 and passage 18 both contained the GFP gene. The PCR product was authentic GFP gene after sequencing verification of these cloned PCR product(data not shown).

2.3 The EGFP positive clone maintenance and morphological characteristics

We cloned the EGFP positive colonies by picking out the stable and strong EGFP expression clones. We further screened the picked clones by propagation up to 20 passages to choose those showed strong green fluorescence in each generation. The undifferentiated positive clones were maintained on MEF feeder layer. The EGFP colonies exhibited compact and clear edges in morphology. The EGFP positive clones grew on MEF feeder layer in the absence of G418 showed no apparent changes in proliferative capacity and in the fluorescent phenotype after long-term culture (Fig. 2A).

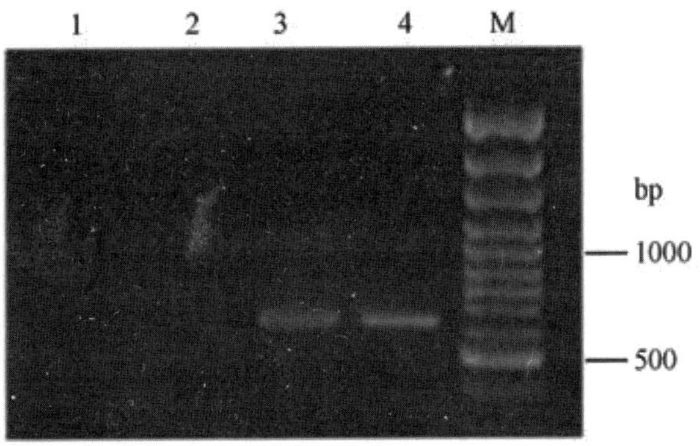

Fig. 1 GFP DNA of EGFP ES cells was amplified

1: control; 2: negative control, templates from the untransfected ES cells; 3 and 4: template from undifferentiated EGFP ES cells passage 8 and 18. The marker was GeneRuler™ 100bp DNA ladder plus (Shanghai Sangon Co Ltd), indicated to the right

2.4 Alkaline phosphatase assay

In mice, alkaline phosphatase staining is commonly used to evaluate pluripotency of ES cell. In the present study, similar staining was performed for characterizing EGFP positive ES cells clones. Most of the cells at passage 8 and 18 exhibited strong alkaline phosphatase (AP) activity. Whereas, those located at the edge of the ES colonies with larger sizes and differentiated morphologies showed weak activity (Fig. 2F).

2.5 Differentiation in vitro

In vitro experiment showed that the EGFP positive clones maintained pluripotency and differentiation capacity. The ES cell clones were trypsinized to single cell or smaller clumps and seeded in culture dishes for 3 day s in proper culture medium. These cells aggregated to form simple embryoid bodies (sEB) (Fig. 2B). Continued culture of the sEB for 5 days longer, the two germ layers and cavities were formed and visible under the inverted microscope. These were mature EBs (maEBs). sEBs and maEBs still showed strong green fluorescence observable under the fluorescence microscope (Fig. 2C).

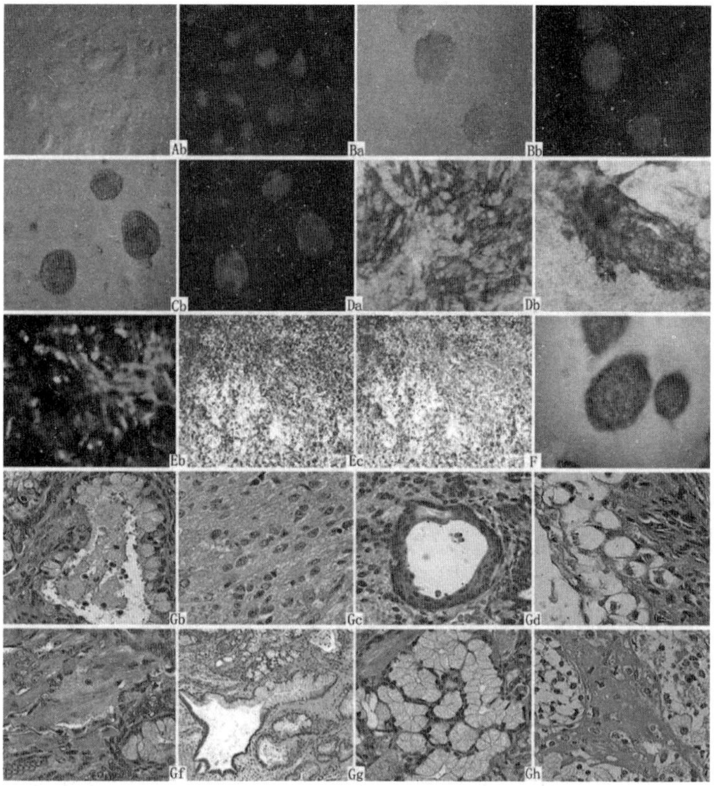

Fig. 2 Identification of EGFP ES cells and the differentiation of cardiomyocyte - like cells

A: Pasage18, murine ES cells underwent stable transfection with EGFP. a. Bright field under the inverted microscope; b. Dark field under the fluorescence microscope. B、C: The transfect ed ES cells and their differentiated cell derivatives are shown: simple embryoid body (sEB). and mature embryoid bodies (maEBs). The B. a, C. a and B. b, C. b are bright and dark fields, respectively. Note that the undifferentiated cells, simple EBs and mature EBs are fluorescent. D: Characterization of immunostaining. Primary culture cardiomyocytes (D. a) and cardiomyocytes derived from EB (D. b) stained with antibodies against cTnT. E: Myocytes characterization under the confocal laser scanning microscopy. The cardiomyocytes derived from EB yielded bright green fluorescence under the confocal laser scanning microscopy at the wave length of 488 nm (E. a) and the morphology of the bright field (E. b). E. c is the overlay of the E. a and E. b photos. F: The transfected EGFP ES cells exhibited strong alkaline phosphatase (AP) activity. G: Teratomas formed by the EGFP positive ES cell lines in male Balb/c nu mice. The EGFP positive ES cells after about 4 months of culture (passages15) from about 60 % ~ 70 % confluent flask were injected subcutaneously into the oxter of 4-week-old male Balb/c nu mice (two or more mice). Six to eight weeks after injection, the resulting teratomas were examined histologically. (G. a) Ciliated columnar epithelium. (G. b) Neurocyte. (G. c) Typical-stratified squamous epithelium. (G. d) Adipose cells. (G. e) Striatedmuscle. (G. f) Gutlike structures. (G. g) Serous gland. (G. h) Cartilage

2.6 Differentiation in vivo

EGFP positive ES cells were injected subcutaneously into the oxter of the Balbc/nu male mouse. Tumors were formed with a diameter of 1 ~ 2cm after about 30 ~ 50 days. Paraffin embedded sections were prepared from midplane of the tumor in about 3 ~ 5μm of thickness and stained with hem atoxylin and eosin. Those sections indicated that the teratom a contained different types of tissues and cells of three germ layers. Squamous epithelium and neurocyte from ectoderm; cartilage, adipose cells, and striated muscle from mesoderm; and gutlike structure, ciliated columnar epithelium and serous gland from endoderm were morphologically identified (Fig. 2G).

2.7 Acquiring the cardiomyocytes-like cells and immunocytochemistry test

Cardiac differentiation was initiated by inducing EB formation from undifferentiated EGFP ES cells. In order to monitor the presence of beating cells in individual EBs, EBs were seeded at low density after 7 days in suspension culture, and the locations of EBs in each well were recorded. The EBs attached and continued to proliferate and differentiate into a heterogeneous population of cells including beating cardiomyocytes. Spontaneously contracting cells appeared as clusters and were identified in approximately 6% of the individual EBs on differentiation day17 and increased to as many as 30% ~ 40% of the EBs by day 22. The percentage was lower compared with 60% ~ 70% of the control (Fig. 3). The commencing time of beating was 3 days later than the control sample. Generally 1 ~ 2 beating areas appeared in one EB. The beating areas were at the most of four in one EB. The beating frequencies range from 30 to 130 per minute. Under confocal laser scanning microscopy, the contracting cell clusters showed bright green fluorescence (Fig. 2F).

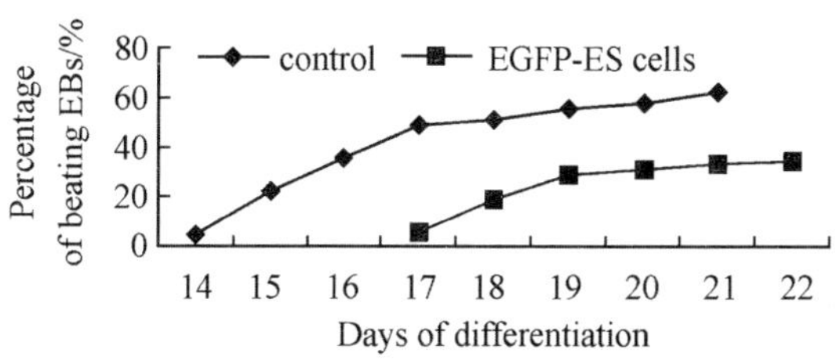

Fig. 3 Percentage of beating EBs with the days of differentiation

The cardiac-specific troponinT (cTnT), a subunit of the troponin complex, is a special marker for the cardiomyocytes. The monoclonal antibody against troponin T was used to detect the expression of the cTnT in the cardiomyocytes derived from the EGFP positive ES cells. Immunostaining showed that cTnT was present only in the beating area of the culture, but not in undifferentiated EGFP ES cells or differentiated cultures without evidence of contraction (Fig. 2. Eb). Similar positive result of immunostaining was also obtained from the primary culture of mouse cardiomyocytes (Fig. 2. Ea). The presence of cTnT in the differentiated cells derived from the EGFP positive ES clones indicated they were indeed cardiomyocytes.

3 Discusion

From the experiment, we found that subculture of the EGFP ES cell lines can be carried out on the MEF without the presence of the G418, which was different to other reports[5-8]. PCR analysis showed that different passages EGFP ES cells cultured on the MEF still carried the GFP gene after 10 passages (from 8 to 18 passages). It seems that the transfected ES cells in our study expressed the green fluorescence protein without the selecting pressure indicating that the gene may integrate into the ES genome. The present study demonstrated that EGFP positive ES cells possessed strong alkaline phosphatase (AP) activity, most of the cells exhibited intensive alkaline phosphatase staining which is similar to other reports[2,13,14].

The in vitro and in vivo differentiation ability is the other important criteria for evaluating the pluripotency of ES cells. The EGFP labeled cells exhibited the potency to undergo differentiation in vitro by forming the embryoid body with various cell types and structures in steps from the sEB to the maEB. When maEBs were seeded in the gelatin coated plate, subsequent culture resulted in spontaneously contracting cells. Independent of the stages of development, sEB, maEB, or the contracting cells all emitted bright green fluorescence. Immunocytochemistry staining indicated that the contracting cells were the cardiomyocytes since the cardiac specific cTnT marker was present on these cells while the surrounding non-beating cells were negative to cTnT. The staining area was comparable to the newborn mouse cardionmyocytes primary culture. The percentage of cardiomyocytes differentiation of the EGFP ES cells was lower than that of control. And the commencing time of beating was not as early as the control. What was the mechanism underline these differences were currently unknown. We also found that both embryoid body formation and differentiation capacity of ES cells were closely related to the growth status of the ES cells and the serum in the culture medium. Stable differentiation percentage could be obtained only when proper serum was used. Furthermore, no difference was found in the beating frequency of the myocytes derived from EGFP ES cells and non-transfected ES cells. The beating frequency became slower when the temperature and freshness of the medium dropped. Histological examination of teratoma derived from the EGFP ES cells in nude mice revealed that the solid areas of teratoma contained tissues representative of all three germ layers. Differentiated tissues including squamous epithelium, neurocyte, cartilage, adipose cells, striated muscle, gutlike structure, ciliated columnar epithelium, serous gl and were observed. This was similar to some related report[6,8,13,14].

To sum up, the above results completely revealed that the EGFP ES cells possess the typical characteristics of undifferentiated embryonic stem cell. And green fluorescence existed in the differentiated cells. The availability of EGFP ES cell lines provided a tractable model to study cell differentiation and tissue generation in vivo and in vitro. Transplantation of GFP-positive ES cells to murine embryos and injury mouse models are under way, which may pave ways to some new findings.

Acknowledgments We thank Dr. Xu Zheng-Ping and Doctoral candidate Zhou Qing-Jun for technical assistance. We are also grateful for colleagues who help us.

References

[1] Evans MJ, Kaufman MH. Establishment in culture of pluripotential cells from mouse embryos. Nature, 1981, 292: 154 – 156.

[2] Martin GR. Isolation of a pluripotent cell line from early mouse embryos cultured in medium conditioned by teratocarcinoma stem cells. PNAS of the USA, 1981, 78: 7634 – 7638.

[3] Gossler A, Joyner AL, Rossant J. Mouse embryonic stem cells and reporter constructs to detect developmentally regulated genes. Science, 1989, 244: 463 – 465.

[4] Klug MG, Soonpaa MH, Koh GY et al. Genetically selected cardiomyocytes from differentiating embronic stem cells form stable intracardiac grafts. The Journal of Clinical Investigation, 1996, 98: 216 – 224.

[5] EigesR, Schuldiner M, DrukkerM. Establishment of human embryonic stem cell – transfected clones carrying a marker for undifferentiated cells. CurentBiology, 2001, 11: 514 – 518.

[6] Zhao WN(赵文宁), MengGL(孟国良), XueYF(薛友纺) et al. Labeling of three different mouse ES celllines with the green fluorescence protein. Chinese Journal of Genetics(遗传学报), 2003, 30 (8): 743 – 749.

[7] Teng L(滕路), MengGL(孟国良), XingY(邢阳) et al. Labeling embryonic stem cells

with enhance green fluorescent protein on the hypoxanthi neguaninephos phoribosyl transferase locus. Chinese Medical Journal(中华医学杂志),2003,116(2):267 -272.

[8]Shen G(沈干),Cong XQ(丛笑倩),Wu Z(吴铮)et al. Establishment of cell line of mouse embryonic stem cells and its label with GFP. The Journalof South east University (Medical Science Edtion)[东南大学学报(医学版)],2003,22 (2):7 1 -74.

[9]Meng GL(孟国良),ShangKG(尚克刚). Improvement of preparing method of primary mouse embryonic fibroblast cells. Biotechnology(生物技术),1997,7(2):38 -39.

[10]Sambrook J,Rusel l DW et al. Molecular Cloning:A Laboratory Manual. 3rd ed,New York:Cold Spring Harbor Laboratory Press,2001

[11]Mathur A,Hong Y,Kemp BK et al. Evaluation of fluorescent dyes for the detection of mitochondrial membrane potential changes in cultured cardiomyocytes. Cardiovascular Research, 2000,46:126 -138.

[12]Mummery CD,Oostwaard W,Doevendans P et al. Differentiation of human embryonic stem cells to cardiomyocytes role of coculture with visceral endoderm -like cells. Circulation,2003, 107:2733 -2740.

[13]Thomson JA,Itskovitz -Eldor J,Shapiro SS et al. Embryonic stem cell lines derived from human blastocysts. Science,1998,282:1145 -1147.

[14]Reubinoff BE,Pera MF,Fong CY et al. Embryonic stem cell lines from human blastocysts:somatic differentiation in vitro. Nature Biotechnology,2000,18:399 -404.

注:本文曾发表于2004年20卷第6期《生物工程学报》。

The role of humic substances in the anaerobic reductive decholorination of 2,4 – dichlorophenoxyacetic acid by Comamonas Koreensis strain CY01

Yibo Wang　Chunyuan Wu　Xiaojing Wang　Shungui Zhou*

摘　要:以 Comamonas Koreensis CY01 为研究对象,研究了厌氧条件下 HS 的存在对微生物介导的2,4 – D(2,4 – 二氯苯氧乙酸)还原脱氯过程的影响。实验结果表明,(i)CY01 菌株不但具有 Fe(III)/HS 还原特性,而且还具有使 2,4 – D 进行脱氯降解的能力,随着降解过程中 Cl – 产量的增加,2,4 – D 和电子供体葡萄糖的浓度均有降低,说明 CY01 引起的2,4 – D 的降解过程是伴随着电子供体葡萄糖的氧化同时进行的;(ii)反应体系中加入 AQDS,会促进 2,4 – D 的降解过程;(iii)当反应体系中产生生物源的 AH2QDS(还原型 AQDS)后,再除去菌体,仍然会引起2,4 – D 的降解,这说明 AH2QDS 对 2,4 – D 的还原作用,即电子从 AH2QDS 向2,4 – D 的传递过程是一个非生物的纯化学过程;(iv)随着 AH2QDS 将电子传递给2,4 – D,AH2QDS 被重新氧化成 AQDS,AQDS 再次作为电子受体接受来自 CY01 传递的电子,使之在 Fe(III)/HS 还原菌和 2,4 – D 之间充当电子穿梭体。综合以上实验结果,说明 CY01 菌株引起的2,4 – D 的还原脱氯过程是一个生物化学过程,CY01 首先氧化葡萄糖产生电子,然后 AQDS 将产生的电子加速传递至 2,4 – D,使2,4 – D 发生还原脱氯降解,这可能也代表了自然界厌氧环境下一条很重要的电子流动方式。因此我们提出了 Fe(III)/HS 还原菌介导的 2,4 – D 还原脱氯机制假说,以期揭示厌氧条件下 2,4 – D 加速降解的内在本质,为建立基于 Fe(III)/HS 呼吸的土壤有机氯污染原位修复技术提供科学依据。

* 作者简介:王弋博(1973—　),女,甘肃天水人,现为天水师范学院生物工程与技术学院教授,博士,主要从事环境生物学研究。

The role of the humic model compound, anthraquinone – 2, 6 – disulfonate (AQDS), in the anaerobic reductive dechlorination of 2,4 – Dichlorophenoxyacetic acid (2,4 – D) by the Fe(III) – and humic substances HS – reducing bacterium, *Comamonas Koreensis* strain CY01 was investigated. The results taken as a whole indicate that (i) strain CY01 can couple glucose oxidation to 2,4 – D reductive dechlorination; (ii) reductive dechlorination of 2,4 – D by strain CY01 is greatly stimulated by the addition of AQDS; (iii) the transfer of electrons from biogenic AH_2QDS to 2,4 – D is an abiotic process which can take place in the absence of microorganisms; (iv) AH_2QDS was re – oxidized during the chemical reaction, AQDS can serve again as electron acceptor for microorganisms, thus acting as electron shuttles. All the results suggested that 2,4 – D reductive dechlorination by CY01 strain is a biochemical process that oxidizes the electron donors and transfers the electron to the acceptors through redox mediator, AQDS. We proposed the possible mechanism for the HS dependent reduction of 2,4 – D. Our results suggested that microbial reduction of HS and subsequent chemical reduction of organic pollutants represent an important path of electron flow in anoxic natural environments. This work is a necessary preliminary step for better understanding the biodegradation of 2,4 – D in subsurface soils.

1. Introduction

2,4 – Dichlorophenoxyacetic acid (2,4 – D) is the third – most widely used herbicide in North America and the most widely used herbicide in the world[1]. It is the active ingredient in several formulations of herbicides recommended for the control of broadleaf weeds. It has major uses in agriculture crops, forestry, turf, non – crop and aquatic weeds [2]. Continuous use and improper disposal may cause soil percolation and groundwater contamination. It causes toxicity in receiving waters and inhibition of biological treatment systems even at low concentrations. The central nervous system is a target organ for the effects of this herbicide in different animal species [3]. Even though human exposure to these herbicides has been associated with numerous clinical manifestations, such as nervous system, liver and kidney damage[4], their continued use is still widely practiced. In situations where these compounds serve as environmental hazards, bioremediation appears to be a potentially powerful clean – up tool[5].

2,4 – D does not persist for long in the environment, because it is readily susceptible to microbial degradation[6]. Microbial degradation of 2,4 – D has been extensively studied, the degradative pathway for this herbicide has been elucidated and the enzymes involved have been characterized. As a result of this considerable attention, a number of bacterial genera, such as *Arthrobacter*, *Alcaligenes*, *Achromobacter*, *Bacillus*, *Bortedella*, *Burkhoderia*, *Pseudomonas*, *Ralstonia*, *Sphingomonas*, *Sarcina*, *Comamonas* (*Delftia*) and *Sporocytophaga*, are known to degrade 2,4 – D, both in mixed and pure cultures[6-9]. Although some studies described 2,4 – D degradation by selected bacteria, there have been no reports, to date, on the degradation of 2,4 – D improved by humic substances (HS) by *Comamonas* strain. Mechanisms of microbially catalyzed reductive dechlorination of 2,4 – D is not well understood and may be species – dependent [10].

Aerobic biotransformation of 2,4 – D has been observed in both pure and mixed microbial cultures[11]. However, anaerobic biodegradation of 2,4 – D has not been as thoroughly elucidated. It is known that the biological removal of halogen atoms from halogenated compounds under anaerobic conditions occurs by reductive dehalogenation[12]. Reductive dehalogenation reactions catalyzed by anaerobic bacteria are either co – metabolic processes or linked to respiration; a process termed dehalorespiration[13]. In the process of dehalorespiration, an anaerobic bacterium utilizes a halogenated compound as a terminal electron acceptor, the reduction of which is coupled to ATP production [14]. Dechlorination is the most commonly reported reductive dehalogenation process, due to the widespread pollution of chlorinated compounds. In the cases of 2,4 – D, reductive dechlorination has been observed in sewage sludge [15], pond sediment[16] and a methanogenic Aquifer [17].

HS and quinines [e. g. anthraquinone – 2,6 – disulfonate (AQDS)], representative of structural moieties in humus, have also been shown to mediate the abiotic reductive dechlorination of polychlorinated pollutants by inorganic electron donors[18]. HS has been also reported to act as an electron mediator to enhance the reduction efficiency of chlorinated aliphatic compounds in aqueous solutions containing bulk reductant [19] and to facilitate the microbially mediated anaerobic dechlorination of chlorinated hydrocarbons under iron – reducing conditions[20]. Although many abiotic experiments have shown that HS catalyzed reductive dechlorination reactions, there has been much less evidence that HS is involved in biological dechlorination [21]. It has been recognized that HS may play an important role in the anaerobic biodegradation and biotransformation of organic as well as inorganic compounds[22]. An increasingly popular approach for the remediation of chloridized environmental pollutants is biological reductive dechlorination. As little studies concerning the roles of HS in anaerobic biodegradation of 2,4 – D by bacteria, it is still not clear so far whether HS are involved in the dechlorination processes of 2,4 – D by *Comamonas*. Therefore, knowledge of the role of HS in the dechlorination of 2,4 – D is thus required.

Although a capacity forbiodegradation of 2,4 – D has been found in *Comamonas* microorganisms, the precise mechanism of anaerobic reductive dechlorination is still unknown. In this work, we report 2,4 – D reductive dechlorination mediated via HS by *Comamonas Koreensis*. A facultative anaerobic bacterium, designated CY01, was isolated successfully from an ancient forest sample collected from Guangzhou, China. It was identified as a strain of *Comamonas Koreensis* CY01 had been tested for its ability to degrade 2,4 – D and the knetics of 2,4 – D degradation by *Comamonas Koreensis* wasexplored. The results indicated that Fe(III) – and HS – reducing bacterium, *Comamonas Koreensis* CY01, may have a more important role in conversion of 2,4 – D in the presence of AQDS than previous considered.

2. Material and methods

2.1 Bacterial strains

An efficient Fe(III) and HS-reducting bacterium, *C. Koreensis* strain CY01, used in the study was isolated from the submerged forest sediment in Sihui City, China. Strain CY01 was tested for a number of key characteristics by using standard procedures [23] and was determined according to Bergey's Manual of Determinative Bacteriology (9th edition, 1994).

2.2 Culture conditions

The strain CY01 was cultured aerobically at 30℃ with shaking at 180 rpm in Luria – Bertani (LB) medium or anaerobically in a defined medium (per liter of distilled water: 2.5g $NaHCO_3$, 0.25g NH_4Cl, 0.6 g NaH_2PO_4, 0.1g KCl, 0.2g yeast extract, vitamin solution, and mineral solution). After autoclaving and cooling under an atmosphere of N_2/CO_2 (80/20, vol/vol), 5mmol l^{-1} glucose and 1 mmol l^{-1} AQDS as electron donor and acceptor were added. into the defined medium. Strict anaerobic techniques were used throughout the study [24]. All gases were passed through a filter prior to use.

Microbial reduction of 2,4 – D, cultures of CY01 were grown overnight in LB medium. The cells were harvested under aerobic conditions and centrifuged for 10 min at

8000rpm at 4℃, washed twice, and resuspended in ultrapure water. The microbial reduction of 2,4 – D was conducted in 20 ml solutions of 2,4 – D(20 ppm), glucose(5 mmol l^{-1}), AQDS(3 mmol l^{-1}), and 1 ml of the cell solution. The solutions(initial pH, 6.5) were dispensed into 25 – mL serum vials, bubbled with N_2/CO_2(80:20) and filtered(0.22 μm filters) before incubation. Then the serum vials were stoppered with butyl rubber bungs and crimped with aluminium caps at a constant temperature(30℃) in an anaerobic station(Manufactured In USA By Sheldon Manufacturing, Inc 300N, 26Th Cornelius, or 97113). The control experiments were performed in the same manner except that no electron donor, glucose were added to the 2,4 – D solutions [32].

2.3 Reagents

Methanol and acetic acid of high-performance-liquid chnmatography(HPLC) – gradient grade was purchased from Shanghai Reagent Co., China. Chemical reagents such as AQDS, 2,4 – D, 2,4 – dichlorophenol(2,4 – DCP) and D(+) – glucose used in the experiment were of analytical grade and used without further purification. Standards of these reagents were purchased from Sigma – Aldrich. Solutions were prepared using ultrapure water and NaOH or HCl in proper amounts was used to get the suitable pH value.

2.4 Analysis techniques

All the samples were centrifuged at 4000rpm for 15min and filtered through 0.22 μm syringe filter; the filtrates were stored for all immediate analysis.

The concentration of Cl^- was determined by ion chromatography(Dionex ICS – 90) with an ion column(IonPac AS14A 4mm × 250mm). A mobile phase consisting of Na_2CO_3(8.0 mM) and $NaHCO_3$(1.0 mM) solution was operated at a flow rate of 1.0 ml min^{-1}.

Glucose was determined spectrophotometrically using the method described by Miller [26].

The concentrations of 2,4 – D and main aromatic intermediates, resulting from the degradation of 2,4 – D were analyzed using a high – performance – liquid chromatography apparatus(waters 1527/2487, made in USA). HPLC conditions were: injector volume, 1 ml; mobile phase flow rate, 1 ml min^{-1}; UV detector wavelength 285 nm; reverse – phase C_{18} column(4.6mm × 250 mm) and an isocratic mobile phase(vol/vol) (methanol: ultrapure water: acetic acid = 60:38:2).

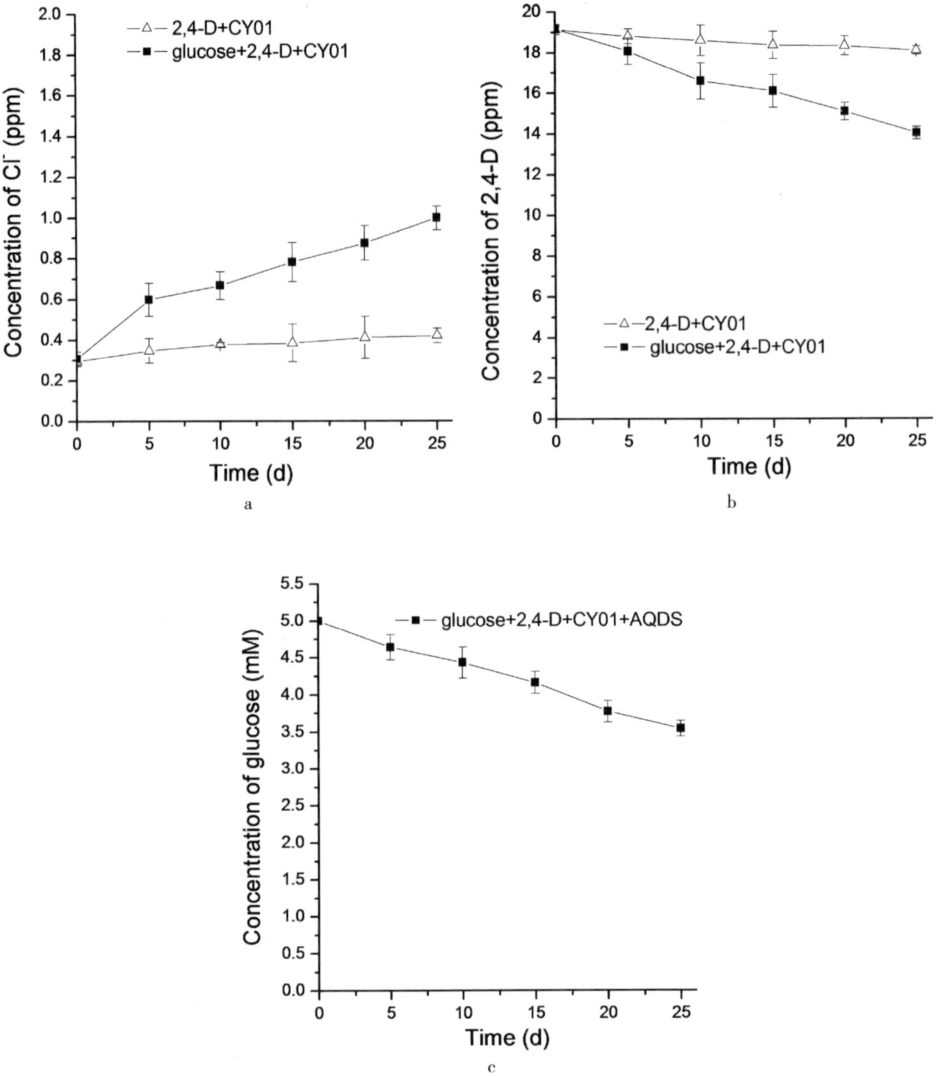

Fig. 1. The effect of *Comamonas Koreensis* CY01 on the production of Cl^- (a) and the biodegradation of 2,4 – D (b) and glucose (c). The experiments were performed at 30℃ by addition of 5 mmol l^{-1} glucose, 20 ppm 2,4 – D and 1 ml cell solution (1.3 $\times 10^8$). Bars represent standard deviations of three independent experiments with three replicates; values are means of triplicate ± S. E.

All the experiments run were conducted in duplicate and analysis of each parameter was done triplicate for each run. Comparison analyse was performed by SPSS 10.0 statistical package.

3 Results and discussion

3.1 Isolation and identification of *C. Koreensis* CY01

In anoxic habitats, ferric - iron and humics - reducing microorganisms presumably play an important role in the oxidation of organic matter[27]. Given the abundance of HS in some soils and sediments, electron transfer to HS might also be important if a diversity of microorganisms is capable of this form of respiration[28]. Further understanding of the potential importance of microbial HS reduction requires knowledge about the distribution and diversity of microorganisms that might be responsible for this reduction in sedimentary environments[29].

In this study, a facultative anaerobic bacterium, designated CY01, was isolated from the submerged forest sediment in Sihui City, Guangdong Province, China. The isolate was identified as a strain of *C. Koreensis* based on its biochemical, physiological and morphological characteristics as well as the analysis of 16S rDNA sequence and the DNA G + C content. CY01 strain was a Gram - negative, non - motile, non - flagella, rod - shaped(1.2 - 1.5 μm long, 0.3 - 0.4 μm wide), oxidase - and catalase - positive(under aerobic growth conditions) bacterium. The G + C content of the DNA was 64.8%. Analysis of 16S rDNA of CY01 indicated that the isolate formed a monophyletic clade with the members of the genus *Comamonas*. The closest phylogenetic relative among the valid species was *C. Koreensis*, with 98% 16S rDNA similarity[30]. The optimal temperature and pH value for cultivation was 30 - 32℃ and 6.5 - 7.0, respectively. Considering the results reported here, all the experiments were conducted at 30℃ and pH 6.5.

Generally, a wide phylogenetic diversity of microorganisms capable of Fe(III) reduction is also able to reduce HS[28]. Strain CY01 showed the highest Fe(III) and AQDS reduction activity of all the strains isolated in this study. AQDS is a model compound for quinine moieties in humic substances(it is a functional analogue, not a structural analogue or a model humic acid as claimed frequently in the literature). Therefore, AQDS has been used to research HS - reducing instead of HS[27]. CY01 could not only reduce Fe(III) and AQDS, but also 2,4 - D with glycerol, glucose, citric acid and sucrose as electron donors under anaerobic growth conditions.

3.2 Reductive dechlorination of 2,4 – D by *Comamonas Koreensis* CY01 under anaerobic condition

During the 25 – day cultivation period, under anaerobic condition, the addition of CY01 to the defined medium, respectively, resulted in a slight degradation of 2,4 – D and glucose as well as the production of Cl^- in triplicate samples (the slight levels in the presence of CY01 was to be statistically significant, $P < 0.05$), whereas nearly no degradation of 2,4 – D (Fig. 1b), consumption of glucose (Fig. 1c) and production of Cl^- (Fig. 1a) were observed of the control. The dechlorination rate of (YD) was 13.3%. The results indicated that (i) CY01 strain has the ability to reduce 2,4 – D directly, whereas the 2,4 – D – reducing ability of the microorganism is relatively weak; and (ii) the reduction of 2,4 – D by CY01 strain must depend on the presence of glucose, when glucose was used as the electron donor. Under anaerobic or reducing conditions, several chlorinated organic compounds have been shown to be dechlorinated by anaerobic and facultative microorganisms [5]. Our results also demonstrated that 2,4 – D could be microbially reduced under anaerobic condition.

3.3 Effect of AQDS on the biodegradation of 2,4 – D by CY01 strain

HS are known to stimulate chlorinated organic compounds reduction by serving as electron shuttles between the cells and chlorinated organic compounds[18]. Previous studies demonstrated that *Comamonas* was involved in the process of 2,4 – D degradation, but the mechanisms involved were not elucidated during this process [29]. Therefore the role of HS in the anaerobic reductive dechlorination of 2,4 – D was nest examined. As can be seen from Fig. 2, under anaerobic condition, 2,4 – D was reduced significantly by CY01 strain. Concomitant with the decrease of glucose, the amount of Cl^- increased up to ~3 ppm, because of the decomposition of 2,4 – D. Glucose was oxidized using 2,4 – D as electron acceptor by CY01. The positive correlation between the oxidation of glucose and the reduction of 2,4 – D. The results suggested that (i) the higher efficiency dechlorination of 2,4 – D by the CY01 strain was dependent on the introduction of AQDS which were reduced during the assay. (ii) CY01 can couple glucose oxidation to 2,4 – D reductive dechlorination, with AQDS to mediate the course. The stimulation of 2,4 – D reductive dechlorination caused by CY01 in the presence of AQDS clearly demonstrated that this particular quinine can serve as an electron shuttle between CY01 and 2,4 – D. Quinones in humus, once microbially reduced, can transfer their reducing equivalents to chlorinated compounds [31].

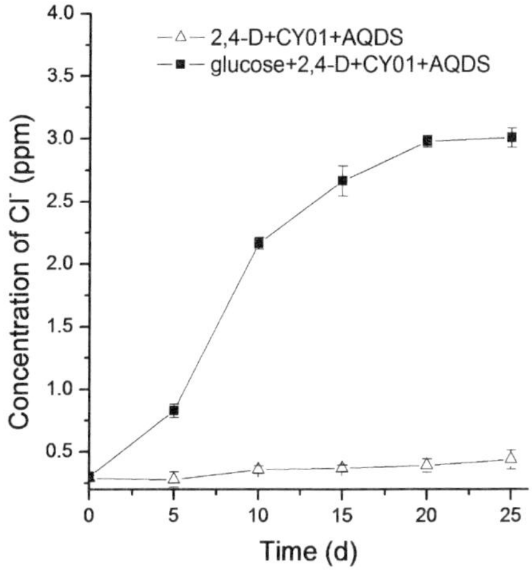

a

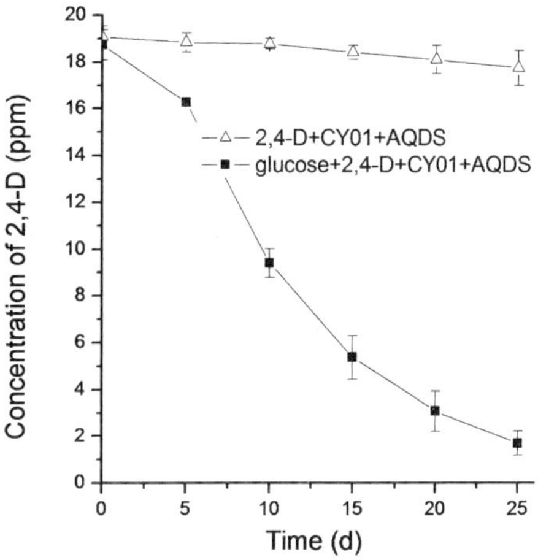

b

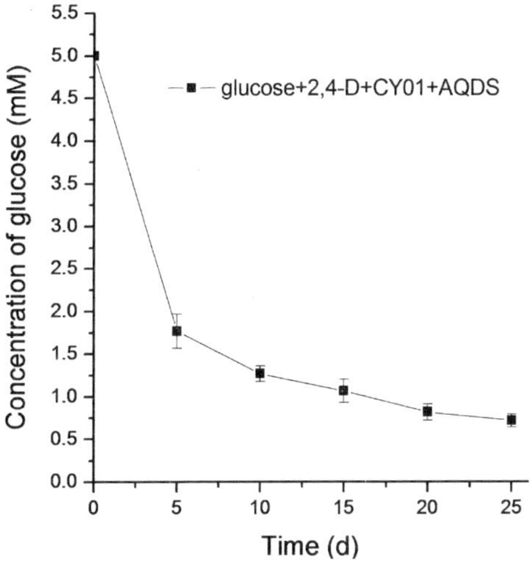

c

Fig. 2 The effect of AQDS on the the production of Cl^- (a) and the biodegradation of 2,4 – dichlorophenoxyacetic acid (b) and glucose (c) in the presence of *Comamonas Koreensis* CY01. The experiments were performed at 30℃ by addition of 5 mmol l^{-1} glucose, 20 ppm 2,4 – D, 3 mmol l^{-1} AQDS and 1 ml cell solution (1.3×10^8). Bars represent standard deviations of three independent experiments with three replicates; values are means of triplicate ± S. E.

It is accounted that the complete dechlorination of 20 ppm 2,4 – D could produce 6 ppm Cl^-. However, during the 25 – day incubation, in the presence of AQDS, about 85% of 2,4 – D was reduced and 80% of glucose was consumed; only about 3ppm Cl^- was produced. Considerably less dechlorination occurred in incubations lacking glucose. In the presence of AQDS, the removal of Cl^- was less than 50% after 25 day of cultivation, but the biodegradation rate of 2,4 – D and glucose was ~80 – 85%. This was an indication that some of the intermediates derived from 2,4 – D decomposition remained, mainly as chlorinated organic compounds, such as 2,4 – dichlorophenol, 2,4 – dichloroanisol (2,4 – DCA) and 4,6 – dichloroanisolresorcinol, *et al.* [1]. Within a 25 – day period, it appears that the dechlorination of 2,4 – D is the result of CY01 action and that the addition of AQDS enhances dechlorination activity. *C. Koreensis* CY01 and 2,4 – D via HS accelerates the microbial reductive dechlorination of 2,4 – D.

AQDS reduction is a good predict of the capacity for HS reducing[28]. Using an intermediate is an effective strategy to accelerate degradation of some environmental pollutants [32]. It was generally considered that HS can serve as redox mediators for microbial reductive dechlorination. This is because reduced HS can directly transfer the electrons gained from microbial reduction to chlorinated compounds. HS are reoxidized during this chemical reaction; HS can serve again as electron acceptors for microorganisms, thus acting as electron shuttles[33 – 35].

3.4 Effect of AQDS on 2, 4 – D reductive dechlorination without CY01 strain

Because the addition of AQDS enhanced the conversation of 2,4 – D by increasing the rate and extent of dechlorination of this pollutant by *C. Koreensis* CY01, we next examined if enhanced 2,4 – D reductive dechlorination was the result of abiotic process of AQDS. The results showed the higher background concentration of Cl^- of controls with dead CY01 strain (Fig. 3d) than that of controls without CY01 (Fig. 3a). Cl^- concentration of controls with dead CY01 strain was about 1.1 – 1.2 ppm, whereas that of controls without CY01 strain was about 0.2 – 0.3 ppm. The observations can be explained as below: when cells were killed by autoclaving, the component of cytoplasm spilled from the dead cells (including Cl^-), thus Cl^- concentration increased in the controls with dead CY01 strain.

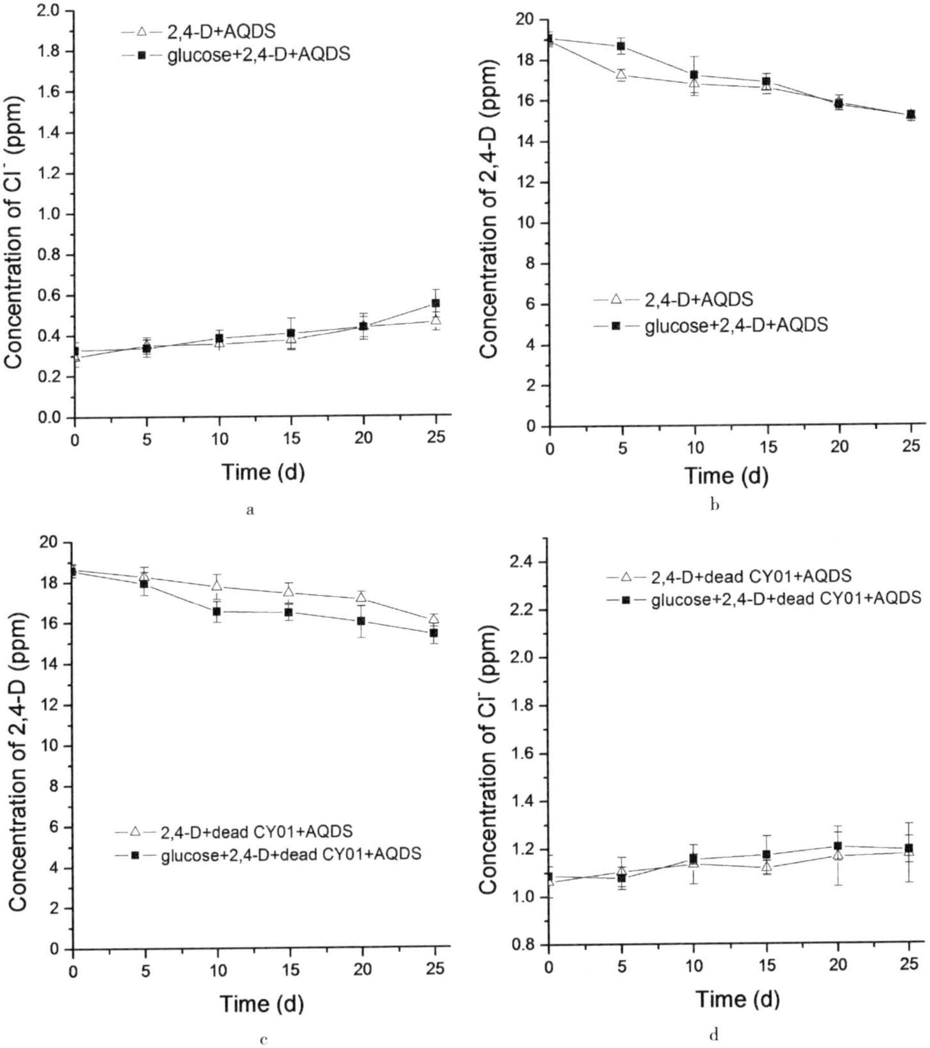

Fig. 3 The production of Cl^- (a, c) and the biodegradation of 2,4 – D (b, d) without *Comamonas Koreensis* CY01 or with dead *Comamonas Koreensis* CY01 cell in the presence of AQDS. The experiments were performed at 30 ℃ by addition of 5 mmol l^{-1} glucose, 20 ppm 2,4 – D, 3 mmol l^{-1} AQDS or 1 ml autoclaved cell solution (1.3×10^8). Bars represent standard deviations of three independent experiments with three replicates: values are means of triplicate ± S. E.

As also seen from Fig. 3, without CY01 cells or with dead CY01 cells, the addition of AQDS hardly affected the production of Cl^- (Fig. 3 a and d) and the decreasing of 2, 4 - D(Fig. 3b and c). The negligible dechlorination achieved in no cells the dechlorination rate was about 0.63% or autoclaved cell samples (the dechlorination rate was about 0.34%) in the presence of AQDS indicated that 2,4 - D reductive dechlorination was a biological process and not a chemical reaction as reduced AQDS was not able to reduce 2,4 - D in the absence of CY01 strain, or when cells were killed by autoclaving. The absence of *C. Koreensis* CY01 strain cancelled the stimulating effect of AQDS on 2, 4 - D - dechlorination in anaerobic condition.

3.5 Effect of the biologically reduced AQDS on reductive dechlorination of 2,4 - D

The above results provided evidence that *C. Koreensis* CY01 and AQDS were responsible for the enhanced conversion of 2,4 - D in anaerobic condition. To further investigate the possible mechanism of microbially catalyzed reductive dechlorination process of 2,4 - D in the presence of AQDS, the further research on the effect of biogenic AH_2QDS may be necessary. By autoclaving to exclude the dechlorination activity of strain CY01, and only that of biologically reduced AQDS was investigated. As shown from Fig. 4, controls with dead CY01 strain served as calibration standard, the net production of Cl^- (Fig. 4a) can be calculated as about 2.3 - 2.7 ppm (the dechlorination rate was about 38.3 - 45%) during 25 day. The biodegradation rate of 2,4 - D and glucose(Fig. 4b and c) almost reached the level of samples without autoclaving (Fig. 2b and c). The biogenic AH_2QDS was shown to directly cause the chemical reduction of 2, 4 - D when the microorganisms were moved by autoclaving from samples. Once oxidized by 2,4 - D, AQDS may again accept electrons from humics and Fe(III) - reducing strain, CY01. The results suggested that 2, 4 - D reductive dechlorination by CY01 strain is a biochemical process that oxidizes the electron donors and transfers the electron to the acceptors through redox mediator, AQDS. Quinones in humus, once microbially reduced, can transfer their reducing equivalents to chlorinated compounds [21]. Thus, in environments, even small quantities of humics could be important in electron transfer as they could be recycled as electron acceptors numerous times[20].

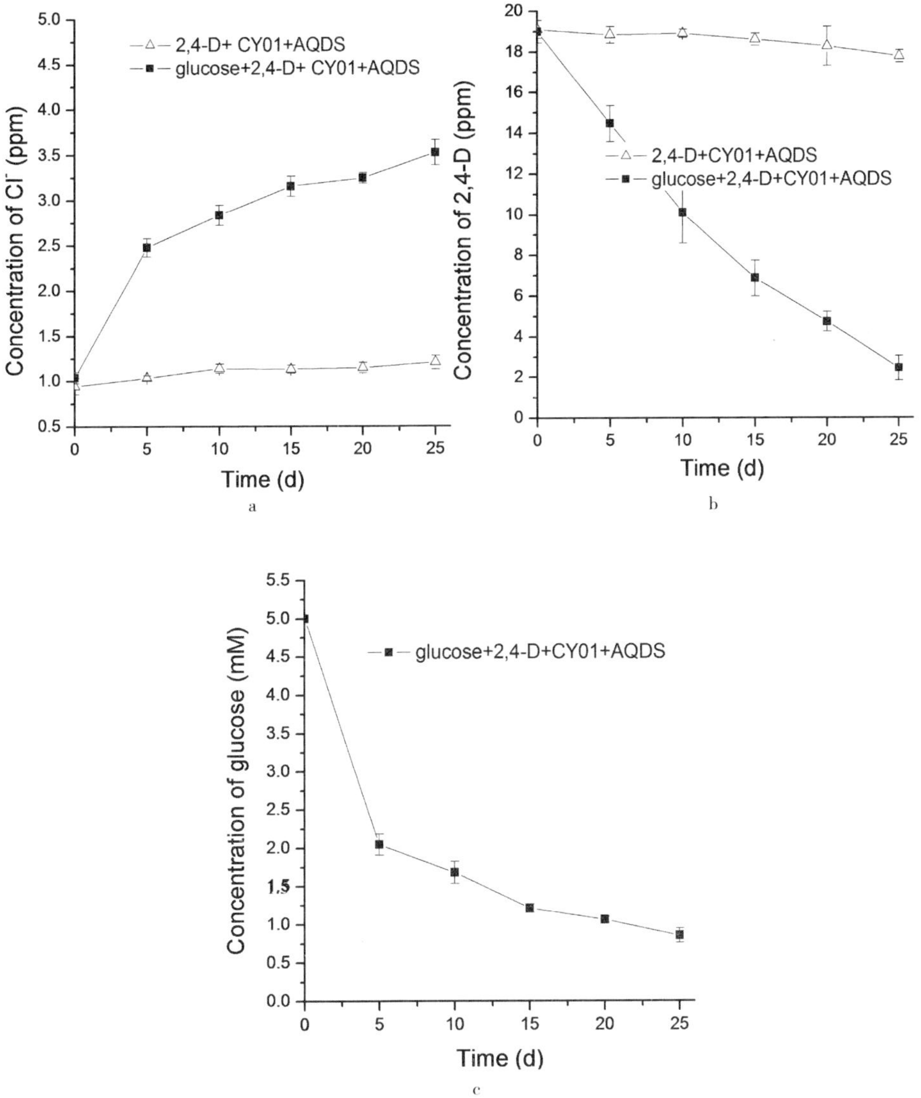

Fig. 4 The effect of biogenic AH_2QDS on the production of Cl^- (a) and the biodegradation of 2,4 – D(b) and glucose(c). The experiments were performed at 30℃ by addition of 5 mmol l^{-1} glucose, 20 ppm 2,4 – D, 3 mmol l^{-1} AQDS and 1 ml cell solution(1.3×10^8). Bars represent standard deviations of three independent experiments with three replicates; values are means of triplicate ± S. E.

The proposed mechanism for the HS dependent reduction of 2,4 – D is: AQDS is reduced to the corresponding hydroquinone(AH_2QDS) by CY01 strain, then the biogenic AH_2QDS can transfer electrons directly to 2,4 – D in a purely chemical reaction, however, there may be biological interaction between microbial anaerobic humic reduction and 2,4 – D reductive dechlorination, which was little known. Because AH_2QDS were reoxidized during the chemical reaction, AQDS can serve again as electron acceptors for microorganisms, thus acting as electron shuttles. This allows an indirect reducing of 2,4 – D without direct contact to the bacterial cells in the presence of AQDS, thus enhancing the rates of microbial reduction of 2,4 – D. This was proved indirectly by Kappler et at al. [36] and Lovely.

3. Conclusions

In this study, the influence of HS on the dechlorination of 2,4 – D by *C. Koreensis* CY01 was examined. The results indicated that HS and Fe(Ⅲ) – reducing bacterium may have more important role in the conversion of 2,4 – D in the presence of HS than previously considered. To our knowledge, the present study constitutes the first report for the anaerobic biodegradation of 2,4 – D linked to HS reduction.

Strain CY01 has the ability to reduce 2,4 – D directly, whereas the 2,4 – D – reducing ability of the microorganism is relatirely weaks. However, when the humic model compound, AQDS was introduced, the dechlorination activity of strain CY01 was enhanced under the same conditions. The reduction of 2,4 – D by CY01 strain with or without AQDS must depend on the presence of glucose, when glucose was used as the electron donor. These results suggested that strain CY01 can couple glucose oxidation to 2,4 – D reduction, with AQDS to mediate the course. Electron shuttling between *C. Koreensis* CY01 and 2,4 – D via HS accelerates the anaerobic biodegradation of 2,4 – D. It was believed that CY01 may transfer electrons via HS to 2,4 – D. The negligible dechlorination achieved in no cells or autoclaved cell samples in the presence of AQDS indicated that 2,4 – D reductive dechlorination was a biological process in anaerobic condition. The biogenic AH_2QDS was shown to directly cause the chemical reduction of 2,4 – D when the microorganisms were moved by autoclaving from samples. Higher efficiency of, 2,4 – D reductive dechlorination depends on the combiation of the biologically reducd AQDS and chemical reducton of 2,4 – D. AQDS can serre as electron shuttle.

Acknowledgements

The research work was funded by the Nature Science Foundation of China (NO

30570165,40601043,20777013)

REFERENCES

[1]C. Badellino, C. A. Rodrigues, R. Bertazzoli, Oxidation of pesticides by in situ electrogenerated hydrogen peroxide: study for the degradation of 2,4 – dichlorophenoxyacetic acid, J. Hazar. Mater. 137(2006)856 – 864.

[2]J. S. Busa, L. E. Hammond, Regulatory progress, toxicology, and public concerns with 2,4 – D: Where do we stand after two decades?, Crop Protection. 26(2007)266 – 269.

[3]A. A. Bortolozzi, A. M. E. Duffard, R. Duffard, M. C. Antonelli, Effects of 2,4 – dichlorophenoxyacetic acid exposure on dopamine D2 – like receptors in rat brain, Neurotoxicol Teratol. 26(4)(2004)599 – 605.

[4]I. R. Gonza′lez, E. R. Leal, R. F. Cerrato, F. E. Garcı′a, N. R. Seijas, H. M. P. Varaldo, Bioremediation of a mineral soil with high contents of clay and organic matter contaminated with herbicide 2,4 – dichlorophenoxyacetic acid using slurry bioreactors: Effect of electron acceptor and supplementation with an organic carbon source, Process Biochem. 41(2006)1951 – 1960.

[5]J. G. Leahy, R. R. Colwell, Microbial degradation of hydrocarbons in the environment, Microbiol. Mol. Biol. Rev. 54(1990)305 – 315.

[6]T. Vroumsia, R. Steiman, F. S. Murandi, J. L. B. Guyod, Fungal bioconversion of 2,4 – dichlorophenoxyacetic acid (2,4 – D) and 2,4 – dichlorophenol (2,4 – DCP), Chemosphere. 60(2005)1471 – 1480.

[7]J. O. Ka, W. E. Holben, J. M. Tiedje, Genetic and phenotypic diversity of2,4 – dichlorophenoxyacetic acid(2,4 – D) degrading bacteria isolated from 2,4 – D – treated field soils, Appl. Environ. Microbio. 60(1994)1106 – 1115.

[8]W. Dejonghe, J. Goris, S. E. Fantroussi, M. Hofte, P. D. Vos, W. Verstraete, Effect of dissemination of 2,4 – dichlorophenoxyacetic acid(2,4 – D) degradation plasmids on 2,4 – D degradation and on bacterial community structure in two different soil horizons, Appl. Environ. Microbiol. 66(2000)3297 – 3304.

[9]E. M. Montiel, N. R. Ordaz, C. R. Granados, C. J. Ramı′rez, C. J. G. Mayer, 2,4 – D – degrading bacterial consortium isolation, kinetic characterization in batch and continuous culture and application for bioaugmenting an activated sludge microbial community, Process Biochem. 41(2006)1521 – 1528.

[10]F. W. Picardal, R. G. Arnold, H. Couch, A. M. Little, M. E. Smith, Involvement of Cytochromes in the anaerobic biotransformation of tetrachloromethane by*Shewanella putrefaciens* 200, Appl. Environ. Microbiol. 59(1993)3763 – 3770.

[11]J. W. Neilson, K. L. Josephson, S. D. Pillai, I. L. Pepper, Polymerase chain reaction and gene probe detection of the 2,4 – dichlorophenoxyacetic acid degradation plasmid, pJP4, Appl. En-

viron. Microbiol. 58(1992)1271 - 1275.

[12]B. V. Chang, J. Y. Liu, S. Y. Yuan, Dechlorination of 2,4 - dichlorophenoxyacetic acid and 2,4,5 - trichlorophenoxyacetic acid insoil, The Sci. of the Total Environ. 215(1998)1 - 8.

[13]C. Holliger, W. Schumacher, Reductive dehalogenation as respiratory process, Antonie van Leeuwenhoek 66(1994)239 - 246.

[14]A. W. Boyle, V. K. Knight, M. M. Häggblom, L. Y. Young, Transformation of 2,4 - dichlorophenoxyacetic acid in four different marine and estuarine sediments: effects of sulfate, hydrogen and acetate on dehalogenation and side - chain cleavage, FEMS Microbiol. Ecology 29(1999)105 - 113.

[15]M. D. Mikesell, S. A. Boyd, Reductive dechlorination of the pesticides 2,4 - D and 2,4,5 - T and pentachlorophenol in anaerobic sludges, J. Environ. Qual. 14(1985)337 - 340.

[16]S. A. Gibson, J. M. Suflita, Extrapolation of biodegradation results to groundwater aquifers: Reductive dehalogenation of aromatic compounds, Appl. Environ. Microbiol. 52(1986)681 - 688.

[17]S. A. Gibson, J. M. Suflita, Anaerobic biodegradation of 2,4,5 - trichlorophenoxyacetic acid in samples from a methanogenic aquifer: simulation by short - chain organic acids and alcohols, Appl. Environ. Microbiol. 56(1990)1825 - 1832.

[18]J. A. Field, F. J. Cervantes, Z. F. Van der, G. Lettinga, Role of quinones in the biodegradation of priority pollutants: a review, Water Sci. Technol. 42(2000)215 - 222.

[19]G. P. Curtis, M. Reinhard, Reductive dehalogenation of hexachloroethane, carbon tetrachloride, and bromoform by anthrahydroquinone disulfate and humic acid, Environ. Sci. Technol. 28(1994)2393 - 2401.

[20]D. R. Lovley, J. D. Coates, E. L. Blunt - Harris, E. J. P. Phillips, J. C. Woodward, Humic substances as electron acceptors for microbial respiration. Nature 382(1996)445 - 448.

[21]F. J. Cervantes, L. Vu - Thi - Thu. G. Lettinga, J. A. Field, Quinone - respiratio improves dechlorination of carbon tetrachloride by anaerobic sludge, Appl. Microbiol. Biotechnol. 64(2004)702 - 711.

[22]Y. G. Hong, J. Guo, Z. C. Xu, M. Y. Xu, G. P. S, Humic substances act as electron acceptor znd redox mediator for microbialdissimilatory azoreduction by *Shewanella decolorationis* S12, J. Microbiol. Biotechnol. 17(2007)428 - 437.

[23]P. Gerhardt, R. G. E. Murray, W. A. Wood, N. R. Krieg, Methods for General and Molecular Bacteriology, Washington, DC: American Society for Microbiology, 1994.

[24]J. M. Vargas, K. Kashefi, E. L. Blunt - Harris, D. R. Lovely, Microbiological evidence for Fe(Ⅲ)reduction on early Earth, Nature 395(1998)65 - 67.

[25]L. Apper, D. Mcknight, J. Robinfulton, E. B. Harris, K. Nevin, D. lovely, P. Hatcher, Fulvic acid oxidation statedetectionusing fluorescence spectroscopy, Environ. Sci. Technol. 36(2002)

3170 –3175.

[26] G. L. Miller, Use of dinitrosalicylic acid reagent for determination of reducing sugar, Analytical Chem. 31(1959)426 –428

[27] L. S. Kristina, K. Andreas, S. Bernhard, Enrichment and isolation of ferric – iron – and humic – acid – reducing bacteria, Methods in enzymology 397(2005)58 –70

[28] D. R. Lovley, J. L. Frage, E. L. Blunt – Harris, L. A. Hayes, E. J. P. Phillips, J. D. Coates, Humic Substances as a mediator for microbially catalyzed metal reduction, Acta hydrochim. hydrobiol. 26(1998)152 –157.

[29] J. D. Coates, D. J. Ellis, E. L. Blunt – Harris, C. V. Gaw, E. E. Roden, D. R. Lovley, Recovery of humic – reducing bacteria from a diversity of environments, Appl. Environ. Microb. 64 (1998)1504 –1509.

[30] Y. H. Chang, J. I. Han, J. Chun, K. C. Lee, M. S. Rhee, Y. B. Kim, K. S. Bae, *Comamonas Koreensis* sp. nov., a non – motile species from wetland in Woopo, Korea, Inter. J. System. Evolu. Microb. 52(2002)377 –381.

[31] F. J. Cervantes, L. Vuthith, G. Lettinga, J. A. Field, Quinone – respiration improves dechlorination of carbon tetrachloride by anaerobic sludge, Environm. Biotechnol. 64 (2004) 702 –711.

[32] M. L. Fultz, R. A. Durst, Mediator compounds for the electrochemical study of biological redox systems. Anal. Chim. Acta 140(1982)1 –18.

[33] D. R. Lovley, Microbial reduction of iron. Manganese and other metals. Adv. Agron. 54 (1995)175 –231.

[34] D. R. Lovley, Bioremediation of organic and metal contaminants with dissimilatory metal reduction, J. Industr. MicrobioI. 14(1995)85 –93.

[35] D. R. Lovley, I. D. Coates, D. A. Saffarini, D. Lonergan, Dissimilatory Iron Reduction. In: Winkelman G., Carrano C. l(Ed.): Iron and Related Transition Metals in Microbial Metabolism. Harwood Academic Publishers, Switzerland, 1997, pp. 187 –215.

[36] A. Kappler, M. Benz, A. Brune, B. Schink, Electron shuttling via humic acids in microbial iron(III) reduction in a freshwater sediment, FEMS Microbiol. Ecol. 47(2004)85 –92.

注:本文曾发表在2009年第164卷的Journal of Hazardous Materials杂志上

Effect of Biological Soil Crusts on Microbial Activity in Soils of the Tengger Desert(China)

Yanmei Liu　Zisheng Xing　HangyuYang *

摘　要:土壤微生物作为土壤的重要生物成分,在土壤的形成和修复过程中起着重要的作用。本研究旨在分析生物土壤结皮对荒漠生态系统土壤微生物活性的影响。分别于2013年4月、7月、10月和2014年1月,在腾格沙漠人工植被固沙区和天然植被区采集生物土壤结皮下沙丘土壤,测定其微生物活性的变化。研究表明:人工植被固沙区和天然植被区的藻-地衣结皮和藓类结皮均可显著提高土壤理化性质、土壤基础呼吸、土壤碱性磷酸酶、蛋白酶和纤维素酶的活性和,降低代谢熵($p < 0.05$);生物土壤结皮对土壤微生物活性的影响也受结皮发育阶段和固沙年限的影响。发育晚期的藓类结皮下土壤基础呼吸和土壤酶的活性显著高于发育早期的藻-地衣结皮下土壤酶的活性和基础呼吸($p < 0.05$),而代谢熵则相反;固沙年限也显著影响土壤微生物活性指标,土壤基础呼吸和土壤酶的活性与固沙年限存在显著的线性正相关关系($p < 0.05$),而代谢熵与固沙年限存在显著的线性负相关关系。生物土壤结皮下土壤基础呼吸和土壤酶的活性表现明显的季节变化,夏季最高,秋季次之,春季和冬季最低,而代谢熵与之相反。因此,腾格里沙漠东南缘的植被固沙区生物土壤结皮可提高该区土壤质量,有利于荒漠生态系统的修复。

Soil microbes, as an important biological component of soils, have a function in the formation of soils and soil - remediation processes. This paper aims to analyze effects of biocrusts on soil microbial activities in desert ecosystems. Two sets of samples were collected under biocrusts in April, July, October, 2013, and January, 2014, in natural and

* 作者简介:刘艳梅(1978—　),女,甘肃天水人,现为天水师范学院生物工程与技术学院教授,博士,主要从事荒漠区土壤生物修复研究。

revegetated areas of the Tengger Desert. The results showed that biocrusts significantly improved soil physicochemical properties, basal respiration and the quantity of soil alkaline phosphatase, protease, and cellulose, and decreased qCO_2 in vegetated areas. Impact of biocrusts on soil microbial activities also varied, depending on the successional stage of crusts and the restoration age. Soil basal respiration and enzyme activity were obviously higher, but qCO_2 were significantly lower in moss – dominated crusts than those dominated by cyanobacteria – lichen. Soil basal respiration and enzyme activity positively correlated with the restoration age, but qCO_2 negatively correlated with the restoration age. Soil basal respiration and enzyme activity were the highest in summer, followed by autumn, and the lowest in spring and winter; whereas, qCO_2 displayed an opposite trend. The study suggests that biocrusts have the ability to improve soil quality and promote soil recovery in vegetated areas of the Tengger Desert.

1. Introduction

Arid and semiarid lands account for as much as 33 – 40% of Earth' sterrestrial surface, and are expanding rapidly (Billings et al., 2003). Vegetation cover in these areas is patchy and discontinuous due to combined impacts of harsh environmental factors such as prolonged drought, high temperatures and high soil erosion rates. Nevertheless, biocrusts are able to adapt successfully to these adverse environments, colonize the bare disaggregated geological substrate (Li, 2012) and cover as much as 70% of the interspaces between the sparse vegetation in these areas (Steven et al., 2014). Numerous studies have examined the ecosystem functions of biocrusts in desert ecosystems, such as fixing of C and N, improving soil structure, enhancing soil stability, modifying soil temperature, moisture and local hydrology, reinforcing plant colonization and promoting soil invertebrate and microbial diversity (Belnap and Lange, 2003; Bowker et al., 2013; Darby et al., 2010; Liu et al., 2013; Neher et al., 2009).

Soil microbes, as an important biological component of soils, have a function in soil formation and soil remediation processes, through the decomposition of organic matter, formation of humus, and nutrient cycling. Soil basal respiration is mostly associated with activity of microbes, therefore it indicates the potential mineralization rate of soil organic matter by soil microbes from a desert ecosystem (Pell et al., 2006). The ratio of basal respiration to microbial biomass carbon (metabolic quotient: qCO_2), provides a method to relate both the amount and activity of soil microbes (Anderson and Domsch, 1990).

Soil basal respiration, qCO_2 and enzyme activity provide a measure of microbial activity, which is a sensitive indicator of soil quality changes in response to environmental changes (Creamer et al., 2014; Raiesi and Beheshti, 2015; Wardle and Ghani, 1995). Recent studies have examined the relationship between biocrusts and soil microbial activity. For example, Bastida et al. (2014) found that the values of soil basal respiration and enzyme activities were higher in biocrusts than in soil beneath biocrusts in south – east Spain. Miralles et al. (2012a, 2012b) found that the values of soil basal respiration and the activities of arylsulphatase, β – glucosidase, casein – protease, cellulase and phosphomonoesterase were higher in soil beneath biocrusts than in bare substrate of the Tabernas Desert. Yu and Steinberger (2012) reported that the values of soil basal respiration were two – fold higher in biocrusts – covered interdune than in playa in the western Negev Desert. Studies by Zhang et al. (2012) and Liu et al. (2014) showed that soil catalase, urease, dehydrogenase and sucrase activities were high beneath biocrusts compared with bare soil without crust. It is also suggested that the successional stages of crusts could influence soil basal respirationand enzyme activities such as catalase, urease, dehydrogenases, sucrase and nitrogenase (Liu et al., 2014; Miralles et al., 2012b; Wu et al., 2009).

Although the influence of biocrusts on soil basal respiration, qCO_2 and enzyme activity has been studied, few studies have specifically examined how the soil microbial activity parameters respond to differences in successional stages of crusts, such as those in restored sand dunes in the Tengger Desert. Moreover, less is known about the effects of biocrusts on qCO_2 and enzyme activity at different soil depths and in different seasons. The present study aims to clarify functions of biocrusts in soil processes based on soil basal respiration, qCO_2 and enzymatic activities. Firstly, we determined the impacts of biocrusts on soil basal respiration, qCO_2 and enzyme activities in the study area and secondly, showed how these parameters varied with successional stages of crusts. Thirdly, we examined how these parameters changed after trampling disturbance. Finally, we studied the spatial and temporal variations of soil basal respiration, qCO_2 and enzyme activities under biocrusts in the study area.

2. Material and methods

2.1 Study area

The field sites (i. e., revegetated and natural vegetated) were located in the desert steppe region at the Shapotou Desert Research and Experiment Station, bordering the Tengger Desert, northern China (37°32′N, 105°02′E) at an elevation of 1300 m a. s. l. The sites are representative of the transition zone between desertified steppe and sand dunes. The climate is tropical dry with an average precipitation and potential evaporation of approximately 186 and 3000 mm yr^{-1}, respectively. Average annual temperature is 10.0 ℃, with minimum and maximum temperatures of -6.9℃ in January and 24.3 ℃ in July, respectively. The nutrient - poor soils are dry with moisture content consistently in the range of 3% - 4%, unconsolidated and movable. The soils can be divided into typical sierozem and aeolian sandy soils.

The revegetated areas were covered by sand dunes with a zero - irrigation vegetation system established in 1956, 1964, 1981 and 1991 using straw - checkerboards sand barriers to fix shifting dunes and subsequently plant xerophytic shrubs to protect the Baotou - Lanzhou railway from sand burial (Li et al., 2012). Once sand - binding vegetation was established, biocrusts then colonized and gradually developed with an coverage of 95%, 89%, 65% and 57% of the surface revegetated in 1956, 1964, 1981 and 1991, respectively (Li et al., 2011; Liu et al., 2013). The natural predominant plants are *Helianthemum scoparium* and *Agriophyllum squarrosum* with a coverage of approximately 1% in revegetated areas (Jia et al., 2008).

The natural vegetation areas were located at Hongwei, a vegetated protective system of the railway and characterized by undisturbed natural vegetation, 22 km from the revegetated areas. This site has the same landscape and soil type as the revegetated areas. The dominant plant species include *Ceratoides lateens*, *Artemisia ordosica*, *Caragana korshinskii*, *Oxytropis aciphylla* and *Artemisia capillaries* with a cover of approximately 20 - 45%. Currently, major biocrusts were dominated by 2 - 3.5mm thick cyanobacteria - lichen in the early - successional stage and 8 - 20 mm thick moss - dominated crusts at a late - successional stage in the revegetated and natural vegetation areas.

2.2 Soil sample collection and preparation

Soil samples were collected in July 2013 from natural vegetation areas at Hongwei and the revegetated areas established in 1956, 1964, 1981 and 1991, respectively. Thus, soil samples from the revegetated areas had 57,49,32 and 22 years of development, respectively in 2013. (1) We selected five 10 m × 10 m quadrats to include both cyanobacteria – lichen and moss crusts, with 20 – m spacing between two adjoining quadrats in revegetated and natural vegetation areas. In addition, five 10 m × 10 m quadrats without biocrusts were also established as reference sites in the mobile sand dunes. Within each quadrat, soil samples covered by cyanobacteria – lichen and moss crusts as well as from the mobile sand dunes were collected with a core sampler in the 0 – 10, 10 – 20 and 20 – 30 cm soil layers. A composite soil sample comprising five soil samples was collected from five quadrats for each soil layer. (2) Soil samples beneath disturbed biocrusts were also collected to study the effect of disturbance on soil microbial activity. The disturbance was performed by two people (about 60 kg each) wearing lug – soled boots trampling the biocrusts resulting in three levels of trampling: severe (< 35% untrampled), medium (35% – 65% untrampled) and non – trampling (> 65% untrampled) in revegetated and natural vegetation areas, respectively. Fifteen 1 m × 1 m quadrats containing biocrusts with 5 – m spacing between two adjoining quadrats were set up in July 2013. In each quadrat, soil samples under trampling were collected in the 0 – 5 and 5 – 15 cm soil depths in July 2014. (3) All soil samples were used to analyze soil basal respiration, qCO_2 and enzyme activities.

Soil samples were collected in April, July and October 2013 and January 2014 from the revegetated areas of 1956 and natural vegetation areas. Five 10 m × 10 m quadrats were selected, which included both cyanobacteria – lichen and moss crusts with 20 – m spacing between two adjoining quadrats established in the two areas. All soil samples under biocrusts from four seasons were used to determine spatial and temporal variability of soil basal respiration, qCO_2 and enzyme activities.

Each soil sample was placed in a sealed plastic bag and brought back to the laboratory. After removing large plant parts and stones, the sample was sieved with a 2 – mm mesh and divided into two parts. One part of the composite sample was air – dried for the estimation of soil physicochemical parameters and the other fresh part was stored at 4 ℃ for analysis of soil basal respiration, qCO_2 and enzyme activities.

2.3 Analyses of soil physicochemical properties

Soil water content was determined gravimetrically by drying 30g of fresh soil at 105℃ for 24 h. Soil texture (percentage content of sand, silt and clay) was determined by the Malvern – S Laser Radiation method. Soil pH wasmeasured in a 1 :2.5 soil/water extract. Soil organic C was determined by the wet combustion method using a mixture of potassium dichromateand sulfuric acid(H_2SO_4) under heating(Yeomans and Bremner, 1998). Total N was measured using a Kjeltec digestion, and available N was determined by the alkaline diffusion method. Total phosphorus (P) was determined by spectrophotometer after H_2SO_4 – perchloric acid digestion(Olsen and Sommers, 1982) and available P was determined according to the molybdenumblue method after sodium bicarbonate extraction(Olsen et al. , 1954).

2.4 Soil basal respiration and qCO_2 determination

To assess soil basal respiration, 50 g of fresh soil adjusted to 60% water holding capacity was weighed into 500 – ml glass containers with hermetic lids. Then a smaller flask containing 20 ml of 0.1M sodium hydroxide(NaOH) was placed in the bottom of glass containers and incubated for 7 d at 25℃ in darkness. The carbon dioxide evolved was determined by back – titration of the excess NaOH using 0.1 M hydrochloric acid (HCl) after addition of 5 ml of a saturated barium chloride solution, utilizing phenolphthale in as an indicator. Soil microbial biomass carbon(MBC) was determined by the chloroform fumigation – extraction method and calculated from the equation: MBC = 2.64(C in fumigated soil extract - C in unfumigated soil extract) (Vance et al. , 1987). The qCO_2 was calculated as the soil basal respiration rate divided by MBC (Anderson and Domsch, 1990).

2.5 Determination of soil enzyme activities

Alkaline phosphatase activity was analyzed on the basis of the method of Tabatabai and Bremner(1969) as described in Miralles et al. (2012a). Two grams of fresh soil samples was kept at 37℃ for 1 h with 4 ml of Modified Universal Buffer(pH 11.0) and 1 ml of 16 mM *p* – nitrophenyl phosphate as decomposed substrate solution before blending with 0.25 ml of methylbenzene. After 1 h of suspension, 4 ml of 0.5 M calcium chloride was added and the *p* – nitrophenolreleased was extracted with 4 ml of 0.5 M NaOH. The enzymatic activity was quantified by reference to calibration curves corresponding to *p* – nitrophenol standards. The alkaline phosphatase activity was measured using an UV/visible spectrophotometer at 400 nm and expressed as μg *p* – nitrophenol g^{-1} h^{-1}.

Protease activity was determined as described by Nannipieri et al. (1979). Five grams of fresh soil sample was kept at 50 ℃ for 2 h with 25 ml of 2% casein and 20 ml of 0.05 M Tris – HCl buffer(pH 8.1), and 25 ml of 15% trichloroacetic acid was added. The mixtures were immediately centrifuged and 7.5 ml of alkaline reagent and 5 ml of 33% Folin – Ciocalteu phenol reagent was added to the 5 ml of supernatant in the tube. The mixtures were immediately filtered and the supernatant measured using an UV/visible spectrophotometerat 700 nm. Protease activity was expressed as μg tyrosine g^{-1} h^{-1}.

Cellulase activity was assayed on the basis of the method of Xu and Zheng(1986) with the following modification: 10 g of fresh soil sampleswerekept at 37 ℃ for 72 h with 20 ml of 1% carboxymethyl – cellulose solution and 5 ml of phosphate buffer(pH 5.5) before blending with 1.5 ml of methylbenzene. The mixtures were immediately filtered and 3 ml of 3,5 – dinitrosalicylic acid solution and 5 ml of distilled water was added to 2 ml of supernatant in the tube. All tubes were bathed in boiling water for 5 min and then cooled to room temperature. The supernatant was measured using an UV/visible spectrophotometer at 508 nm and cellulase activity expressed as μg glucose g^{-1} h^{-1}.

2.6 Data analysis

All statistical analyses were performed with SPSS 16.0 statistical and Canoco 4.5 software. Influence of biocrusts, stage of crust formation, restoration age, and trampling disturbance on soil physicochemical properties, basal respiration, qCO_2 and enzyme activities were analyzed using one – way analysis of variance (ANOVA) with Duncan's multiple range test. Three – way ANOVA was used to evaluate the effect of the interactions among the successional stage of crusts, the restoration age and soil depth on soil physicochemical properties, basal respiration, qCO_2 and enzyme activities. Two – way ANOVA was performed to test the influence of the interactions between trampling intensity and the successional stage of crusts on soil basal respiration, qCO_2 and enzyme activities. Pearson's correlation test was used to determine the correlations of the restoration age, soil depth and trampling intensity with soil microbial activity parameters. Significance was defined at $P < 0.05$. The associations among soil physicochemical properties and soil microbial activity parameters were evaluated by redundancy analysis (RDA) using Canoco 4.5 software.

3 Results

3.1 Effects of biocrusts on soil physicochemical properties

Both cyanobacteria – lichen and moss crusts affected soil texture, resulting in an increase in the content of soil clay and silt and reduction in the content of sand in 0 – 30 cm soil in comparison with bare soil(Table 1). Soil organic C, total N, availableN, total P, available P and pH were higher beneath biocrusts than in bare soil at the depth of 0 – 30 cm(Table 1). These soil physicochemical properties markedly varied with the successional stage of crusts, the restoration ageand soil depth(Table 2). Moss crusts had higher in contents of soil clay and silt, organic C, total N, availableN, total P and available P, pH, and lower in soils and contents than cyanobacteria – lichen crusts(Table 1). The content of soil clay and silt, organic C, total N, availableN, total P and available P and pH value positively correlated with the restoration age with all r values being >0.90($P<0.05$), and negatively correlated with contents of sand($R^2=0.95$). The content of clay and silt, organic C, total N, available N, total P and available P and pH value under biocrusts declined with soil depth($P<0.01$), whereass sand contents under biocrusts increased with soil depth($P<0.01$).

3.2 Effects of biocrusts on soil basal respiration, $q\mathrm{CO_2}$ and enzyme activities

Soil basal respiration and the activities of soil alkaline phosphatase, protease and cellulase were markedly higher beneath cyanobacteria – lichen(Fig. 1; A_1, A_3, A_4 and A_5) and moss crusts(Fig. 1; B_1, B_3, B_4 and B_5) than in the bare soil at the depth of 0 – 20 cm($P<0.05$), varying significantly with the successional stage of crusts, the restoration age and soil depth(Table 3, $P<0.01$). Soil basal respiration and the activities of soil alkaline phosphatase, protease and cellulase activities under biocrusts increased with the increase of restoration age($r>0.95$; $P<0.05$). Soil basal respiration and the activities studied enzyme were significantly greater beneath moss than cyanobacteria – lichen crusts(Table 3, $P<0.001$). Moreover, soil basal respiration and the activities studied enzyme both under cyanobacteria – lichen and moss – dominated crusts were clearly negatively correlated with soil depth with all R^2 values >0.93($P<0.05$).

Table 1 Soil physicochemical properties under cyanobacteria – lichen and moss crusts (depth 0 – 30 cm) in vegetated areas

Biocrusts	Years since sand fixed	Sand (%)	Silt (%)	Clay (%)	Organic carbon (g kg^{-1})	Total nitrogen (g kg^{-1})	Available nitrogen (mg kg^{-1})	Total phosphorous (g kg^{-1})	Available phosphorous (mg kg^{-1})	pH
Cyanobacteria – lichen crusts	NVA	90.57d	5.90a	3.53a	3.97a	0.19a	23.33a	1.369a	29.43a	8.12a
	57	94.57c	4.15b	1.31b	2.97b	0.14b	18.28b	0.607b	18.74b	8.05b
	49	95.10c	3.73c	1.17b	2.84b	0.13bc	15.94bc	0.570c	14.31c	8.04b
	32	95.80b	3.20d	1.00b	2.62c	0.12cd	13.61cd	0.565c	12.43d	8.03bc
	22	96.20b	3.08d	0.71bc	2.48c	0.11d	12.44d	0.473d	5.72e	7.99c
	0	99.03a	0.70e	0.27c	0.58d	0.07e	7.39e	0.401e	3.76f	7.81d
Moss crusts	NVA	89.00f	6.86a	4.06a	4.61a	0.21a	26.07a	1.52a	29.73a	8.15a
	57	92.57e	5.40b	2.07b	3.15b	0.17b	21.00b	0.66b	21.83b	8.05b
	49	93.87d	4.74c	1.39c	2.98bc	0.14c	18.30c	0.60bc	15.44c	8.05b
	32	94.40c	4.41c	1.21c	2.85c	0.14c	15.93cd	0.58bc	13.40d	8.04b
	22	95.40b	3.80d	0.80d	2.58d	0.12d	14.38d	0.55c	8.41e	7.99c
	0	99.03a	0.70e	0.27e	0.58e	0.08e	7.39e	0.40d	3.76f	7.81d

NVA = natural vegetation areas. Different letters following property values indicate significant differences at the $P < 0.05$ level.

Table 2 Statistical analysis of soil physicochemical properties under cyanobacteria – lichen and moss crusts

	df	Sand (%)	Silt (%)	Clay (%)	Organic C	Total N	Available N	Total P	Available P	pH
The successional stage of crusts	1	.00***	.00***	.00 ns	1.26***	.00***	109.00***	.08***	54.81**	.00 ns
Soil depth	2	.03***	.02***	.00***	17.2***	.00***	483.31***	.28***	1691.02***	.15***
The restoration age	5	.02***	.01***	.00**	26.1***	.00***	619.96***	2.56***	1555.81***	.22***
The successional stage of crusts * Soil depth	2	.00***	.00***	8.88E-5ns	.01 ns	1.81E-5***	4.20 ns	.00ns	2.66ns	.00 ns
The successional stage of crusts * The restoration age	5	.00*	9.66E-5**	5.74E-5ns	.22***	2.38E-5***	4.74ns	.01**	8.27ns	.00 ns
Soil depth * The restoration age	10	.00***	.00***	.00*	1.06***	6.19E-5***	19.17**	.05***	85.11***	.02***
The successional stage of crusts * Soil depth * The restoration age	10	9.11E-5ns	3.98E-5 ns	2.92E-5 ns	.03 ns	8.41E-6***	1.20 ns	.00 ns	6.87 ns	3.98E-5 ns
Error	72									

df = degrees of freedom; *** = $P<0.001$, ** = $P<0.01$, * = $P<0.05$, ns = not significant.

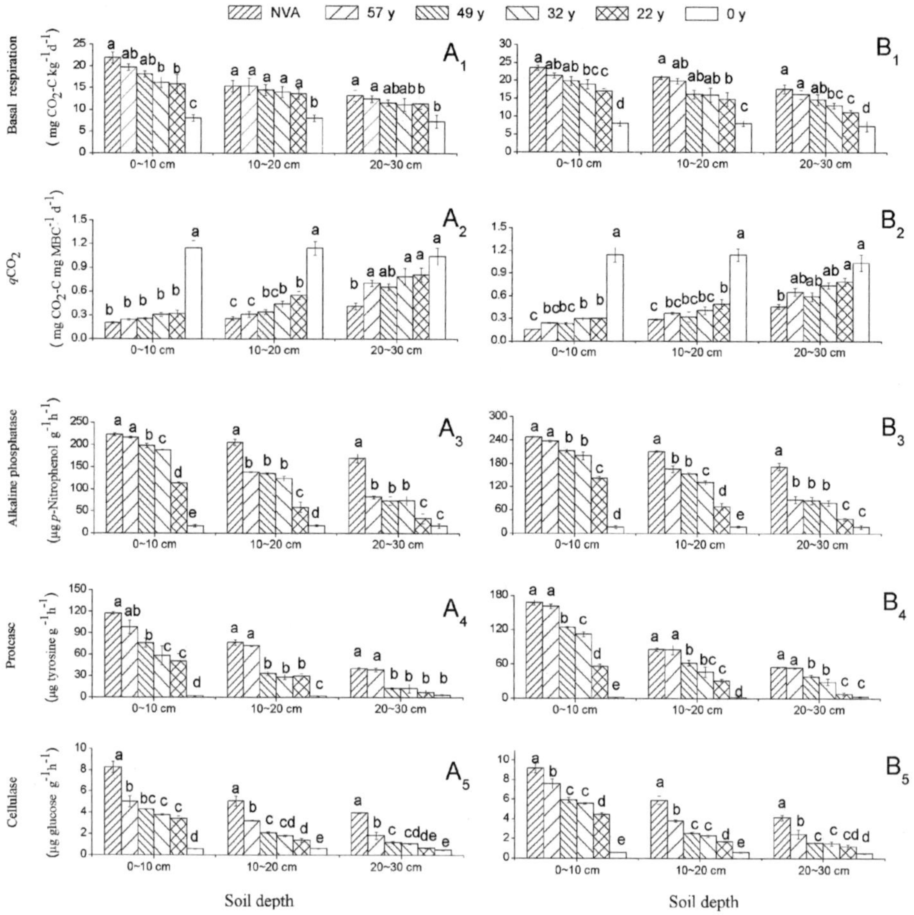

Fig. 1 Soil basal respiration, qCO_2 and enzyme activities under cyanobacteria - lichen (A_{1-5}) and moss crusts (B_{1-5}) in vegetated areas of the Tengger Desert. NVA refers to natural vegetation areas. 57, 49, 32, 22 and 0 y refer to the restoration age. Different letters indicate significant differences at the $P < 0.05$ level. Error bars show SE, n = 3.

The soil qCO_2 was significantly lower beneath cyanobacteria - lichen and moss crusts than in the bare soil at the depth of 0 - 20 cm ($P < 0.05$) (Fig. 1; A_2 and B_2). The qCO_2 varied with the successional stage of crusts, the restoration age and soil depth (Table 3, $P < 0.01$). Moss crusts had lower qCO_2 than cyanobacteria - lichen crusts (Table 3, $P < 0.01$). The qCO_2 was negatively correlated with restoration age ($R^2 = 0.794$, $P < 0.05$) and the impacts of the restoration age on qCO_2 also varied with soil depth (Table 3, $P < 0.001$). The qCO_2 under cyanobacteria - lichen and moss crusts increased significantly with soil depth (Table 3, $P < 0.001$) with all $r > 0.97$.

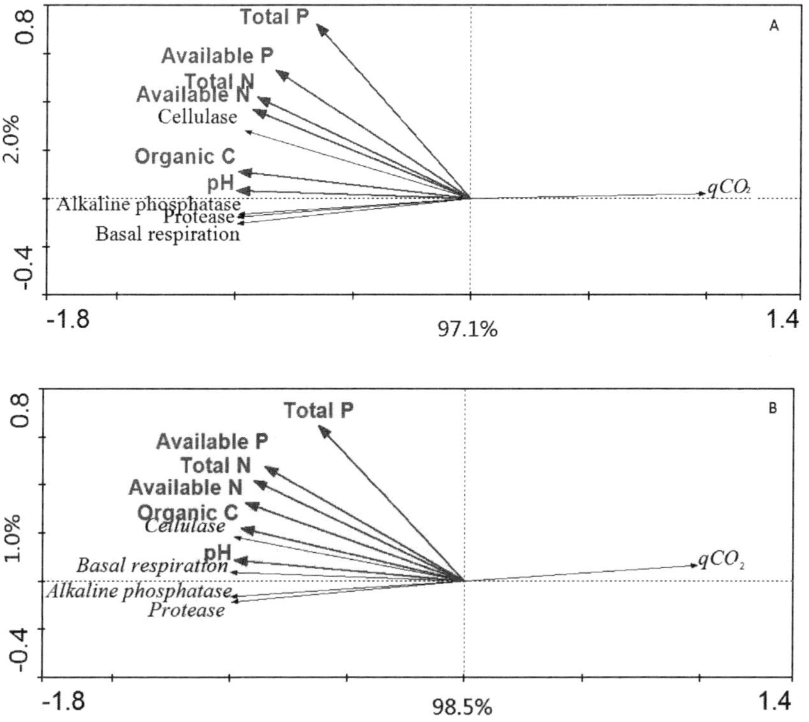

Fig. 2 Redundancy analyses(RDA) on the correlations between soil physicochemical properties and soil microbial activity parameters under cyanobacteria – lichen(A) and moss crusts(B). Red arrow lines represent soil physicochemical properties and blue arrow lines indicate soil microbial activity parameters.

The RDA ordination biplots showed the correlations between soil physicochemical properties and soil microbial activity parameters under biocrusts(Fig. 2;A and B). Axis 1 and axis 2 in Fig. 2A were significant and accounted for 97.1% and 2.0 % of the variation under cyanobacteria – lichen crusts, respectively. Axis 1 in Fig. 2A was positively correlated with soil pH, organic C, available N and total N, with correlation coefficients of 0.99, 0.99, 0.93 and 0.91, respectively. Similarly, Axes 1 and 2 in Fig. 2B were significant and explained 98.5% and 1.0% of the total variance under moss crusts, respectively. Axis 1 in Fig. 2B was highly positively correlated with soil pH, organic C, available N and total N, with correlation coefficients of 0.98, 0.95, 0.94 and 0.90, respectively. With a Monte Carlo permutation test, it was found that soil pH, organic C, available N and total N were the major factors that explain the variations in soil microbial activity parameters beneath cyanobacteria – lichen and moss crusts in the study areas. Soil basal respiration and enzyme activities also showed positive correlations with soil pH, organic C, availableN and total N, whereas qCO_2 was negatively correlated with these soil physicochemical properties. This finding indicated that soil microbial activity was enhanced with improved soil conditions.

Table 3 Statistical analysis of soil basal respiration, qCO_2 and enzyme activities under cyanobacteria – lichen and moss crusts

	df	Soil basal respiration	qCO_2	Soil alkaline phosphatase activities	Soil protease activities	Soil cellulase activities
The successional stage of crusts	1	.00 ***	.01 **	3209.7 ***	13.5 ***	11060.4 ***
Soil depth	2	.00 ***	.97 ***	73952.6 ***	96.9 ***	33851.8 ***
The restoration age	5	.00 ***	1.51 ***	77795.1 ***	62.3 ***	19808.8 ***
The successional stage of crusts * Soil depth	2	2.77E – 6 ns	.00ns	342.9 ns	2.6 ***	1888.7 ***
The successional stage of crusts * The restoration age	5	1.29E – 5 ns	.00 ns	183.0 ns	0.8 **	998.5 ***
Soil depth * The restoration age	10	6.51E – 6 **	.08 ***	4269.1 ***	4.2 ***	1912.7 ***
The successional stage of crusts * Soil depth * The restoration age	10	2.58E – 6 ns	.00ns	70.5 ns	0.3 ns	197.9 **
Error	72					

df = degrees of freedom; * * * = $P < 0.001$, * * = $P < 0.01$, ns = not significant.

3.3 Impacts of tramping biocrusts on soil basal respiration, qCO_2 and enzyme activities

There was no significant effect of medium trampling on soil basal respiration and activities of soil alkaline phosphatase, protease and cellulase in the 0 – 5 cm and 5 – 15 cm soil layers in vegetated areas (Figs. 3 and 4). However, severe trampling of biocrusts markedly reduced these measures compared with non – trampling biocrusts in 0 – 5 cm

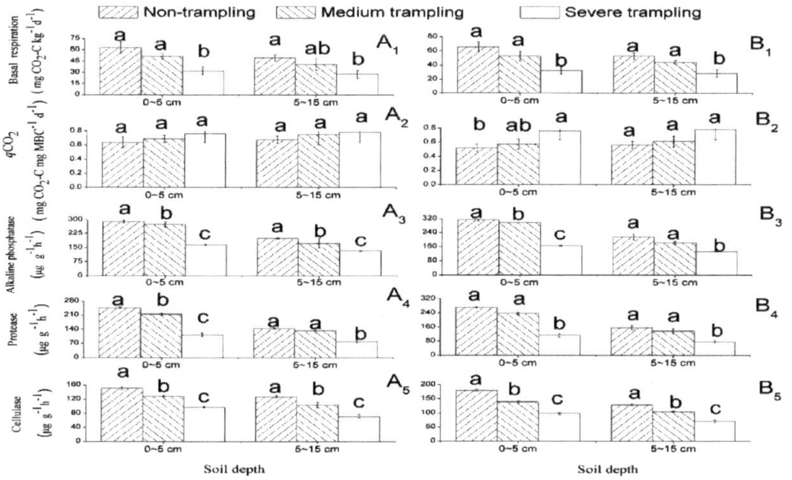

Fig. 3 Impacts of trampling disturbance of cyanobacteria – lichen($A_{1\sim5}$) and moss crusts($B_{1\sim5}$) on soil basal respiration, q CO_2 and enzyme activities in revegetated areas of the Tengger Desert. Different letters indicate significant differences at the $P < 0.05$ level. Error bars show SE, n = 3

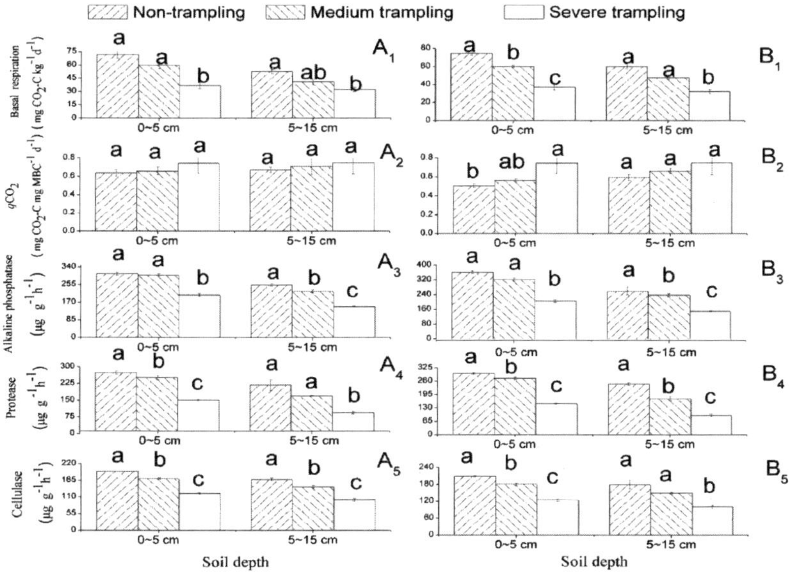

Fig. 4 Impacts of trampling disturbance of cyanobacteria – lichen($A_{1\sim5}$) and moss crusts($B_{1\sim5}$) on soil basal respiration, q CO_2 and enzyme activities in naturally vegetated areas of the Tengger Desert. Different letters indicate significant differences at the $P < 0.05$ level. Error bars show SE, n = 3.

and 5 – 15 cm soil layers(Figs. 3 and 4, $P < 0.05$). There was a decreasing trend in qCO_2 values from severe trampling to non – trampling crusts, although this was not significant for 0 – 15 cm soil layers in two study sites(Figs. 3 and 4). A simple linear regression showed that soil basal respiration and three studied enzyme activities were all negatively correlated with trampling intensity(all $R^2 > 0.85$, $P < 0.05$), whereas qCO_2 was positively correlated with trampling intensity(all $r > 0.94$, $P < 0.05$). Two – way ANOVA also showed that soil basal respiration, qCO_2 and enzyme activities beneath disturbed crusts obviously varied with the successional stage of crusts and trampling intensity(Table 4). For the same trampling intensity, soil basal respiration and three enzyme activities beneath disturbed moss crusts were greater than underneath disturbed cyanobacteria – lichen crusts(Figs. 3 and 4; Table 4; $P < 0.05$), whereas qCO_2 was lower underneath disturbed moss than disturbed cyanobacteria – lichen crusts(Figs. 3 and 4; Table 4; $P < 0.05$). However, there were no significant interactions on soil basal respiration, qCO_2 and three studied enzyme activities between the successional stage of crusts and trampling intensity of biocrusts(Table 4).

3.4 Seasonal variations in soil basal respiration, qCO_2 and enzyme activities

Soil basal respiration and qCO_2 under biocrusts showed obvious seasonal variations in both revegetated and natural vegetation areas. Soil basal respiration under cyanobacteria – lichen and moss crusts had maximum values in summer(July), followed by autumn(October), spring(April) and winter(January) in the 0 – 30 cm soil layer in revegetated and natural vegetation areas(Figs. 5 and 6; A_1 and B_1). In contrast, qCO_2 increased from minimum to maximum values from summer(July), autumn(October) and spring(April) to winter(January) in the 0 – 30 cm soil layer in revegetated and natural vegetation areas(Figs. 5 and 6; A_2 and B_2).

The activities of soil alkaline phosphatase, protease and cellulase under cyanobacteria – lichen and moss crusts exhibited a clear seasonal pattern in the vegetated areas(Figs. 5 and 6). The activities of soil protease and cellulase under cyanobacteria – lichen and moss crusts ranged from the highest to lowest values in summer(July), autumn(October), spring (April) and winter(January) in 0 – 30 cm soil layer in natural and revegetated areas(Figs. 5 and 6; A_4, B_4, A_5 and B_5). Soil alkaline phosphatase activity under biocrusts also showed a seasonal trend, increasing from minimum values in spring(April) to maximum values in summer(July), then declining in autumn(October) and further declining in winter(January) in natural and revegetated areas(Figs. 5 and 6; A_3 and B_3).

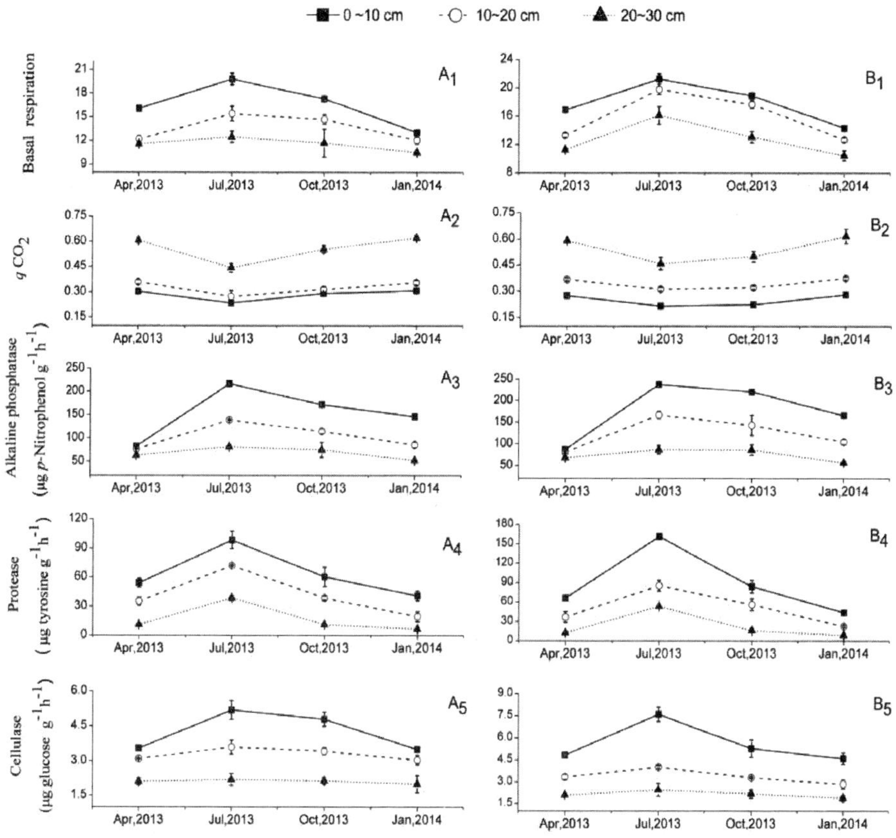

Fig. 5 Seasonal variations of soil basal respiration(mg CO_2 – C $kg^{-1}d^{-1}$), qCO_2 (mg CO_2 – C mg MBC^{-1} d^{-1}) and enzyme activities under cyanobacteria – lichen($A_{1\sim5}$) and moss crusts($B_{1\sim5}$) in revegetated areas. Error bars show SE, n = 3.

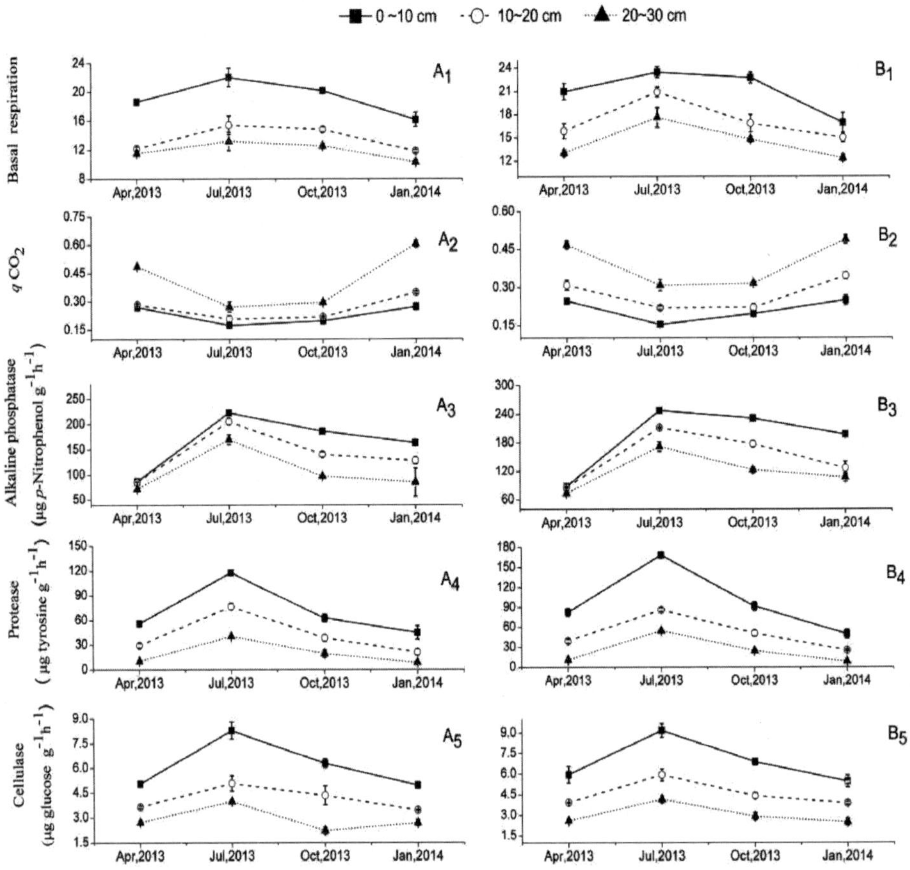

Fig. 6 Seasonal variations of soil basal respiration(mg CO_2 – C $kg^{-1}d^{-1}$), qCO_2 (mg CO_2 – C mg MBC^{-1} d^{-1}) and enzyme activities under cyanobacteria – lichen(A_{1-5}) and moss crusts(B_{1-5}) in naturally vegetated areas. Error bars show SE, n = 3.

4 Discussion

4.1 Effects of biocrusts on soil basal respiration, q CO_2 and enzyme activities

The biocrusts in vegetated areas enhanced soil physicochemical properties, basal respiration and the activity of soil alkaline phosphatase, protease and cellulase, and reduced qCO_2. These results are similar to the findings of Miralles et al. (2012a, 2012b) and Bastidaet al. (2014), who found that soil basal respiration and several hydrolase enzyme activities were higher in biocrusts than in bare substrata in south – east Spain.

High soil basal respiration under biocrusts indicates higher metabolic capacity of soil microbes and higher microbial biomass under biocrusts than in bare soil in vegetated areas. In contrast, a sharp increase in qCO_2 indicates reduced microbial efficiency, which appears to be related to stress and resulted from low substrate quality and carbon – use efficiency of soil microbes (Agnelli et al., 2001; Anderson and Domsch, 1990). Disturbed soil microbial communities have high respiration in relation to their biomass and results in high qCO_2 (Araújo et al., 2013). The 3 – 4 fold lower qCO_2 under biocrusts than in bare soils (Fig. 1) reveals the comparatively high metabolic efficiency of soil microbial communities under biocrusts. High soil basal respiration and enzyme activities as well as the low qCO_2 under biocrustsis most likely attributed to biocrusts which contribute significant amounts of nutrients to soils and provide suitable soil temperature and moisture for soil microbes, improving soil microbial activity (Wardle and Ghani, 1995) in desert areas. Therefore, enhanced soil microbial activity under biocrusts could indicate the beneficial effects of biocrusts in vegetated areas.

The study showed that severe trampling of biocrusts caused a drastic decline in soil basal respiration but an increase in qCO_2. Lower soil basal respiration and higher qCO_2 beneath severely trampled biocrusts compared with untrampled biocrusts suggests that the trampling of biocrusts could down – size metabolic capacity of soil microbes. Thus, our study suggests that biocrusts were sensitive to trampling as reported in previous study (Williams et al., 2008). Under severe trampling on biocrusts, the ability of biocrusts to deposit nutrients and create stable microsites may have been reduced (Eldridge and Whitford, 2009), which led to a decline of soil microbial activity. The enhanced soil basal respiration and reduced qCO_2 under biocrusts as well as low soil basal respiration and high qCO_2 following severe trampling of biocrusts clearly indicate that biocrusts enhance soil microbial activities in the vegetated areas.

Soil cellulase, protease and alkaline phosphatase play key roles in soil C, N and P cycling, respectively (Heinze et al., 2014; Miralles et al., 2012a); thus, their higher activities under biocrusts than in bare soils indicate that biocrusts could accelerate soil C, N and P cycling in desert areas. This may be attributed to the following: (i) Biocrusts can increase the content of soil C, N and P in desert ecosystems as reported by other authors (Chamizo et al., 2012; Li et al., 2009; Yu et al., 2012) and reinforced by the research. (ii) The relatively abundant nutrients in the soil under biocrusts (i. e. higher organic C, N, P; Table 1) provide rich food resources and a favorable environment (i. e.

pH and soil structure) for soil microbes, stimulate soil microbial activity (Bastida et al., 2014; Liu et al., 2013) and lead to considerable increases in soil basal respiration and enzyme activities as well as decreases in qCO_2 (Belnap and Lange, 2003; Brockett et al., 2012; Li et al., 2011). Soil pH, organic C, available N and total N were found to be mainly responsible for the changes in soil basal respiration, qCO_2 and enzyme activities in the desert areas. Soil nutrients (i. e., organic C and N) and pH may impose physiological constraints on the survival and growth of soil microbes (Zhang et al., 2016), and thereby directly alter soil microbial activity. In addition, we found that severe trampling of biocrusts led to a considerable decline in activities of soil alkaline phosphatase, protease and cellulase compared with non – trampled biocrusts, which may be attributed to the disturbance of trampling on soil C, N and P turnover in the biocrusts. Therefore, trampling could be concluded to cause a loss of soil nutrients and accelerate soil degeneration in vegetated areas. In summary, the improved soil enzyme activities and basal respiration and decreased qCO_2 under biocrusts, as well as reduced soil enzyme activities after severe trampling, suggest that biocrusts enhanced soil quality and had contributed to soil recovery in vegetated areas of the Tengger Desert.

4.2 Association of soil basal respiration, qCO_2 and enzyme activities with successional stage of crusts

Late – stage moss crusts had higher soil basal respiration and activities of soil alkaline phosphatase, protease and cellulase, and lower qCO_2, than early – stage cyanobacteria – lichen crusts. Similarly, Yu and Steinberger (2012) and Miralles et al. (2012a) reported that late – stage crusts led to greater soil basal respiration and hydrolytic enzyme activities, respectively, than early – stage crusts. The field measurements of Zhao et al. (2016) suggested that soil respiration increased from early – stage to late – stage crusts in the Tengger Desert. Furthermore, density increases in soil microbes (i. e., fungal and bacteria) from early – stage to late – stage crusts (Grishkan et al., 2015; Liu et al., 2013) may also have been responsible for the higher soil basal respiration and lower qCO_2 under late – stage moss crusts than early – stage cyanobacteria – lichen crusts in vegetated areas. The explanation is that late – stage crusts could provide more abundant nutrients (organic C, total N, available N, total P and available P) and suitable environments (improved soil texture and appropriate pH) for soil microbes than early – stage crusts (Table1). The high activities of soil cellulase, protease and alkaline phosphatase indicates that late – stage moss crusts had greater C, N and P turnover than the early –

stage cyanobacteria – lichen crusts in vegetated areas, as also described by other authors (Li et al., 2009, 2012; Miralles et al., 2012a). Late – stage crusts contain dark – colored mosses, lichens and cyanobacteria, which can fix more nutrients than the light – colored early – stage crusts dominated by cyanobacteria (Veluci et al., 2006; Zelikova et al., 2012). Moreover, soil basal respiration, qCO_2 and enzyme activities significantly differed between cyanobacteria – lichen and moss crusts following the same trampling intensity, with higher soil basal respiration and enzyme activities, and lower qCO_2 for moss compared to cyanobacteria – lichen crusts. This suggests that (i) early – stage cyanobacteria – lichen crusts are more susceptible to trampling disturbance than late – stage moss crusts in desert areas and (ii) the ability of biocrusts to improve soil quality increased with the successional stage of crusts in desert areas.

4.3 Association of soil basal respiration, qCO_2 and enzyme activities with the restoration age

The linear relationships of soil basal respiration, qCO_2 and enzyme activities with the restoration age suggest cumulative changes in soil microbial activity and soil C, N and P turnover under biocrusts over time following sand dune stabilization in revegetated areas. Both the increased soil microbial activity and improved turnover of soil nutrients demonstrated improved soil quality with sand fixation. Positive correlations of soil basal respiration, qCO_2 and the three studied enzyme activities with the restoration age were the first found for a desert ecosystem, although similar trends were reported for other ecosystems. For example, soil basal respiration increased with the succession row of forest (Susyan et al., 2011) and qCO_2 decreased with increasing succession age of regrowth vegetation (Frouz and Nováková, 2005; Susyan et al., 2011). Some enzyme activities increased with succession age in coal mine spoil heaps (Baldrian et al., 2008). In the present study, the thickening of biocrusts increased with the sand dune restoration age. During their thickening, biocrusts accumulate soil nutrients, as shown in Table 1, and improved the environment (e. g. soil texture, soil water content and temperature), which promotes the activities of soil microbes and soil enzymes (Hamman et al., 2007). Therefore, the increased thickness of biocrusts may be a major factor in enhancing soil basal respiration and enzyme activities, and decreasing qCO_2, with the time since sand was fixed.

4.4 Spatio - temporal variations in soil basal respiration, qCO_2 and enzyme activities under biocrusts

Soil basal respiration and enzyme activities declined with soil depth, whereas qCO_2 increased with soil depths, as also reported by other authors (Enowashu et al., 2009; Heinze et al., 2014; Miralles et al., 2012b). Lower soil basal respiration and enzyme activities as well as higher qCO_2 were recorded in deeper compared to upper layers under biocrusts, indicating that soil microbes experienced certain environment stresses in deeper soil layers, such as low nutrient levels. The significant differences in soil basal respiration, qCO_2 and enzyme activities under biocrusts and bare soils in the 0 – 20 cm soil layer indicated that biocrusts significantly promoted soil microbial activity and soil C, N and P turnover at this depth. Moreover, the cumulative effect of sand – fixing and soil depth on soil basal respiration, qCO_2 and enzyme activities may imply that biocrusts enhance soil microbial activity and soil nutrient turnover in the 0 – 20 cm soil layer since sand was fixed, whereas soil microbial activity below 20 cm would be improved with a longer period of sand being fixed. Thus, biocrusts gradually improved soil quality in a downward direction to deeper soil over time following the fixing of sand in the study area.

The seasonal differences in soil basal respiration, qCO_2 and enzyme activities under biocrusts may be induced by climatic factors such as soil temperature and moisture, which are the main factors driving the studied seasonal differences (Uvarovet al., 2006). While biocrusts growth could be affected by temperature and moisture, the seasonal variations in soil basal respiration, qCO_2 and enzyme activities under biocrusts may be indirectly influenced by the biocrusts growth. The long and intensive sunlight with relatively plentiful precipitation in summer ensures a stronger growth of biocrusts, leading to higher above – ground biomass and cover of biocrusts than other seasons. It is the exuberant growth of biocrusts that may have improved soil moisture and temperature, creating a more favourable habitat for soil microbial communities and promoting the observed soil microbial activities (e. g. high soil basal respiration and enzyme activities, and low qCO_2) under biocrusts, more in summer than other seasons. Biocrusts growth gradually slowed down in autumn due to reduced illumination, which restricted soil microbial activities. In spring and winter, soil microbial activities were greatly discouraged by the slow/inactive growth of biocrusts due to very little precipitation and low illumination, which reduced soil basal respiration and enzyme activities, and increased qCO_2.

Moreover, seasonal responses in soil basal respiration and enzyme activities under biocrusts were more sensitive in the topsoil layer than in deeper soil, as also described by some previous reports(Susyanet al. ,2011).

5 Conclusions

Biocrusts strongly increased soil basal respiration and activities of soil alkaline phosphatase, protease and cellulase, and reduced qCO_2 in vegetated areas of the Tengger Desert. Severe trampling of biocrusts markedly reduced soil basal respiration and activities of soil alkaline phosphatase, protease and cellulase, and enhanced qCO_2 in 0 – 15 cm soil layer in the study areas. The successional stage of crusts and the restoration age also highly affected soil basal respiration, qCO_2 and enzyme activities. Vigorous soil basal respiration and the three studied enzyme activities, and reduced qCO_2, were observed beneath moss compared to cyanobacteria – lichen crusts, which is a consistent indication of relatively greater soil microbial activities and soil nutrient under late – stage moss crusts. The sand dune restoration age positively influenced soil basal respiration and enzyme activities, and negatively influenced qCO_2, which illustrates the continuous improvement of soil quality along with restoration age in revegetated areas. Moreover, soil basal respiration, qCO_2 and enzyme activities under biocrusts also changed with soil depth and season in vegetated areas. Biocrusts strongly enhanced soil basal respiration and enzyme activities, and reduced qCO_2, in 0 – 20 cm soil depth by improving soil microbial activities in 0 – 20 cm soil layer. Soil basal respiration and enzyme activities showed clear seasonal patterns: the maximum values in summer coupled with the vigorous growth of biocrusts, followed by autumn with the medium values and minimum values in spring and winter. However, the seasonal pattern of qCO_2 was completely opposite to soil basal respiration and activities of the three enzymes. Thus, soil microbial activities under biocrusts were highest in summer, followed by autumn and were lowest in spring and winter in vegetated areas. This study also concluded that soil pH, organic C and available N and total N were the major factors that affect soil basal respiration, qCO_2 and the three enzyme activities beneath biocrusts in the study areas.

This study showed that biocrusts clearly improved soil microbial activity and ultimately enhanced soil quality in desert areas. Moreover, severe disturbance of biocrusts by human can reduce soil quality and consequently cause the deterioration of desert ecosystems.

Acknowledgements

This research was funded by China National Funds for Regional Science(grant No. 41261014) and the China National Funds for Young Scientists(grant No. 41401341).

REFERENCES

[1] Agnelli, A., Ugolini, F. C., Corti, G., Pietramellara, G., 2001. Microbial biomass - C and basal respiration of fine earth and highly altered rock fragments of two forest soils. Soil Biol. Biochem. 33, 613 - 620.

[2] Anderson, T. H., Domsch, K. H., 1990. Application of eco - physiological quotients(qCO_2 and qD) on microbial biomasses from soils of different cropping histories. Soil Biol. Biochem. 22, 251 - 255.

[3] Araújo, A. S. F., Cesarz, S., Leite, L. F. C., Borges, C. D., Tsai, S. M., Eisenhauer, N., 2013. Soil microbial properties and temporal stability in degraded and restored lands of Northeast Brazil. Soil Biol. Biochem. 66, 175 - 181.

[4] Baldrian, P., Trögl, J., Frouz, J., Šnajdr, J., Valášková, V., Merhautová, V., Cajthaml, T., Herinková, J., 2008. Enzyme activities and microbial biomass in topsoil layer during spontaneous succession in spoil heaps after brown coal mining. Soil Biol. Biochem. 40, 2107 - 2115.

[5] Bastida, F., Jehmlich, N., Ondoño, S., Bergen, M., García, C., Moreno, J. L., 2014. Characterization of the microbial community in biological soil crusts dominated by *Fulgensia desertorum* (Tomin) Poelt and *Squamarinacartilaginea*(With.) P. James and in the underlying soil. Soil Biol. Biochem. 76, 70 - 79.

[6] Belnap, J., Lange O. L., 2003. Biological Soil Crust: Structure, Function and Management. Springer - Verlag, Berlin, pp 3 - 30.

[7] Billings S. A., Schaeffer, S. M., Evans, R. D., 2003. Nitrogen fixation by biological soil crusts and heterotrophic bacteriain an intact Mojave Desert ecosystem with elevated CO_2 and added soil carbon. Soil Biol. Biochem. 35, 643 - 649.

[8] Bowker, M. A., Eldridge, D. J., Val, J., and Soliveres, S., 2013. Hydrology in a patterned landscape is co - engineered by soil - disturbing animals and biological crusts. Soil Biol. Biochem. 61, 14 - 22.

[9] Brockett, B. F. T., Prescott, C. E., Grayston, S. J., 2012. Soil moisture is the major factor influencing microbial community structure and enzyme activities across seven biogeoclimatic zones in western Canada. Soil Biol. Biochem. 44, 9 - 20.

[10] Chamizo, S., Cantón, Y., Miralles, I., Domingo, F., 2012. Biological soil crust development affects physicochemical characteristics of soil surface in semiarid ecosystems. Soil Biol. Bio-

chem. 49,96 – 105.

[11] Creamer, R. E., Schulte, R. P. O., Stone, D., Gal, A., Krogh, P. H., Lo Papa, G., Murray, P. J., Pérès, G., Foerster, B., Rutgers, M., Sousa, J. P., Winding, A., 2014. Measuring basal soil respiration across Europe: Do incubation temperature and incubation period matter? Ecol. Indic. 36,409 – 418.

[12] Darby, B. J., Neher, D. A., Belnap, J., 2010. Impact of biological soil crusts and desert plants on soil microfaunal community composition. Plant Soil 328,421 – 431.

[13] Eldridge, D. J., Whitford, W. G., 2009. Soil disturbance by native animals along grazing gradients in an arid grassland. J. Arid Environ. 73,1144 – 1148.

[14] Enowashu, E., Poll, C., Lamersdorf, N., Kandeler, E., 2009. Microbial biomass and enzyme activities under reduced nitrogen deposition in a spruce forest soil. Appl. Soil Ecol. 43, 11 – 21.

[15] Frouz, J., Nova'kova', A., 2005. Development of soil microbial properties in topsoil layer during spontaneous succession in heaps after brown coal mining in relation to humus microstructure development. Geoderma 129,54 – 64.

[16] Grishkan, J., Jia, R. L., Kidron, G. J., Li, X. R., 2015. Cultivable microfungal mommunities inhabiting biological soil crusts in the Tengger Desert, China. Pedosphere25(3),351 – 363.

[17] Hamman, S. T., Burke, I. C., Stromberger, M. E., 2007. Relationships between microbialcommunity structure and soil environmental conditions in a recently burnedsystem. Soil Biol. Biochem. 39,1703 – 1711.

[18] Heinze, S., Chen, Y., El – Nahhal, Y., Hadar, Y., Jung, R., Safi, J., Safi, M., Tarchitzky, J., Marschner, B., 2014. Small scale stratification of microbial activity parameters in Mediterranean soils under freshwater and treated wastewater Irrigation. Soil Biol. Biochem. 70,193 – 204.

[19] Jia, R. L., Li, X. R., Liu, L. C., Gao, Y. H., Li, X. J., 2008. Responses of biological soil cruststo sand burial in a revegetated area of the Tengger Desert, Northern China. Soil Biol. Biochem. 40,2827 – 2834.

[20] Liu, Y. M., Li, X. R., Xing, Z. S., Zhao, X., Pan, Y. X., 2013. Responses of soil microbial biomass and community composition to biological soil crusts in the revegetated areas of the Tengger Desert. Appl. Soil Ecol. 65,52 – 59.

[21] Liu, Y. M., Yang, H. Y., Li, X. R., Xing, Z. S., 2014. Effects of biological soil crusts on soil enzyme activities in revegetated areas of the Tengger Desert, China. Appl. Soil Ecol. 80,6 – 14.

[22] Li, X. R., 2012. Eco – hydrology of biological soil crusts in desert regions of China. China Higher Education Press, Beijing, pp 3.

[23] Li, X. R., Jia, R. L., Chen, Y. W., Huang, L., Zhang, P., 2011. Association of ant nests with successional stages of biological soil crusts in the Tengger Desert, Northern China. Appl. Soil Ecol. 47,59 – 66.

[24] Li, X. R., Zhang, P., Su, Y. G., Jia, R. L., 2012. Carbon fixation by biological soil crusts following revegetation of sand dunes in arid desert regions of China: A four - year field study. Catena 97, 119 - 126.

[25] Li, X. R., Zhang, Y. M., Zhao, Y. G., 2009. A study of biological soil crusts: recent development, trend and prospect. Adv. Earth Sci. 24(1), 11 - 24(in Chinese).

[26] Miralles, I., Domingo, F., Cantón, Y., Trasar - Cepeda, C., Leirós, M. C., Gil - Sotres, F., 2012a. Hydrolase enzyme activities in a successional gradient of biological soil crusts in arid and semi - arid zones. Soil Biol. Biochem. 53, 124 - 132.

[27] Miralles, I., Domingo, F., García - Campos, E., Trasar - Cepeda, C., Leirós, M. C., Gil - Sotres, F., 2012b. Biological and microbial activity in biological soil crusts from the Tabernas desert, a sub - arid zone in SE Spain. Soil Biol. Biochem. 55, 113 - 121.

[28] Nannipieri, P., Pedrazzini, F., Arcara, P. G., Piovanelli, C., 1979. Changes in amino acids, enzyme activities, and biomasses during microbial growth. Soil Sci. 127, 26 - 34.

[29] Neher, D. A., Lewins, S. A., Weicht, T. R., Darby, B. J., 2009. Microarthropod communities associated with biological soil crusts in the Colorado Plateau and Chihuahuan deserts. J. Arid Environ. 73, 672 - 677.

[30] Olsen, S. R., Cole, C. V., Watanabe, F. S., and Dean, L. A., 1954. Estimation of Available Phosphorus in Soils by Extraction with Sodium Bicarbonate. United States Department of Agriculture, Circular 939. United States Government Printing Office, Washington, DC, USA, pp 18 - 19.

[31] Olsen, S. R., Sommers, L. E., 1982. Phosphorus. In: Page AL, Miller RH, Keeney DR (Eds.), Methods of Soil Analysis. American Society of Agronomy, Madison, Wisconsin, pp 403 - 427.

[32] Pell, M., Stenström, J., Granhall, U., 2006. Soil Respiration. In: Bloem, J., Hopkins, D. W., Benedetti, A. (Eds.), Microbial Methods for Assessing Soil Quality. CAB International, Wallingford, Oxfordshire, UK.

[33] Raiesi, F., Beheshti, A., 2015. Microbiological indicators of soil quality and degradation following conversion of native forests to continuous croplands. Ecolo. Indic. 50, 173 - 185.

[34] Steven, B., Gallegos - Graves, L. V., Yeager, C., Belnap, J., Kuske, C. R., 2014. Common and distinguishing features of the bacterial and fungal communities in biological soil crusts and shrub root zone soils. Soil Biol. Biochem. 69, 302 - 312.

[35] Susyan, E. A., Wirth, S., Ananyeva, N. D., Stolnikova, E. V., 2011. Forest succession on abandoned arable soils in European Russia - Impacts on microbial biomass, fungal - bacterial ratio, and basal CO_2 respiration activity. Eur. J. Soil Biol. 47, 169 - 174.

[36] Tabatabai, M. A., Bremner, J. M., 1969. Use of P - nitrophenyl phosphate for assay of soil phosphatase activity. Soil Biol. Biochem. 1, 301 - 307.

[37] Uvarov, A. V., Tiunov, A. V., Scheu, S., 2006. Long - term effects of seasonal and diur-

nal temperature fluctuations on carbon dioxide efflux from a forest soil. Soil Biol. Biochem. 38, 3387 – 3397.

[38]Vance, E. D., Brookes, P. C., Jenkinson, D. S., 1987. An extraction method for measuring soil microbial biomass carbon. Soil Biol. Biochem. 19, 703 – 707.

[39]Veluci, R. M, Neher, D. A., Weicht, T. R., 2006. Fixation and leaching of nitrogen by biological soil crust communities in mesic temperate soils. Microb. Ecol. 51, 189 – 196.

[40]Wardle, D. A., Ghani, A. A., 1995. Critique of the microbial metabolic quotient(qCO_2) as a bioindicator of disturbance and ecosystem development. Soil Biol. Biochem. 27, 1601 – 1610.

[41]Williams, W. J., Eldridge, D. J., Alchin, B. M., 2008. Grazing and drought reduce cyanobacterial soil crusts in an Australian Acacia woodland. J. Arid Environ. 72, 1064 – 1075.

[42]Wu, N., Zhang, Y. M., Downing, A., 2009. Comparative study of nitrogenase activity in different types of biological soil crusts in the Gurbantunggut Desert, Northwestern China. J. Arid Environ. 73, 828 – 833.

[43]Xu, G. H., Zheng, H. Y., 1986. Handbook of Analysis of Soil Microorganism. AgriculturePress, Beijing, pp 249 – 291(in Chinese).

[44]Yeomans, J. C., Bremner, J. M., 1998. A rapid and precise method for routine determination of organic carbon in soil. Commun. Soil Sci. Plan. 19, 1467 – 1476.

[45]Yu, J., Kidron, G. J., Pen – Mouratov, S., Wasserstrom, H., Barness, G., Steinberger, Y., 2012. Do development stages of biological soil crusts determine activity and functional diversity in a sand – dune ecosystem? Soil Biol. Biochem. 51, 66 – 72.

[46] Yu, J., Steinberger, Y., 2012. Spatiotemporal changes in abiotic properties, microbial CO_2 evolution, and biomass in playa and crust – covered interdune soils in a sand – dune desert ecosystem. Eur. J. Soil Biol. 50, 7 – 14.

[47]Zelikova, T. J., Housman, D. C., Grote, E. E., Neher, D. A., Belnap, J., 2012. Warming and increased precipitation frequency on the Colorado Plateau: Implications for biological soil crusts and soil processes. Plant Soil 355, 265 – 282.

[48]Zhang, T., Wang, N. F., Liu, H. Y., Zhang, Y. Q., Yu, L. F., 2012. Soil pH is a key determinant of soil fungal community composition in the Ny – Alesund Region, Svalbard(High Arctic). Front. Microbiol. 7, 227.

[49]Zhang, W., Zhang, G. S., Liu, G. X., Dong, Z. B., Chen, T., Zhang, M. X., Dyson, P. J., An, L. Z., 2012. Bacterial diversity and distribution in the southeast edge of the Tengger Desert and their correlation with soil enzyme activities. J. Environ. Sci. 24(11), 2004 – 2011.

[50]Zhao, Y., Zhang, Z. S., Hu, Y. G., Chen, Y. L., 2016. The Seasonal and Successional Variations of Carbon Release from Biological Soil Crust – Covered Soil. J. Arid Environ. 127, 148 – 153.

Effects of Biological Soil Crusts on Soil Enzyme Activities in Revegetated Areas of the Tengger Desert, China

Yanmei Liu　Hangyu Yang　Xinrong Li　Zisheng Xing *

摘　要:生物土壤结皮作为干旱半干旱荒漠区地表景观的重要组分,占荒漠区地表活体生物面积的70%以上,对沙丘的固定起到了重要的作用。土壤酶活性能敏感的指示土壤的恢复程度,是衡量沙区生态恢复与健康的重要生物学属性,而目前关于生物土壤结皮与土壤酶活性的关系研究很少。为探明生物土壤结皮对土壤酶活性的影响,以腾格里沙漠东南缘的植被固沙区生物土壤结皮覆盖的沙丘土壤为研究对象,2011 年根据固沙时间的不同将样地分为 4 个区(1954、1964、1981 和 1991y 固沙区)进行采样,以流沙区(0 y)为对照。研究表明:植被固沙区的藻－地衣结皮和藓类结皮均可显著提高土壤脲酶、蔗糖酶、过氧化氢和脱氢酶的活性($p < 0.05$);结皮类型显著影响土壤酶的活性,发育晚期的藓类结皮下土壤脲酶、蔗糖酶、过氧化氢和脱氢酶的活性显著高于发育早期的藻－地衣结皮下土壤酶的活性($p < 0.05$);固沙年限显著影响土壤脲酶、蔗糖酶、过氧化氢和脱氢酶的活性,且与这几种土壤酶活性均存在显著的线性正相关关系($p < 0.05$);目前,生物土壤结皮可显著提高 0 ~ 20 cm 土层脲酶、蔗糖酶、过氧化氢和脱氢酶的活性($p < 0.05$),且这种影响随土层的增加而减弱。而且,生物土壤结皮下土壤脲酶、蔗糖酶、过氧化氢和脱氢酶的活性表现明显的季节变化,表现为夏季 > 秋季 > 春季和冬季。因此,腾格里沙漠东南缘的植被固沙区生物土壤结皮的存在与演替提高了土壤酶的活性,这指示了生物土壤结皮有利于该区土壤及其相应生态系统的恢复。

---

* 作者简介:刘艳梅(1978—　),女,甘肃天水人,现为天水师范学院生物工程与技术学院教授,博士,主要从事荒漠区土壤生物修复研究。

Biological soil crusts(BSCs) cover up to 70% of the sparsely - vegetated areas in arid and semiarid regions throughout the world and play a vital role in the dune stabilization in desert ecosystems. Soil enzyme activities could be used as significant bioindicators for soil recovery from sand burial. However, the relationship between BSCs and soil enzyme activities is little known at present. The objective of this paper was to determine whether BSCs could affect soil enzyme activities in revegetated areas of the Tengger Desert. The results showed that BSCs significantly promoted the activities of soil urease, invertase, catalase and dehydrogenase. The effects also varied with crust type and the elapsed time since sand dune stabilization. All soil enzyme activities tested in this study were greater under moss crusts than under cyanobacteria - lichen crusts. The elapsed time since sand dune stabilization was correlated positively with four enzyme activities in our study. The studied enzyme activities varied with soil depth and season regardless of crust types. Cyanobacteria - lichen and moss crusts significantly enhanced all test enzyme activities in the 0 - 20 cm soil layer, negatively correlating with soil depth. All test enzyme activities were greater in summer and autumn with the vigorous growth of crusts than in spring and winter. The study demonstrated that the colonization and development of BSCs could improve soil quality and promote soil recovery in the degraded areas of the Tengger Desert.

1 Introduction

Biological soil crusts(BSCs), being a sub - ecosystem or a microcosm(Castillo - Monroy et al., 2011), cover up to 70% of the interspaces between sparse vegetations in semiarid and arid regions throughout the world(Belnap, 1995). They are a complex mosaic of soil, green algae, lichens, mosses, micro - fungi, cyanobacteria and other bacteria (Belnap and Lange, 2003), varying from 2 mm thick, relatively homogeneous cyanobacteria crusts(Zaady and Bouskila, 2002; Li et al., 2011), to complex crusts dominated by lichens and mosses up to 30 mm thick on the surface of stabilized sand dunes(Li et al., 2002). The succession of BSCs generally starts with the colonization of large filamentous cyanobacteria(such as *Microcoleus sp.*) which are able to adhere to soil particles with a secretion of gelatinous polysaccharide materials(Belnap, 2006; Zaady et al., 2010). Smaller pigmented cyanobacteria(such as *Nostoc spp.* and *Scytonema spp.*) and green algae then invade within the filamentous cyanobacteria. Subsequently, lichens and mosses may grow and colonize with increased soil surface stability and/or improved moisture availability (Eldridge and Greene, 1994; Kidron et al., 2008; Yu et al., 2012). BSCs have been found to play numbers of important ecological roles in desert ecosystems, including enhancing soil aggregation and stability (Belnap, 1996; Guo et al., 2008), adjusting soil temperature and moisture (Belnap, 1995; George et al., 2003), improving soil aeration and porosity (Harper and Marble, 1988; Belnap et al., 2006), adjusting local hydrology (Evans and Johansen, 1999; Belnap and Lange, 2003), promoting vascular plant colonization (Zhao et al., 2011), improving soil invertebrate and microbial diversity (Darby et al., 2007, 2010; Neher et al., 2009; Liu et al., 2011; Liu et al., 2013).

Soil enzymatic activity changes quickly with soil conditions, thereby being sensitive indicators for soil quality variation (Puglisiet al., 2006; Trasar – Cepeda et al., 2008; Lebrun et al., 2012). Moreover, it is relatively easy to determine the activity of soil enzymes. Therefore, soil enzymatic activity has been widely used to study the impacts of environmental changes and human impacts on soil quality in recent years (Lebrun et al., 2012). It is not surprised that the close relationship of BSCs with soil enzyme activities has also attracted increasing attention in recent years. For example, Wu et al. (2009) and Zhao et al. (2010) suggested that nitrogenase activity was influenced by the successional stages of BSCs. Zhang et al. (2012) reported that the activities of catalase, urease, dehydrogenase and sucrase were greater in BSCs than in the bare soil without crust. Miralles et al. (2012) found that the activities of several hydrolase enzymes (i.e., arylsulphatase, ß – glucosidase, casein – protease, cellulase and phosphomonoesterase) were greater in BSCs than in the bare substrate of the Tabernas Desert. Zelikova et al. (2012) measured the relationship of extracellular enzyme activities with BSCs in warming and increased precipitation. Although the relationship of BSCs with soil enzyme activities were studied, relatively little is known about the impacts of BSCs, crust type, and the elapsed time since sand dune stabilization on soil enzyme activities in revegetated areas of the Tengger Desert. In particular, in the Tengger Desert where periods of high temperature are often accompanied by temporary droughts, the spatio – temporal variations in soil enzyme activities under BSCs are rarely known. Understanding the impact of BSCs on below – ground soil enzyme activities and spatio – temporal dynamics could provide important information for enhancing BSC's function in the recovery of degraded desert areas.

The aim of this study was to investigate the BSC's function in soil process through soil enzyme activities in the recovery of degraded desertareas. In a first step, we evaluated the impact of BSCs on soil enzyme activities in revegetated areas of the Tengger Desert in northern China. In a second step, we quantified the variation patterns of soil enzyme activities under BSCs with season and soil depth.

2 Material and methods

2.1 Site description

The study area is located at the southeast fringe of the Tengger Desert in northern China (37°32′N and 105°02′E) with an altitude of 1,339 m. This area is a typical transitional zone from desertified steppe to sand dunes, delineating oasis and desert. The mean annual precipitation is approximately 186 mm, 80% of which falls between May and September. The mean annual potential evaporation is about 3000 mm. The mean annual temperature is 10.0 ℃ with the coldest mean monthly temperature of -6.9 ℃ in January and warmest of 24.3 ℃ in July. A northwest wind is prevailing with an average wind velocity of 3.5 m s^{-1}. The mean annual number of days with dust storms caused by wind is approximately 59 (Li et al., 2011). The soil is loose, infertile and mobile and can thus be classified as orthic sierozem and aeolian sandy soil, with very low but constant moisture content of 3 - 4%, and soil organic matter content of 1.5 - 4.0 g kg^{-1} (Li et al., 2007b; Liu et al., 2011). The predominant plants are semishrubs (*Salsola passerine* Bunge, *Oxytropis aciphylla* Ledeb., and *Ceratoides lateens* Reveal et Holmgren), shrubs (*Remuriasoongorica* Maximand *Caragana korshinskii* Kom.), forbs (*Carex stenophylloides* Krecz), and grasses (*Stipa breviflora* Griseb and *Cleistogenes songorica* Ohwj) (Li et al., 2004).

To protect the natural desertified steppe from sand burial due to the constant expansion of sand dunes in Shapotou region, sand - binding vegetation was first established along the Baotou - Lanzhou railway in 1956 by erecting 1 m × 1 m straw checkerboard sand barriers on shifting sandy surfaces. Parallel stabilized areas were expanded one by one along the railway line in 1964, 1981 and 1991. After the shifting sandy surfaces were stabilized, xerophytic shrubs (*Artemisia ordosia* Krasch, *C. korshinskii* Kom. and *Caragana microphylla* Lam.) were planted within the checkerboard, growing onlywith rain inputs. BSCs colonized and developed once the no - irrigation vegetation was established. The initial cyanobacteria - dominated crusts remain the dominant crust type in revegetated areas of 1991 and the late successional lichen - and moss - dominated crusts are the dominant crust type in revegetated areas of 1956 and 1964. In the study areas, moss crusts were 8 – 20 mm thick and green under wet conditions, while cyanobacteria – lichen crusts were 2 – 3.5 mm thick and dark brown or black with a clear surface microtopography. Total coverage of crust is more than 80% of the revegetated areas at present (Jia et al., 2008). The BSCs on sand surface altered topsoil physical and chemical properties, resulting in increases in the content of soil clay and silt, soil pH, electric conductivity (EC), organic C, total N, total phosphorous as well as the thickness of soil crust and subsoil, and decreases in soil bulk density in revegetated areas as compared to mobile sand dunes (Table 1; Li et al., 2007a; Liu et al., 2012).

Table 1 The physico – chemical properties of topsoil (0 – 10 cm) in revegetated areas, natural vegetation areas and mobile sand dunes

Soil properties	natural vegetation areas	Years after revegetation (y)				
		1956	1964	1981	1991	0 (mobile sand dunes)
Sand (%)	13.54	66.39	68.28	71.54	78.87	99.67
Silt (%)	72.00	22.59	24.79	23.59	15.60	0.12
Clay (%)	14.45	11.01	6.93	4.87	4.45	0.21
Organic carbon ($g\ kg^{-1}$)	20.54	7.74	7.59	4.32	1.65	0.37
Total nitrogen ($g\ kg^{-1}$)	2.07	1.02	0.74	0.54	0.39	0.17
Total phosphorous ($g\ kg^{-1}$)	1.38	0.77	0.75	0.71	0.44	0.4
Soil water content (%)	3.79	1.96	1.87	1.61	1.55	1.2
pH	8.28	7.99	7.95	7.90	7.82	7.42
EC ($m\ s^{-1}$)	1.28	0.19	0.17	0.15	0.14	0.09
Bulk density ($g\ cm^{-3}$)	1.13	1.44	1.47	1.5	1.52	1.53
The thickness of soil crust and subsoil (cm)	4.87	2.5	2.2	1.4	0.72	0

2.2 Soil sample collection and preparation

Soil samples were collected in July 2011 from different revegetated areas stabilized in 1956, 1964, 1981 and 1991, respectively. The mobile sand dunes served as the control. Within each revegetated area, five sub – plots (10 m × 10 m) being both cyanobacteria – lichen and moss crusts were established, at least 20 m spacing between two adjacent sub – plots. Meanwhile, five sub – plots (10 m × 10 m) without crusts were established in the mobile sand dunes with at least 20 m spacing between two adjacent sub – plots. Within each sub – plot, soil samples under cyanobacteria – lichen crusts were collected from each soil layer at depths of 0 – 10 cm, 10 – 20 cm and 20 – 30 cm with a 4 cm inner diameter × 10 cm depth core sampler, respectively. Each soil sample consisted of five soil samples under cyanobacteria – lichen crusts which were taken from five sub – plots with the same soil depth. Soil samples under moss crusts and of mobile sand dunes were also collected using the same method as cyanobacteria – lichen crusts. In total, 72 soil samples under crusts and nine sand dune soil samples were used in analysis of soil enzyme activities. To monitor seasonal variations in soil enzyme activities under crusts, the revegetated areas of 1956 and natural vegetation areas at Hongwei of Shapotou area as experimental locations were chosen on behalf of the artificially revegetated areas and natural vegetation areas, respectively. Five sub – plots (10 m × 10 m) being both cyanobacteria – lichen and moss crusts, at least 20 m spacing between two adjacent sub – plots, were established in the revegetated areas of 1956 and natural vegetation areas, respectively. Soil samples under crusts were collected in April, July, September and December 2011 in revegetated areas of 1956 and natural vegetation areas with the upper method, respectively. Total 144 soil samples under crusts from different seasons were used in analysis of season variations of soil enzyme activities.

Each soil sample was placed in individual plastic bag and taken back to the laboratory. A sieve with 2 mm apertures was used to remove large plant parts and stones before samples were stored at 4 ℃ until analysis. Soil moisture content was determined gravimetrically using 20 g of field moist soil sample oven – dried at 105 ℃ for 24 h.

2.3 Soil enzyme activities

Urease (EC 3.5.1.5) activity was assayed according to the method of Yang et al. (2007). 5 g soil sample was incubated 24 h at 37 ℃ with 10 ml of 10% urea and 20 ml of citrate buffer (pH 6.7) before mixing with 1 ml methylbenzene for 15 min. After incubation, the mixtures were immediately filtered and 1 ml supernatant was treated with 10 ml 37 ℃ distilled water, 4 ml sodium phenate (1.35 M) and 3 ml sodium hypochlorite (active chlorine 0.9%) for 20 min. The amount of N – NH_4^+ released from urea hydrolysis was measured in the supernatant at 578 nm. In both cases, urease activity is expressed as μg N – NH_4^+ g^{-1} h^{-1}.

Invertase (EC 3.2.1.26) activity was assayed according to the method of Jin et al. (2009). 5 g soil sample was incubated 24 h at 37 ℃ with 15 ml of 8% sucrose solution and 5 ml phosphate buffer (pH 5.5) before mixing with 0.2 ml methylbenzene for 15 min. The mixtures were immediately filtered and the 1 ml supernatant into tube was treated with 3 ml 3,5 – dinitrosalicylic acid solution and 5 ml distilled water. Place all of the tubes into a boiling water bath for 5 min and then allow them to cool to room temperature. The amount of glucose released from sucrose hydrolysis was measured in the supernatant at 508 nm. Invertase activity is expressed as μg glucose g^{-1} h^{-1}.

Catalase (EC 1.11.1.6) activity was analyzed according to the method of Jin et al. (2009). Catalase activity was analyzed by back – titrating residual H_2O_2 with $KMnO_4$ and was expressed in 0.1mol l^{-1} $KMnO_4$ ml g^{-1} h^{-1}. 5 g soil sample was added to 40 ml distilled water and 5 ml 0.3% hydrogen peroxide solution. The mixture was oscillated at 25 ℃ for 20min. After oscillation, the reaction was stopped by adding 5 ml of 1.5 mol l^{-1} sulfuric acid. The mixture was filtered and titrated using 0.1mol l^{-1} $KMnO_4$.

Dehydrogenase (EC 1.1.1.1) activity was determined as described by Belén Hinojosa et al. (2004) with the following modification: 15 g soil sample mixed with 0.15 g of $CaCO_3$ was incubated with 1 ml 3% 2,3,5 – triphenyltetrazolium chloride (TTC) and 10 ml distilled water at 37℃ in darkness. After 24 h, 10 ml methanol was added, and the suspension homogenized, filtered and washed with methanol until the reddish color caused by the reduced TTC (triphenylformazan, TPF) had disappeared from soil. The optical density at 485 nm was compared to those of TPF standards. Dehydrogenase activity is expressed as μg TPF g^{-1} h^{-1}. Enzyme assays were performed in three replicates for each soil sample.

2.4 Statistical analyses

Analysis of variance was used to evaluate if soil enzyme activities differed between crust type, the elapsed time since sand dune stabilization and season. A multivariate analysis of variance was used to evaluate the interactions of crust type, the elapsed time since sand dune stabilization and soil depth on soil enzyme activities. The correlations of the elapsed time since sand dune stabilization and soil depth with soil enzyme activities were examined with Pearson's correlations coefficients. Differences obtained at levels of $p < 0.05$ were considered significant. All data were subjected to statistical analysis using the SPSS 16.0 software.

3 Results

3.1 Effects of BSCs on soil enzyme activities

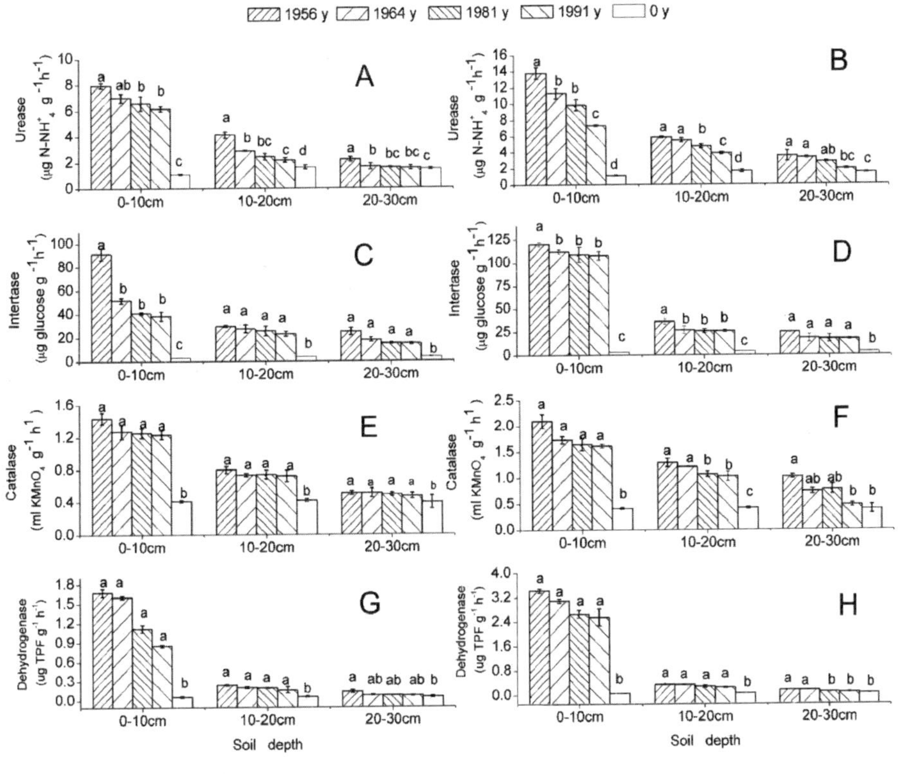

Fig. 1 Soil enzyme activities under cyanobacteria – lichen (A, C, E, and G) and moss crusts (B, D, F, and H) in revegetated areas of the Tengger Desert. 1956, 1964, 1981, 1991 and 0 y refer to the elapsed time since dune fixation. Significant differences ($p < 0.05$) among treatments are indicated by different letters. Error bars show SE ($n = 3$).

BSCs significantly increased the activities of soil urease, invertase, catalase and dehydrogenase in revegetated areas. Soil urease activity was greater under cyanobacteria – lichen and moss crusts than in mobile sand dunes in the 0 – 20 cm soil layers ($p < 0.05$) (Fig. 1, A and B). Soil invertase activity was greater under cyanobacteria – lichen and moss crusts than in mobile sand dunes in the 0 – 30 cm soil layers ($p < 0.05$) (Fig. 1, C and D). Soil urease and invertase activities varied with crust type, the elapsed time since sand dune stabilization and soil depth (Table 2, $p < 0.001$). Moss crusts had greater soil urease and invertase activities than cyanobacteria – lichen crusts and the effects of crust type on them were interactively determined by the elapsed time since sand dune stabilization and soil depth (Table 2, $p < 0.01$). The elapsed time since sand dune stabilization was correlated positively with soil urease and invertase activities ($r = 0.967$ and $r = 0.906$, respectively, $p < 0.05$) and the effects of the elapsed time since sand dune stabilization on them were also dependent upon crust type and soil depth (Table 2, $p < 0.01$). Soil urease activity under cyanobacteria – lichen and moss crusts obviously declined with soil depth (Table 2, $p < 0.001$) with a negative correlation ($r = 0.956$ and $r = 0.970$, respectively, $p < 0.05$). Similarly, soil invertase activity under cyanobacteria – lichen and moss crusts decreased with soil depth (Table 2, $p < 0.001$), showing an obviously negative correlation with soil depth ($r = 0.954$ and $r = 0.909$, respectively, $p < 0.05$).

Soil catalase and dehydrogenase activities (Fig. 1, E and G) were greater beneath cyanobacteria – lichen crusts than in mobile sand dunes in the 0 – 20 cm soil layers ($p < 0.05$). The activities of soil catalase and dehydrogenase were greater beneath moss crusts than in mobile sand dunes in the 0 – 20 cm soil layers ($p < 0.05$) (Fig. 1, F and H). The activities of soil catalase and dehydrogenase varied with crust type, the elapsed time since sand dune stabilization and soil depth (Table 2, $p < 0.05$). The elapsed time since sand dune stabilization correlated positively with soil catalase and dehydrogenase activities ($r = 0.908$ and $r = 0.956$, respectively, $p < 0.05$) and the effect of the elapsed time since sand dune stabilization on them also depended on crust type and soil depth (Table 2, $p < 0.001$). Catalase and dehydrogenase activities under moss crusts were greater than under cyanobacteria – lichen crusts and the effect of crust type on them also depended on the elapsed time since sand dune stabilization (Table 2, $p < 0.001$). The interactions mean that the effect of crust type on soil catalase and dehydrogenase activities varied with the elapsed time since sand dune stabilization. Soil catalase activity under cyanobacteria – lichen and moss crusts obviously declined with soil depth (Table 2, $p < 0.001$), which exhibited an obviously negative correlation with soil depth ($r = 0.979$ and $r = 0.993$, respectively, $p < 0.05$). Similarly, soil dehydrogenase activity under cyanobacteria – lichen and moss crusts declined as soil depth increased

(Table 2, $p < 0.001$), suggesting a negative correlation with soil depth ($r = 0.903$ and $r = 0.893$, respectively, $p < 0.05$).

Table 2 Statistical analysis of soil enzyme activities under cyanobacteria – lichen and moss crusts

Source of variation	df	Mean Square	F
Soil urease activity			
Crust type	1	78.932***	273.854
Soil depth	2	208.340***	722.831
The elapsed time since dune stabilization	4	61.662***	213.935
Crust type * Soil depth	2	7.005***	24.304
Crust type * The elapsed time since dune stabilization	4	7.069***	24.525
Soil depth * The elapsed time since dune stabilization	8	17.331***	60.131
Crust type * Soil depth * The elapsed time since dune stabilization	8	1.711***	5.935
Soil invertase activity			
Crust type	1	5695.419***	57.985
Soil depth	2	23967.210***	244.010
The elapsed time since dunestabilization	4	6495.149***	66.127
Crust type * Soil depth	2	4853.279***	49.411
Crust type * The elapsed time since dune stabilization	4	473.857**	4.824
Soil depth * The elapsed time since dune stabilization	8	1703.419***	17.343
Crust type * Soil depth * The elapsed time since dune stabilization	8	453.324***	4.615
Soil catalase activity			
Crust type	1	.226***	79.986
Soil depth	2	.459***	162.098

续表

Source of variation	df	Mean Square	F
The elapsed time since dune stabilization	4	.177***	62.537
Crust type * Soil depth	2	.007	2.381
Crust type * The elapsed time since dune stabilization	4	.021***	7.469
Soil depth * The elapsed time since dune stabilization	8	.029***	10.317
Crust type * Soil depth * The elapsed time since dune stabilization	8	.001	.477
Soil dehydrogenase activity			
Crust type	1	5.169***	418.025
Soil depth	2	24.799***	2.005E3
The elapsed time since dune stabilization	4	2.532***	204.767
Crust type * Soil depth	2	4.201***	339.728
Crust type * The elapsed time since dune stabilization	4	.330***	26.654
Soil depth * The elapsed time since dune stabilization	8	1.777***	143.745
Crust type * Soil depth * The elapsed time since dune stabilization	8	.265***	21.453

** $p<0.01$　*** $p<0.001$

3.2 Seasonal variations in soil enzyme activities

The activities of soil urease, invertase, catalase and dehyrogenase under cyanobacteria – lichen and moss crusts changed with season in the 0 – 10 cm, 10 – 20 cm and 20 – 30 cm soil layers in both revegetated areas and natural vegetation areas (Fig. 2 and 3). Soil urease and invertase activities under cyanobacteria – lichen and moss crusts increased through time in spring (April) from the lowest values, quickly reached its maximum in summer (July) with significant difference from that in spring and winter ($p < 0.05$), then dropped back and achieved the second highest values in autumn (September) and the third highest values in winter (December) sequentially in revegetated areas (Fig. 2, A, B, C, D) and natural vegetation areas (Fig. 3, A, B, C, D). The activities of soil catalase and dehyrogenase under cyanobacteria – lichen and moss crusts were ranked from the highest to lowest as in summer (July), autumn (September), spring (April), and winter (December) in three soil layers in revegetated areas (Fig. 2, E, F, G, H) and natural vegetation areas (Fig. 3, E, F, G, H).

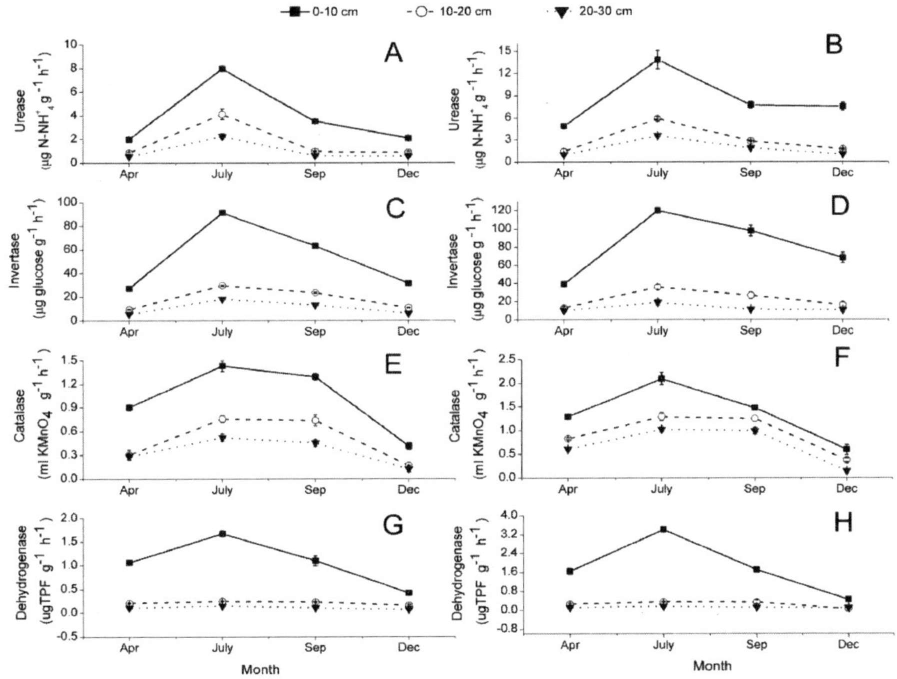

Fig. 2 Seasonal variations of soil enzyme activities under cyanobacteria – lichen crusts in the revegetated areas (A, C, E, and G) and natural vegetation areas (B, D, F, and H). Error bars show SE (n = 3).

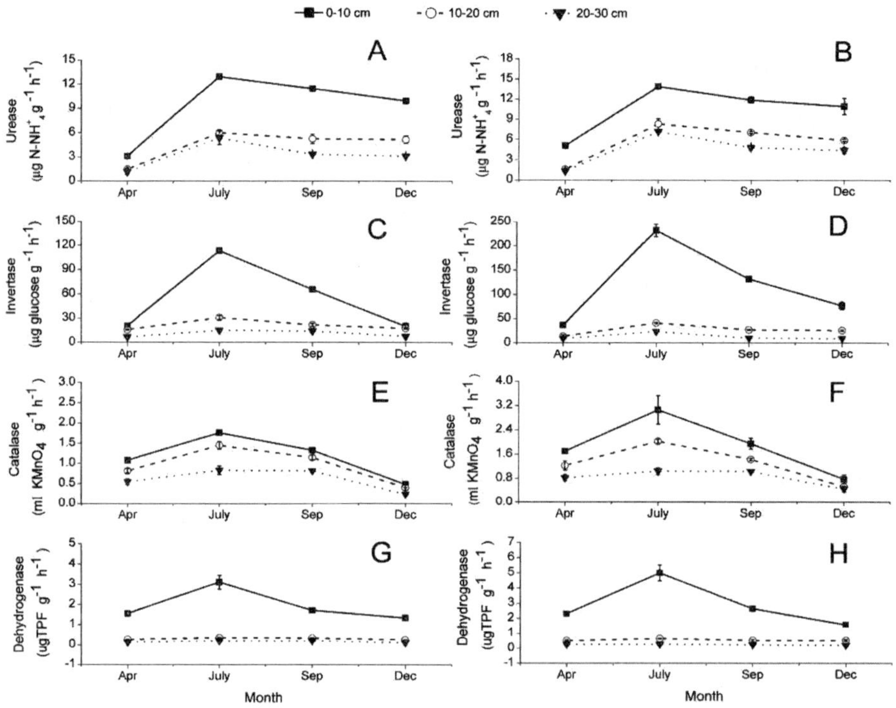

Fig. 3 Seasonal variations of soil enzyme activities under moss crusts in the revegetated areas(A, C, E, and G) and natural vegetation areas(B, D, F, and H). Error bars show SE(n = 3).

4 Discussion

Enzyme activities was found to be suitable soil properties for evaluating soil change in both natural and agroecosystems (Trasar - Cepeda et al., 2000; Puglisi et al., 2006). BSCs obviously enhanced the activities of soil urease, invertase, catalase and dehydrogenasein revegetated areas of the Tengger Desert, similar to Zhang et al. (2012). It may be possible that BSCs could create a favorable environment (suitable soil temperature and moisture) with relatively abundant nutrient (greater soil organic matter and soil nutrient) (Belnap and Lange, 2003; Li et al., 2007a; Darby et al., 2007, 2010; Zhao et al., 2011) for favoring soilenzyme activities (Acosta - Martínez et al., 2007; Brockett et al., 2012). This speculation is supported by our previous finding that soil under BSCs had 11.3 - fold greater organic matter, 10.4 - fold greater total N, 3.25 - fold greater C:N ratio, 1.57 - fold greater soil moisture and 1.13 - fold decreased daytime temperatures than uncrusted soil in the study area(Li et al. ,2011). Soil invertase and urease are crucial to soil C and N cycling, respectively(Ge et al. ,2010). Thus, the greater activities of soil invertase and urease under crusts than mobile sand dunes mean that BSCs could accelerate soil C and N turnover, enhancing soil C and N cycling in desert ecosystem. This is supported by Li et al. (2009), who found that BSCs are dominant components in the input and exchange of C and N in desert ecosystem. Catalase and dehydrogenase were produced mainly by soil microbes and involved in microbial oxidoreductase metabolism(Iyyemperumal and Shi, 2008). Therefore, the increased activities of catalase and dehydrogenase under BSCs in revegetated areas suggest that BSCs enhance soil microbial activities. The improved activities of four enzyme activities-under BSCs in our study may indicate that BSCs played a significant role in improving soil quality and promoting soil recovery in degraded areas of the Tengger Desert.

Greater activities of soil urease and invertase under late – stage moss crusts than early – stage cyanobacteria – lichen crusts indicates that late – stage moss crusts have a greater ability to increase C and N cycling to desert ecosystems compared with the early – stage cyanobacteria – lichen crusts in revegetated areas. This result is in consistent with previous finding that late – stage crusts could cause greater C and N inputs and exchange than early – stage crusts Li et al. (2009) and Darby et al. (2010). Free – living cyanobacteria were ubiquitous on late – stage moss crusts due to the favorable conditions creating by mosses(Veluci et al. ,2006). For example, we observed epiphytic cyanobacteria on late – stage moss crusts in revegetated areas of the Tengger Desert, similar to Veluci et al. (2006). Therefore, late – stage crusts that contain not only dark – colored mosses, lichens, but also cyanobacteria can fix more C and N than light – colored, early – stage crusts, which are dominated by light cyanobacteri(Housman et al. ,2006; Grote et al. ,2010; Zelikova et al. ,2012). Greater activities of soil catalase and dehydrogenase in late – stage moss crusts than early – stage cyanobacteria – lichen crusts suggests that soil microbial activities were greater in the late – stage moss crust compared with the early – stage crusts. Similarly, Miralles et al. (2012) reported that late – stage crusts could lead to greater activity of hydrolytic enzymes than early – stage crusts. These findings can be reinforced by the previous findings that soil under late – stage crusts had 1.26 – fold greater organic matter and 1.88 – fold greater total N than soil under early – stage crusts in revegetated areas(Li et al. ,2011) and the positive correlation between nutrient content and high enzyme activity(Aon and Colaneric, 2001;

Acosta - Martínez et al. ,2007;Zwikel et al. ,2007). In addition, late - stage crusts could provide 1.29 - fold greater soil moisture, 1.15 - fold decreased daytime temperatures and 1.02 - fold decreased pH than early - stage crusts (Li et al. ,2011), which were of great benefit to increase of enzyme activities as soil temperature, moisture and pH were the important factors influencing soil enzyme activities (Brockett et al. , 2012). Veluci et al. (2006) reported similar findings, where in BSCs moderate fluctuations in soil climate, in part by reducing evaporation and absorbing more solar radiation with their dark pigments. Ecologists usually consider that cyanobacteria are primary colonizers which are successively replaced or substituted by moss - lichen crusts. Our results found that the succession of BSCs could be indicated by the changes of some enzyme activities; the late - stage crusts had greater soil enzyme activities compared with early - stage crusts. There is also evidence that early - stage crusts would have greater soil invertase activity than late - stage crusts (Miralles et al. 2012), which contrasts our findings. This may be caused by the differences in species compositions of crusts, climate conditions and geographical factors and so on.

The linear increase of activities of soil urease, invertase, catalase and dehydrogenase with the elapsed time since sand dune stabilization suggested that the cycling of soil C and N and soil microbial activity under crusts may be accumulated with the elapsed time since sand dune stabilization in revegetated areas. Therefore, these benefits to the cycling of soil C and N and soil microbial activity are the proof of the improvement of soil quality. This is the first study to show that four enzyme activities in our study were correlated positively with the elapsed time since sand dune stabilization in desert ecosystem. However, similar trend that some enzyme activities increased with spontaneous succession age was observed in the restored ecosystem in spoil heaps after mining of brown coal (Baldrian et al. ,2008). In 6 - 150 years old alpine glacier foreland soils, enzyme activities generally increased with succession age (Ohtonen et al. ,1999). The improvement of soil enzyme activities with the elapsed time since sand dune stabilization may be attributed mainly to the increased thickness of soil crust and subsoil. The thickening processing of crust with the elapsed time since sand dune stabilization was concomitant with the accumulation of soil nutrient (i. e. , organic C, total N and P; Table 1) and favorable environment (more soil water, suitable temperature and pH) for soil enzyme activities (Li et al. ,2005; Hamman et al. ,2007; Kara et al. ,2008).

The activities of soil urease, invertase, catalase and dehydrogenase under crusts showed a gradual decline with soil depths and this is in agreement with the findings of Niemi et al. (2005) and Enowashu et al. (2009). The significant differences in four enzyme activities in our study between crusted and mobile sand dunes in the 0 – 20 cm soil layers suggested that BSCs could significantly promote the cycling of soil C and N and soil microbial activity in the upper soil layers. In the 20 – 30 cm soil layer, although four enzyme activities in our study under crusts were greater than mobile sand dunes, no significant differences were found in some of them at present. A possible explanation is that greater soil organic matter, nutrient contents, moisture and temperature retention in the top soil (Jia et al., 2007; Liu et al., 2011) can support greater cycling of soil C and N and soil microbial activity, leading to the enhancement of four enzyme activities in our study than the deeper soil layer (Fierer et al., 2003). Meanwhile, the significant interactions between the elapsed time since sand dune stabilization and soil depth on enzyme activities were found. This may imply that BSCs can enhance the cycling of soil C and N and soil microbial activity on the top soil first (in the 0 – 20 cm soil layers) and in the deeper soil layers later as the elapsed time since sand dune stabilization continues toincrease. The result may indicate that BSCs gradually improved soil quality and promoted soil recovery in the vertical direction with increasing of the elapsed time since sand dune stabilization in revegetated areas.

To our current knowledge, this is the first study to determine that soil enzyme activities under crusts changed with season in sandy soils of the Tengger Desert. The seasonal variation in activities of soil urease, invertase, catalase and dehydrogenase under crusts may be caused by climatic factors and soil properties. While climatic factors such as temperature and moisture have been identified as the leading factors for the observed seasonal differences (Criquet et al., 2004; Wittmann et al., 2004), the effects of temperature and moisture on enzyme activities may be indirectly mediated by crust growth. In previous study, Jin et al. (2009) and Yao et al. (2011) reported plant growth can be an important factor regulating seasonal variations on soil enzyme activities. It is the greater sunlight condition (776 h sunshine) and relatively plenty precipitation (78. 0 mm) in summer that ensure more vigorous growth of crusts, leading to greater aboveground biomass, cover and species composition than in autumn, spring and winter. The vigorous growth of crusts improved soil moisture and temperature, created a more suitable habitat for soil microorganisms (Zwikel et al., 2007) and promoted soil enzyme activities under crusts more in summer than other season. The growth of crustss lowed down in autumn due to relatively low precipitation(59.4 mm), low illumination(467 h sunshine), and low soil temperature which discouraged activities of soil microorganism and soil enzyme in fall compared to summer. The very little precipitation and low temperature in spring and winter restricted the growth of crusts, leading to the reduced soil microorganism activities. Thus, soil enzyme activities decreased in spring and winter compared with in summer and autumn. Our finding that enzyme activities under crusts were more sensitive to seasonal variations in the topsoil layer(0 – 10 cm) than in deeper soil layer, similar to the previous findings obtaining from other soils in other places(Aon and Colaneric, 2001; Zwikel et al., 2007). The finding may suggest that enzymatic activities in the topsoil layer could be sensitive indicators to the seasonal change of crusts growth.

5. Conclusion

The present study demonstrated that BSCs significantly increased the activities of soil urease, invertase, catalase and dehydrogenase in revegetated areas of the Tengger Desert. Crust type and the elapsed time since sand dune stabilization could significantly affect the studied enzyme activities. All four enzymes demonstrated consistently greater activities in late – stage moss than early – stage cyanobacteria – lichen crusts due to more suitable soil temperature and moisture, improved soil organic matter content and nutrient content under late – stage moss crusts compared with early – stage cyanobacteria – lichen crusts. The elapsed time since sand dune stabilization was correlated positively with four enzyme activities in our study, illustrating the progressive improvement of soil quality with elapsed time since sand dune stabilization in revegetated areas. In addition, four enzyme activities in our study under crusts also varied with soil depth and season in the study areas. BSCs significantly promoted enzyme activities in the 0 – 20 cm soil layers, improved soil quality of the upper soil layer. Four enzyme activities showed an apparent seasonal pattern, highest in summer, followed by autumn and the lowest in spring and winter in our study. The seasonal patterns of soil enzyme activities mainly due to crusts growth impacted by climatic factors and their interaction with the development of crusts.

The study demonstrated that BSCs significantly enhanced soil enzyme activities in revegetated areas, promote the cycling of soil C and N and soil microbial activity. The colonization and development of BSCs could improve soil quality and benefit the recovery of degraded areas in the Tengger Desert.

Acknowledgements

This research was supported by China National Funds for Regional Science (grant No. 41261014), school project of Tianshui Normal University (grant No. TSA0923) and China National Funds for Distinguished Young Scientists (grant No. 40825001).

REFERENCES

[1] Acosta – Martínez, V., Cruz, L., Sotomayor – Ramírez, D., Pérez – Alegría, L., 2007. Enzyme activities as affected by soil properties and land use in a tropical watershed. Appl. Soil Ecol. 35, 35 – 45.

[2] Aon, M. A., Colaneri, A. C., 2001. II. Temporal and spatial evolution of enzymatic activities and physico – chemical properties in an agricultural soil. Appl. Soil Ecol. 18, 255 – 270.

[3] Baldrian, P., Trögl, J., Frouz, J., Šnajdr, J., Valášková, V., Merhautová, V., Cajthaml, T., Herinková, J., 2008. Enzyme activities and microbial biomass in topsoil layer during spontaneous succession in spoil heaps after brown coal mining. Soil Biol. Biochem. 40, 2107 – 2115.

[4] Belén Hinojosa, M., Carreira, J. A., García – Ruíz, R., Dick, R. P., 2004. Soil moisture pre – treatment effects on enzyme activities as indicators of heavy metal – contaminated and reclaimed soils. Soil Biol. Biochem. 36, 1559 – 1568.

[5] Belnap, J., 1995. Surface disturbances: their role in accelerating desertification. Environ. Monit. Assess. 37, 39 – 57.

[6] Belnap, J., 1996. Soil surface disturbances in cold deserts: effects on nitrogenase activity in cyanobacterial – lichen soil crusts. Biol. Fert. Soils 19, 362 – 367.

[7] Belnap, J., 2006. The potential roles of biological soil crusts in dryland hydrological cycles. Hydrol. Process. 20, 3159 – 3178.

[8] Belnap, J., Lange, O. L., 2003. Biological Soil Crust: Structure, Function and Management. Springer – Verlag, Berlin, 3 – 30.

[9] Belnap, J., Phillips, S. L., Troxler, T., 2006. Soil lichen and moss cover and species richness can be highly dynamic: The effects of invasion by the annual exotic grass *Bromus tectorum*, precipitation, and temperature on biological soil crusts in SE Utah. Appl. Soil Ecol. 32, 63 – 76.

[10] Brockett, B. F. T., Prescott, C. E., Grayston, S. J., 2012. Soil moisture is the major factor influencing microbial community structure and enzyme activities across seven biogeoclimatic zones in western Canada. Soil Biol. Biochem. 44, 9 – 20.

[11] Castillo – Monroy, A. P., Bowker, M. A., Maestre, F. T., Rodríguez – Echeverría, S., Martinez, I., Barraza – Zepeda, C. E., Escolar, C., 2011. Relationships between biological soil

crusts, bacterial diversity and abundance, and ecosystem functioning: insights from a semi – arid Mediterranean environment. J. Veg. Sci. 22, 165 – 174.

[12] Criquet, S., Ferre, E., Farnet, A. M., Le Petit, J., 2004. Annual dynamics of phosphatase activities in an evergreen oak litter: influence of biotic and abiotic factors. Soil Biol. Biochem. 36, 1111 – 1118.

[13] Darby, B. J., Neher, D. A., and Belnap, J. 2007. Soil nematode communities are ecologically more mature beneath late – than early – successional stage biological soil crusts. Applied Soil Ecology 35, 203 – 212

[14] Darby, B. J., Neher, D. A., and Belnap, J. 2010. Impact of biological soil crusts and desert plants on soil microfaunal community composition. Plant and Soil 328, 421 – 431.

[15] Eldridge, D. J., Greene, R. S. B., 1994. Microbiotic soil crusts: A view of their roles in soil and ecological processes in the rangelands of Australia. Aust. J. Soil Res. 32, 389 – 415.

[16] Enowashu, E., Poll, C., Lamersdorf, N., Kandeler, E., 2009. Microbial biomass and enzyme activities under reduced nitrogen deposition in a spruceforest soil. Appl. Soil Ecol. 43, 11 – 21.

[17] Evans, R. D., Johansen, J. R., 1999. Microbiotic crusts and ecosystem processes. Crit. Rev. Plant Sci. 18, 183 – 225.

[18] Fierer, N., Schimel, J. P., Holden, P. A., 2003. Variations in microbial community composition through two soil depth profiles. Soil Biol. Biochem. 35, 167 – 176.

[19] Ge, C. R., Xue, D., Yao, H. Y., 2010. Microbial biomass, community diversity, and enzyme activities in response to urea application in tea orchard soils. Communications in Soil Science and Plant Analysis, 41(7): 797 – 810

[20] George, D. G., Roundy, B. A., St. Clair, L. L., Johansen, J. R., Schaalje, G. B., Webb, B. L., 2003. The effects of microbiotic soil crusts on soil water loss. Arid Land Res. Manag. 17, 113 – 125.

[21] Grote, E., Belnap, J., Housman, D., Sparks, J., 2010. Carbon exchangein biological soil crust communities under differential temperaturesand soil water contents: implications for global change. Glob Chang Biol 16, 2763 – 2774.

[22] Guo, Y. R., Zhao, H. L., Zuo, X. A., Drake, S., Zhao, X. Y., 2008. Biological soil crust development and its topsoil properties in the process of dune stabilization, Inner Mongolia, China. Environ. Geol. 54, 653 – 662.

[23] Hamman, S. T., Burke, I. C., Stromberger, M. E., 2007. Relationships between microbial community structure and soil environmental conditions in a recently burned system. Soil Biol. Biochem. 39, 1703 – 1711.

[24] Harper, K. T., Marble, J. R., 1988. A role for nonvascular plants in management of arid and semiarid rangelands. In: Tueller, P. T. (Ed.), Vegetationa science applications for rangeland a-

nalysis and management. Kluwer Academic Publishers, Dordrecht, Netherlands. pp, 135 – 169.

[25] Housman, D., Powers, H., Collins, A., Belnap, J., 2006. Carbon and nitrogen fixation differ between successional stages of biological soil crusts in the Colorado Plateau and Chuhuahuan Desert. J Arid Environ 66, 620 – 234.

[26] Iyyemperumal, K., Shi, W., 2008. Soil enzyme activities in two forage systems following application of different rates of swine lagoon effluent or ammonium nitrate. Appl. Soil Ecol. 38, 128 – 136.

[27] Jia, R. L., Li, X. R., Liu, L. C., Gao, Y. H., Li, X. J., 2008. Responsesof biological soil cruststo sand burial in a revegetated area of the Tengger Desert, Northern China. Soil Biol. Biochem. 40, 2827 – 2834.

[28] Jia, X. H., Li, X. R., Li, Y. S., 2007. Soil organic carbon and nitroge dynamics during there – vegetation process in the arid desert region. Chin. J. Plant Ecol. 31, 66 – 74(inChinese).

[29] Jin, K., Sleutel, S., Buchan, D., De Neve, S., Cai, D. X., Gabriels, D., Jin, J. Y., 2009. Changes of soil enzyme activities under different tillage practices in the Chinese Loess Plateau. Soil Till. Res. 104, 115 – 120.

[30] Kara, Ö., Bolat, ì., Çakıro ğ lu, K., Öztürk, M., 2008. Plant canopy effects on litter accumulation and soil microbial biomass in two temperate forests. Biol. Fert. Soils 45, 193 – 198.

[31] Kidron, G. J., Vonshak, A., Abeliovich, A., 2008. Recovery rates of microbiotic crusts within a dune ecosystem in the Negev desert. Geomorphology 100, 444 – 452.

[32] Lebrun, J. D., Trinsoutrot – Gattin, I., Vinceslas – Akpa, M., Bailleul, C., Brault, A., Mougin, C., Laval, K., 2012. Assessing impacts of copper on soil enzyme activities in regard to their natural spatiotemporal variation under long – term different land uses. Soil Biol. Biochem. 49, 150 – 156.

[33] Li, X. R., He, M. Z., Duan, Z. H., Xiao, H. L., Jia, X. H., 2007a. Recovery of topsoil physicochemical properties in revegetated sites in the sand – burial ecosystems of the Tengger Desert, northern China. Gemorphology 88, 254 – 265.

[34] Li, X. R., Jia, R. L., Chen, Y. W., Huang, L., Zhang, P., 2011. Association of ant nests with successional stages of biological soil crusts in the Tengger Desert, Northern China. Appl. Soil Ecol. 47, 59 – 66.

[35] Li, X. R., Kong, D. S., Tan, H. J., Wang, X. P., 2007b. Changes in soil and in vegetation following stabilisation of dunes in southeastern fringe of the Tengger Dersert, China. Plant Soil 300, 221 – 231.

[36] Li, X. R., Wang, X. P., Li, T., Zhang, J. G., 2002. Microbiotic crust and its effect on vegetation and habitat on artificially stabilized desert dunes in Tengger Desert, North China. Biol. Fert. Soils 35, 147 – 154.

[37] Li, X. R., Xiao, H. L., Liu, L. C., Zhang, J. G., Wang, X. P., 2005. Long – term effect of

sand – fixed vegetation on restoration of biodiversity in Shapotou region in Tengger Desert. J. Desert. Res. 25(2),173 – 181(in Chinese).

[38] Li, X. R., Zhang, Y. M., Zhao, Y. G., 2009. A study of biological soil crusts: recent development, trend and prospect. Adv. Earth Sci. 24(1), 11 – 24(in Chinese).

[39] Li, X. R., Zhang, Z. S., Zhang, J. G., Wang, X. P., Jia, X. H., 2004. Association between vegetationpatterns and soil properties in the southeastern Tengger Desert, China. Arid Land Res. Manag. 18, 369 – 383.

[40] Liu, Y. M., Li, X. R., Jia, R. L., Huang, L., Zhou, Y. Y., Gao, Y. H., 2011. Effects of biological soil crusts on soil nematode communities following dune stabilization in the Tengger Desert, Northern China. Appl. Soil Ecol. 49, 118 – 124.

[41] Liu, Y. M., Li, X. R., Xing, Z. S., Zhao, X., Pan, Y. X., 2013. Responses of soil microbial biomass and community composition to biological soil crusts in the revegetated areas of the Tengger Desert. Appl. Soil Ecol. 65, 52 – 59.

[42] Miralles, I., Domingo, F., Cantón, Y., Trasar – Cepeda, C., Leirós, M. C., Gil – Sotres, F., 2012. Hydrolase enzyme activities in a successional gradient of biological soil crusts in arid and semi – arid zones. Soil Biol. Biochem. 53, 124 – 132.

[43] Neher, D. A., Lewins, S. A., Weicht, T. R., Darby, B. J., 2009. Microarthropod communities associated with biological soil crusts in the Colorado Plateau and Chihuahuan deserts. J. Arid Environ. 73, 672 – 677.

[44] Niemi, R. M., Vepsäläinen, M., Wallenius, K., Simpanen, S., Alakukku. L., Pietola, L., 2005. Temporal and soil depth – related variation in soil enzyme activities and in root growth of red clover(*Trifolium pratense*) and timothy(*Phleum pratense*) in the field. Appl. Soil Ecol. 30, 113 – 125.

[45] Ohtonen, R., Fritze, H., Pennanen, T., Jumpponen, A., Trappe, J., 1999. Ecosystem properties and microbial community changes in primary succession on a glacier forefront. Oecologia 119, 239 – 246.

[46] Puglisi, E., Del Re, A. A. M., Rao, M. A., Gianfreda, L., 2006. Development and validation of numerical indexes integrating enzyme activities of soils. Soil Biol. Biochem. 38, 1673 – 1681.

[47] Trasar – Cepeda, C., Leirós M. C., Gil – Sotres, F., 2008. Hydrolytic enzyme activities in agricultural and forest soils. Some implications for their use as indicators of soil quality. Soil Biol. Biochem. 40, 2146 – 2155.

[48] Trasar – Cepeda, C., Leirós, M. C., Seoane, S., Gil – Sotres, F., 2000. Limitation of soil enzymes as indicators of soil pollution. Soil Biol. Biochem. 32, 1867 – 1875.

[49] Veluci, R. M., Neher, D. A., Weicht, T. R., 2006. Fixation and leaching of nitrogen by biological soil crust communities in mesic temperate soils. Microbial Ecology 51, 189 – 196

[50] Wittmann, C., Kähkönen, M. A., Ilvesniemi, H., Kurola, J., Salkinoja – Salonen, M. S.,

2004. Areal activities and stratification of hydrolytic enzymes involved in the biochemical cycles of carbon, nitrogen, sulphur and phosphorus in podsolized boreal forest soils. Soil Biol. Biochem. 36, 425 - 433.

[51] Wu, N., Zhang, Y. M., Downing, A., 2009. Comparative study of nitrogenaseactivity in different types of biological soil crusts in the Gurbantunggut Desert, Northwestern China. J. Arid Environ. 73, 828 - 833.

[52] Yang, C. L., Sun, T. H., Zhou, W. X., Chen, S., 2007. Single and joint effects of pesticides and mercury on soil urease. J. Environ. Sci. 19, 210 - 216

[53] Yao, H. Y., Bowman, D., Shi, W., 2011. Seasonal variations of soil microbial biomass and activity in warm - and cool - season turfgrass systems. Soil Biol. Biochem. 43, 1536 - 1543.

[54] Yu, J., Kidron G. J., Pen - Mouratov, S., Wasserstrom, H., Barness, G., Steinberger, Y., 2012. Do development stages of biological soil crusts determine activity and functional diversity in a sand - dune ecosystem? Soil Biol. Biochem. 51, 66 - 72.

[55] Zaady, E., Ben - David, E. A., Sher, Y., Tzirkin, R., Nejidat, A., 2010. Inferring biological soil crust successional stage using combined PLFA, DGGE, physical and biophysiological analyses. Soil Biol. Biochem. 42, 842 - 849.

[56] Zaady, E., Bouskila, A., 2002. Lizard burrows association with successional stages of biological soil crusts in an arid sandy region. J. Arid Environ. 50, 235 - 246.

[57] Zelikova, T. J., Housman, D. C., Grote, E. E., Neher, D. A., and Belnap, J. 2012. Warming and increased precipitation frequency on the Colorado Plateau: Implications for biological soil crusts and soil processes. Plant Soil 355, 265 - 282.

[58] Zhang, W., Zhang, G. S., Liu, G. X., Dong, Z. B., Chen, T., Zhang, M. X., Dyson, P. J., An, L. Z., 2012. Bacterial diversity and distribution in the southeast edge of the Tengger Desert and their correlation with soil enzyme activities. J. Environ. Sci. 24(11), 2004 - 2011.

[59] Zhao, H. L., Guo, Y. R., Zhou, R. L., Drake, S., 2011. The effects of plantation development on biological soil crust and topsoil properties in a desert in northern China. Geoderma 160, 367 - 372.

[60] Zhao, Y., Xua, M., Belnap, J., 2010. Potential nitrogen fixation activity of different aged biological soil crusts from rehabilitated grasslands of the hilly Loess Plateau, China. J. Arid Environ. 74, 1186 - 1191.

[61] Zwikel, S., Lavee, H., Sarah, P., 2007. Temporal dynamics in arylsulfatase enzyme activity in various microenvironments along a climatic transect in Israel. Geoderma 140, 30 - 41.

Advances in Cadaverine Bacterial Production and Its Applications

Weichao Ma　Kequan Chen　Yan Li　Ning Hao
Xin Wang　Pingkai Ouyang *

摘　要:1,5－戊二胺是一种广泛分布于原核和真核生物中的具有多种生物活性的天然多胺,正日益成为一种重要的工业化学品,并在多个领域展现出广泛的应用前景,特别是作为单体用于合成生物基聚酰胺。基于1,5－戊二胺的聚酰胺5X具有优异的性能和环境友好特性,因而有广泛的应用前景。本文总结了近期关于1,5－戊二胺在细菌中的生物合成、代谢及生理学功能,着重介绍了1,5－戊二胺在大肠杆菌的代谢调控机制。文章还综述了微生物发酵法和全细胞催化法生产1,5－戊二胺的进展及1,5－戊二胺分离纯化方法。此外,对1,5－戊二胺在生物基聚酰胺合成中的应用进行了总结。并对利用可再生资源生产1,5－戊二胺进行了展望和对以后的研究提供了建议。

Cadaverine, a natural polyamine with multiple bioactivities that is widely distributed in prokaryotes and eukaryotes, is becoming an important industrial chemical. Cadaverine exhibits broad prospects for various applications, especially as an important monomer for bio－based polyamides. Cadaverine－based PA 5X has broad application prospects owing to its environmentally friendly characteristics and exceptional performance in water absorption and dimensional stability. In this review, we summarize recent findings on the biosynthesis, metabolism, and physiological function of cadaverine in bacteria, with a focus on the regulatory mechanism of cadaverine synthesis in *Escherichia coli* (*E. coli*). We also describe recent developments in bacterial production of cadaverine

* 作者简介:马伟超(1980—　)男,内蒙古赤峰人,现为天水师范学院生物工程与技术学院副教授,博士,主要从事工业微生物分子育种与发酵工程研究。

by direct fermentation and whole - cell bioconversion, and recent approaches for the separation and purification of cadaverine. In addition, we present an overview of the application of cadaverine in the synthesis of completely bio - based polyamides. Finally, we provide an outlook and suggest future developments to advance the production of cadaverine from renewable resources.

1 INTRODUCTION

Cadaverine(1,5 - diaminopentane) is a natural polyamine with multiple bioactivities that is formed through the direct decarboxylation of L - lysine catalyzed by lysine decarboxylase, and that is widely distributed in prokaryotes and eukaryotes. Cadaverine plays an important role in cell survival at acidic pH and protects cells that are starved of inorganic phosphate, Pi, under anaerobic conditions [1,2]. In plants, it is involved in regulating diverse processes such as plant growth and development, cell signaling, stress response, and insect defense [3]. Cadaverine formation is also related to animal growth [4] and tumorigenesis [5,6]. Thus, cadaverine has a wide range of potential applications in agriculture and medicine. Cadaverine has a similar structure to the synthesized petrochemical hexamethylenediamine, and thus can be used to replace hexamethylenediamine in the production of polyamides or polyurethanes [7,8]. Cadaverine is becoming an important industrial chemical because it exhibits broad prospects for various applications, and especially for the synthesis of fully bio - based polyamides by polymerization with dicarboxylic acids derived from renewable sources.

At present, cadaverine can be produced by direct microbial fermentation or whole - cell bioconversion. For direct microbial fermentation, cadaverine - producing strains are mainly engineered from the conventional L - lysine producers *Corynebacterium glutamicum* (*C. glutamicum*)[9] and *Escherichia coli*(*E. coli*)[10], because L - lysine is the direct precursor of cadaverine. For whole - cell bioconversion, the biocatalysts are dominated by *E. coli* overexpressed lysine decarboxylase [11,12]. In view of the above, the present review focuses on the biosynthesis, metabolism, and physiological function of cadaverine in bacteria, recent advances in cadaverine production, and applications of cadaverine in bio - based polyamide PA 5X synthesis.

2 CADAVERINE METABOLISM IN BACTERIA

2.1 Anabolism of cadaverine in bacteria

Cadaverine is formed through the decarboxylation of L – lysine, and thus its biosynthesis depends on L – lysine. Two different pathways evolved separately for L – lysine biosynthesis: the diaminopimelic acid (DAP) route in bacteria and plants; and the α – aminoadipic acid (AAA) pathway in most fungi and some archaea [13,14]. The DAP pathway has three variants for the synthesis of *meso* – diaminopimelate [15 – 17] (Fig. 1). The first, most prevalent variant, in which the intermediates are succinylated, is widely distributed in eubacteria [18], the so – called lower fungi [19], plants [20], and archaea [14,21]. The second variant involves the N – acylation of intermediates, and is only present in *Bacillus* species, such as *Bacillus megaterium* [18,22]. The third variant is the dehydrogenase pathway, in which Δ1 – piperidine – 2,6 – dicarboxylate is converted to *meso* – diaminopimelate by a one – step reaction; this variant has extremely limited occurrence in *C. glutamicum*, *Bacillus sphaericus*, *Pseudomonas* species, *Brevibacterium* species, and some plants such as soybean, maize, and wheat [13,18].

The biosynthesis and metabolism of cadaverine has been extensively studied in *E. coli*. Two types of lysine decarboxylases are involved in cadaverine synthesis: the constitutive LdcC [23] and the inducible CadA [24]. They show about 68% and 69% identity, respectively, in DNA and amino acid sequences; however, optimum LdcC activity occurs at about pH 7.6, and optimum CadA activity occurs at around pH 5.6 [25]. Furthermore, CadA exhibits higher thermal stability and higher decarboxylation activity [25,26].

CadA is a member of the *E. coli* Cad system, which protects cells against acid stress by inducing protein expression under conditions of low pH with excess L – lysine [27]. The *E. coli* Cad system has two principal components: the cadBA operon coding for lysine decarboxylase (CadA) and lysine/cadaverine antiporter (CadB), and the regulatory protein CadC [28,29]. CadC is a multifunctional inner membrane protein, which is composed of a DNA – binding domain with a winged helix – turn – helix motif at the cytoplasmic N – terminal, a transmembrane (TM) domain, and a periplasmic C – terminal [30]. The N – terminal can bind to the Cad1 and Cad2 sites of the upstream region of cadBA to activate cadBA expression [31]. The C – terminal is a pH sensor domain that is responsible for sensing the external H^+ concentration [32]. The TM domain and periplasm – exposed amino acids of CadC are important in mediating and/or stabilizing the association between CadC and LysP through the formation of salt bridges and/or disulfide bonds [30]. LysP is an L – lysine – specific permease, which perceives exogenous L – lysine and regulates CadC levels [30,33,34]. Transcription of the cadBA operon is also modulated by the histone – like DNA – binding protein H – NS, which acts to reduce the expression of CadA and CadB under non – inducing conditions [35,36].

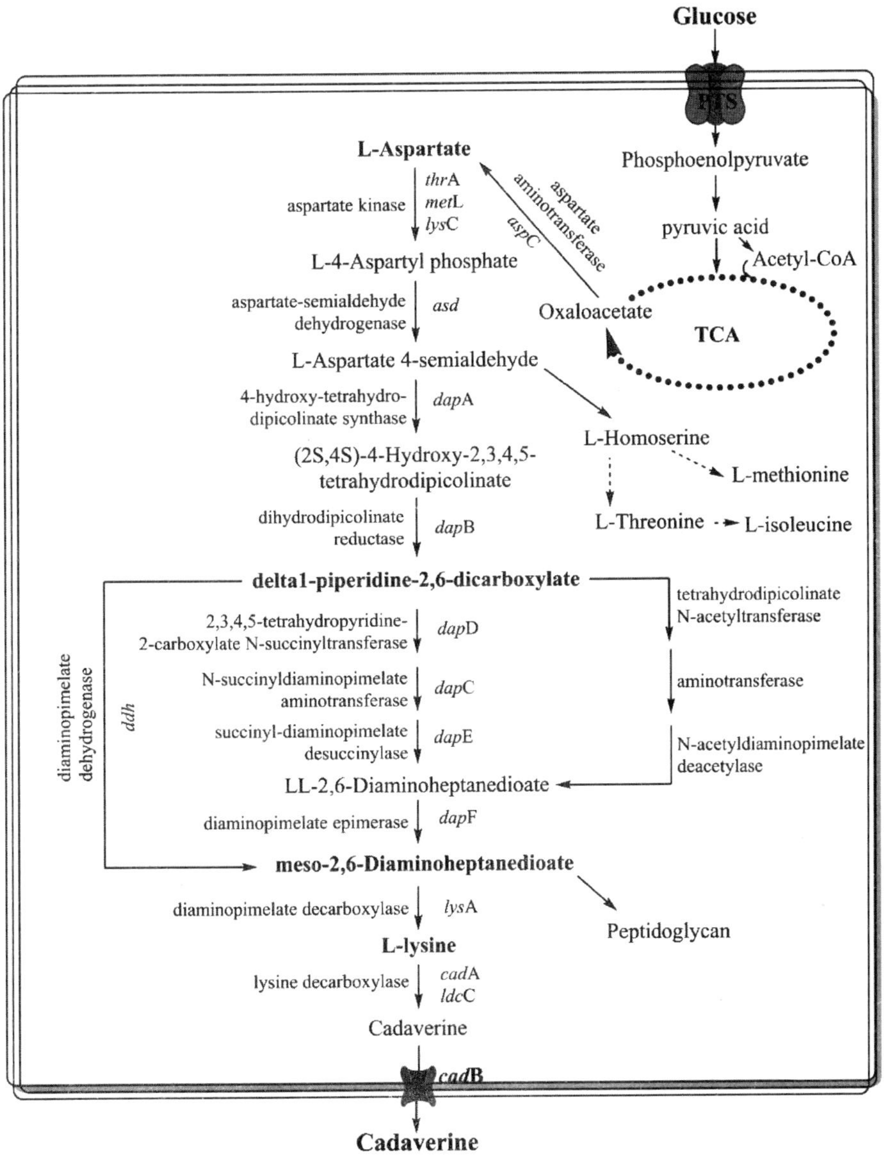

Fig. 1 Schematic representation of the biosynthesis of cadaverine through the diaminopimelic acid (DAP) pathway.

Fig. 2 summarizes the regulatory mechanism of the cadBA operon in *E. coli.* Under normal growth conditions, the expression of the cadBA operon is inhibited by the binding of H – NS proteins to Cad1 and Cad2 sites, which are located at – 144 to – 112 bp and – 89 to – 59 bp of the cadBA upstream region, respectively [31,35,36]. Meanwhile, the transcriptional activator CadC is anchored in the inner membrane and cannot perform the activation function because of the interaction with membrane protein LysP. Thus, cadBA operon is not transcribed, and no cellular cadaverine is produced [37]. When acidic stress occurs ($pH < 6.8$) in a lysine – rich environment (> 5 mM), the interaction between CadC and LysP is weakened due to their pH – dependent conformational transition and/or the opening of the disulfide bridge between the two proteins [32,37]. In addition, the conformation of LysP is further altered upon the binding/translocation of L – lysine, which leads to the disassociation of CadC and LysP [30,38]. The activated CadC then replaces H – NS proteins and binds the Cad1 and Cad2 sites of the cadBA upstream region to activate cadBA operon transcription and expression [36,38]. As cadaverine accumulates, the excess cadaverine (> 235 mM [37]) acts as a negative effector of cadBA expression by deactivation of CadC through binding to the site at the dimerization interface of CadC [28,39]. In addition, CadA activity is inhibited by ppGpp, a stringent response effector that accumulates rapidly in cells that are starved for amino acids, to prevent excessive L – lysine consumption [40].

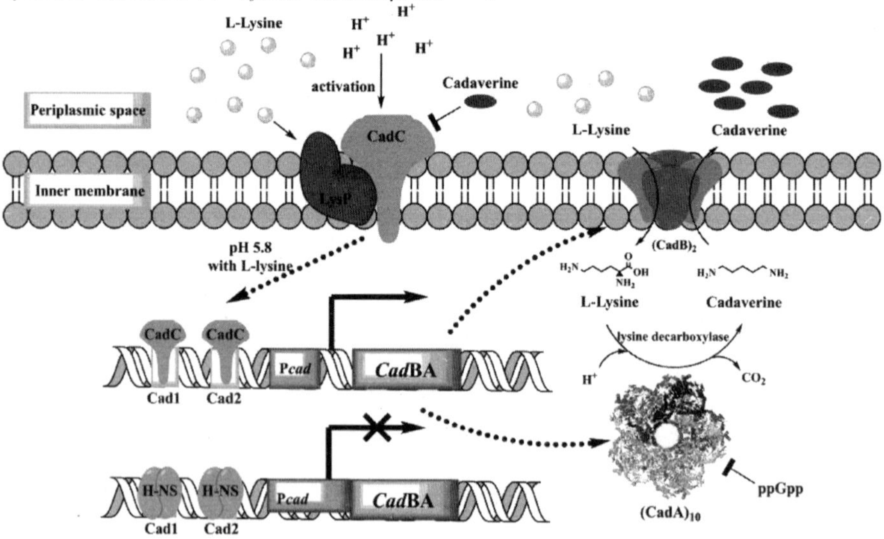

Fig. 2 Regulation of cadaverine biosynthesis in *E. coli.*

In addition to the Cad system of *E. coli*, another cadaverine synthesis system has been identified in *Lactobacillus saerimneri* 30a. This system consists of a specific lysine decarboxylase and a bifunctional antiporter that has an affinity for ornithine and putrescine as well as for lysine and cadaverine [41].

2.2 Intermediary metabolism of cadaverine in bacteria

Although the utilization or degradation of cadaverine has not been observed for *E. coli* thus far [8], several bacteria species are known to use cadaverine as an intermediate. The acetylation of cadaverine has been studied in detail in *C. glutamicum*. In *C. glutamicum*, the encoding gene NCgl1469, which has been identified and functionally assigned as cadaverine acetyltransferase, is responsible for the formation of *N* – acetylcadaverine. Deletion of NCgl1469 led to complete elimination of cadaverine acetylation and a cadaverine yield increase of 11% [42]. Cadaverine can also be aminopropylated by agmatine/cadaverine aminopropyl transferase (ACAPT) to form *N* – (3 – aminopropyl) cadaverine (APC) in *Pyrococcus furiosus*[43]. Another cadaverine catabolic pathway has been confirmed in members of *Pseudomonas*, in which cadaverine is metabolized by transamination to α – piperidine and oxidized to δ – aminovaleric acid (AMV) [44].

Cadaverine has been proven to be an essential constituent of the cell wall, and plays an important role in the structural linkage between the outer membrane and peptidoglycan in some strictly anaerobic bacteria [45]. This role was first discovered in a peptidoglycan of *Selenomonas ruminantium* by Kamio et al. [46,47]; a similar structure was found in *Veillonella alcalescens*[48], *Veillonella parvula*[49], and *Anaerovibrio lipolytica*[50]. The proposed primary structure of the cadaverine – containing peptidoglycan is shown in Fig. 3 [45,50].

Cadaverine is also involved in the biosynthesis of siderophores, which are exported into the local extracellular milieu where they bind Fe^{3+} with high avidity [51,52]. The currently known cadaverine – based siderophores (Fig. 4) include arthrobactin [51,53], bisucaberin B [54], bisucaberin [55], desferrioxamine B [56 – 58], desferrioxamine E [57,59], and desferrioxamine G_1 [60].

3 BIO – BASED PRODUCTION OF CADAVERINE

Cadaverine is synthesized through the direct decarboxylation of L – lysine, which is catalyzed by lysine decarboxylase in living cells. However, little research is available on

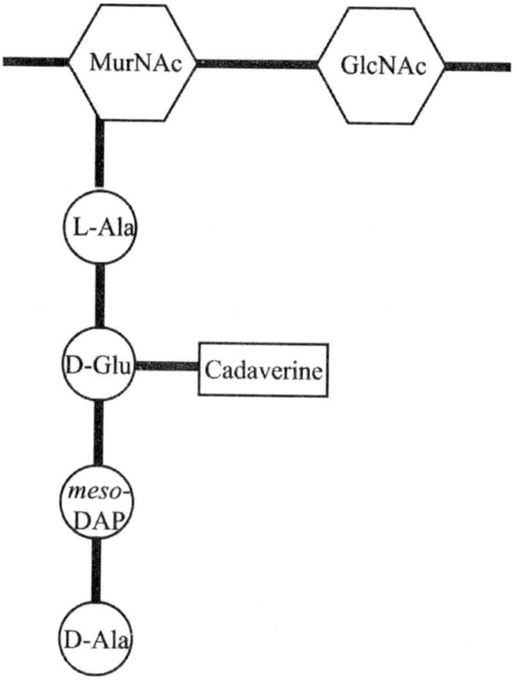

Fig. 3　Chemical structure of the cadaverine - containing peptidoglycan. (MurNAc and GlcNAc refer to *N* - acetylmuramic acid and *N* - acetylglucosamine, respectively.)

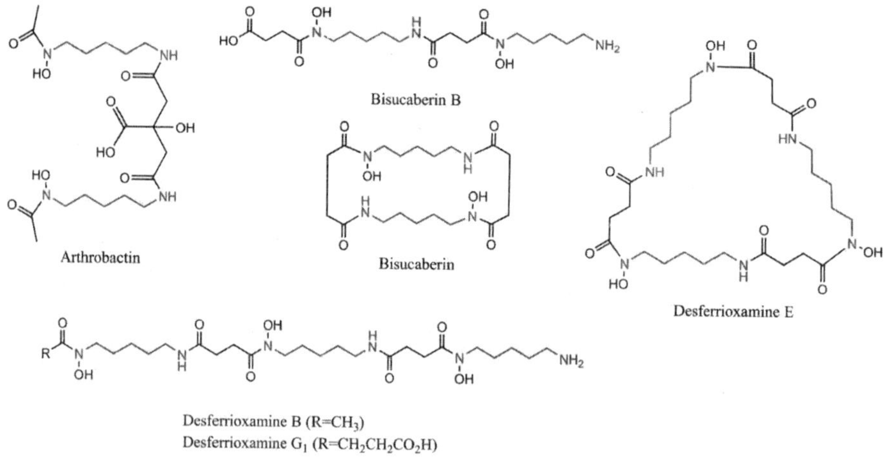

Fig. 4　Cadaverine - based siderophores.

the production of cadaverine by direct enzyme catalysis, due to product inhibition in lysine decarboxylase, which loses 50% activity at 3 g L^{-1} cadaverine [61]. In one recent exception, an attempt was made to catalyze L – lysine decarboxylation using glutaraldehyde – cross – linked lysine decarboxylase; however, the immobilization yield and activity retention were only 30.5% and 8.04%, respectively [62]. Thus, it is still impractical to produce cadaverine by direct enzyme catalysis, whether using a free or an immobilized enzyme. On the other hand, great progress has been achieved in cadaverine production by direct fermentation of metabolic engineering strains or by whole – cell bioconversion from L – lysine.

3.1 Metabolic engineering of cadaverine production

In fermentative cadaverine production, the most commonly used producer strain is engineered from the conventional L – lysine producer*C. glutamicum*, which was first isolated by Kinoshita et al. in 1957 [63], and which has been continuously optimized for L – lysine production for up to 50 years [7]. Especially in recent years, metabolic engineering has strongly contributed to the performance of *C. glutamicum* [64-69], which is able to produce 120 g L^{-1} of L – lysine with a productivity of 4 g L^{-1} h^{-1} and a yield of 0.55 g g^{-1} glucose in fed – batch culture within 30 h [7,70]. The annual global production of L – lysine is currently approximately 2 million tons, using a mutant strain of *C. glutamicum*[71].

Because of its remarkable performance in L – lysine production, *C. glutamicum* was one of the first species engineered for fermentative cadaverine production. In a first proof of principle, Mimitsuka et al. constructed an L – homoserine auxotrophic *C. glutamicum*, in which the L – homoserine dehydrogenase gene (*hom*) was replaced by the lysine decarboxylase gene (*cadA*) of *E. coli*; the resulting strain produced 2.6 g L^{-1} of cadaverine after 18 h fermentation [72]. However, a concentration of 2.3 g L^{-1} L – lysine remained in the fermentation broth, which was presumed to be caused by the cadaverine excretion deficiency in *C. glutamicum*. Hence, lysine decarboxylase activity was inhibited by high concentrations of intracellular cadaverine. This hypothesis was confirmed by Li et al., who constructed an engineered *C. glutamicum* by co – expression of the *E. coli* cadaverine/lysine antiporter CadB and the *Hafnia alvei* lysine decarboxylase (LDC). Results showed that the cadaverine secretion rate increased by 22% and the yield of total cadaverine and extracellular cadaverine increased by 30% and 73%, respectively [73]. A similar study on modulation of cell permeability was carried out by Matsushima et al. through the addition of Tween 40 at the mid – exponential phase, causing a 1.5 – fold increase in cadaverine yield compared with cultivation without Tween

40 addition [74].

The real breakthrough in cadaverine production using *C. glutamicum* was accomplished by Christoph Wittmann and colleagues. Using transcriptome analysis, they identified a major facilitator superfamily permease, cg2893, which is a cadaverine exporter in *C. glutamicum*, and subsequently examined the effect of cg2893 overexpression on cadaverine production. Their results showed a 20% increase in product yield, with 40% enhancement in specific production rate [61].

The cadaverine producer *C. glutamicum* DAP – 16 was then engineered from the L – lysine hyper – producer *C. glutamicum* LYS – 12 [70] using the following process: genome – based expression of a codon – optimized variant of *E. coli* LdcC; deletion of N – acetyltransferase NCgl1469 and lysine exporter LysE; and overexpression of permease cg2893 by replacement of the native promoter by the *sod* promoter. Fed – batch fermentation of DAP – 16 resulted in the production of 88 g L^{-1} of cadaverine in 50 h, with a productivity of 2.2 g L^{-1} h^{-1} and a yield of 0.29 g(cadaverine)(g glucose)$^{-1}$ [75]. Buschke et al. also contributed to improvements in cadaverine production. They targeted the genes that limited cadaverine production performance from five – carbon sugars using a systems – wide comparison of the fluxome and transcriptome on glucose and xylose as the sole carbon sources in the cadaverine producer *C. glutamicum* DAP – Xyl1. Next, they constructed the cadaverine producer *C. glutamicum* DAP – Xyl2 with the following characteristics: Fructose bisphosphatase (Fbp) and the entire *tkt* operon were overexpressed; isocitrate dehydrogenase (Icd) was attenuated; L – lysine exporter (lysE) and N – acetyl transferase (cg1722) were knocked out; and the xylose isomerase (XylA) and xylulokinase (XylB) from *E. coli* were introduced. As a result, *C. glutamicum* DAP – Xyl2 achieved 103 g L^{-1} of cadaverine from xylose with a product yield of 32%, in a fed – batch process [76].

In addition to *C. glutamicum*, *E. coli* is a competitive L – lysine producer and has been engineered for cadaverine production. *E. coli* and *C. glutamicum* have a similar biosynthesis pathway and metabolic regulation of L – lysine, so *E. coli* has been systematically engineered and fermented accordingly to produce L – lysine [77 – 80], resulting in the production of 136 g L^{-1} of L – lysine with a productivity of 2.8 g L^{-1} h^{-1} and a yield of 0.56 g(lysine)(g glucose)$^{-1}$ [81]. However, the use of *E. coli* for cadaverine production has not been studied extensively; the highest titer of cadaverine produced by metabolic engineered *E. coli* is only 9.61 g L^{-1} thus far, with a productivity of 0.32 g

L^{-1} h^{-1} and a yield of 0.12 g(lysine)(g glucose)$^{-1}$[10]. There are two possible reasons for the low bioconversion rate of L – lysine to cadaverine in *E. coli*. The first is that *E. coli* exhibits low tolerance to cadaverine, which inhibits cell growth and causes cell lysis after 8 h exposure to 0.3 – 0.5 M cadaverine [10]; the other is that cadaverine regulates membrane permeability of *E. coli*. Cadaverine has been shown to interact with the bacterial porins OmpC and OmpF, inducing porin closure in a concentration – dependent manner [82,83], and the extent of this effect is greater from the periplasmic side than from the extracellular side [84]. As a result, the uptake of nutrients and the efflux of deleterious metabolites are decreased, which halts the bioconversion of L – lysine to cadaverine. Since *E. coli* possesses considerable capability to produce high amounts of L – lysine and has a competitive advantage in L – lysine decarboxylase activity [12], it is the most promising species for fermentative cadaverine production and is not being used to its full potential.

Several studies have been carried out to extend the substrate range for cadaverine production. Tateno et al. developed an approach to produce cadaverine from soluble starch using engineered *C. glutamicum* overexpressing α – amylase (AmyA) from *Streptococcus bovis* 148 and CadA from *E. coli*, and the cadaverine titer reached 2.24 g L^{-1} after 21 h fermentation [85]. The feasibility of using methanol as a carbon source for cadaverine fermentation was also evaluated in *Bacillus methanolicus* overexpressing *E. coli* CadA, which led to a volume – corrected concentration of 11.3 g L^{-1} of cadaverine by high – cell density fed – batch fermentation [86].

3.2 Cadaverine production by whole – cell biocatalysis

Whole – cell bioconversion is a promising method due to its convenience and high efficiency, as well as its high product concentration and purity, which reduce the cost of product separation. In addition, L – lysine is economically viable as a raw material for cadaverine production, because there is a thriving, million – ton industry for its production [8].

In comparison with fermentative cadaverine, which is mainly produced by *C. glutamicum*, whole – cell bioprocesses are dominated by *E. coli*. The use of whole – cell biocatalysis for cadaverine production was initially reported in a patent by Nishi et al. Nishi et al. reported obtaining 69 g L^{-1} cadaverine after 6 h whole – cell bioconversion using the *E. coli* overexpressed CadA, with a productivity of 11.5 g L^{-1} h^{-1}[87].

In whole – cell bioprocesses, limited permeability is a widespread problem: The natural barrier functions of the cell wall and cell membrane limit the transport of starting material into the cell, as well as the release of products accumulated in the cell [88,89]. Thus, an efficient substrate and product transport system is important in whole - cell bioconversions. There are several ways to reduce the permeability barrier, including chemical (e. g., by adding detergents or solvents) or physical (e. g., temperature shock) treatment [90]. However, these treatments may damage the biocatalysts and pose further difficulty for downstream processes [88]. In comparison, modulating cell permeability through genetic approaches is more promising. For cadaverine transportation in *E. coli*, there is a lysine/cadaverine antiporter (CadB) that facilitates cadaverine excretion and coordinates cadaverine excretion with lysine uptake, which is driven solely by concentration gradients without other energy inputs [29,91]. The effect of CadB overexpression on biocatalyst activity was first investigated by Ma et al. Their results showed that co - overexpression of CadA and CadB with N - terminal fused *pel*B signal peptide from the pectate lyases of *Erwinia carotovora* was effective in reducing the biocatalyst permeability barrier. The resulting strain, BL - DAB, produced cadaverine with a titer of 221 ± 6 g L^{-1} and a molar yield of 92% [11].

Pyridoxal 5′ - phosphate (PLP), the cofactor of lysine decarboxylase, is critical for whole - cell activity. The activity of lysine decarboxylase is positively correlated to the amount of PLP, and the maximum activity occurred at a ratio of 1.0 - 1.2 mol of PLP per mol of protein [24]. Kim et al. reported that when no PLP was added in whole - cell bioconversion, only 20% of the 1 M L - lysine was decarboxylated to cadaverine, whereas the conversion rate reached 80% with the addition of 0.025 mM PLP [92]. PLP can also be self - supplied by enhancing intracellular PLP synthesis in biocatalysis. For example, Ma et al. constructed the whole - cell biocatalyst *E. coli* AST3, in which PLP synthesis was enhanced by introducing the *de novo* PLP synthesis pathway (R5P pathway) of *Bacillus subtilis* into a CadA/CadB overexpressing strain, BL - DAB. This *E. coli* biocatalyst was capable of accumulating intracellular PLP at a concentration of 1051 nmol $g(DCW)^{-1}$ after optimizing the co - expression of proteins and culture conditions. As a result, cadaverine productivity reached 28 g $g(DCW)^{-1}$ h^{-1} without PLP addition, which was 1.2 - fold and 2.9 - fold higher than that of BL - DAB with and without PLP supplementation, respectively [93].

The effect of substrate L - lysine concentration on biocatalyst activity was also investigated. Oh et al. reported that in a bioconversion system with cells concentrated to

OD600 of 10, an increase in the initial L – lysine concentration of up to 150 g L^{-1} resulted in significant decreases in the activity of whole – cell biocatalysts [12]. Fortunately, substrate inhibition can be relieved by optimizing L – lysine feeding strategies [11].

Another research focus is to maintain biocatalyst activity, which is significantly reduced with the increase of cadaverine concentration during the whole – cell bioconversion process. Oh et al. reported that an initial rate of L – lysine consumption of 40.42 mM h^{-1} could be reached, whereas the average reaction ratc was only 10 mM h^{-1}, indicating a significant decrease in biocatalyst activity [12]. Cell lysis may be one of the reasons for this decrease; therefore, the application of immobilized cells for cadaverine production has been investigated recently. Bhatia et al. examined the deactivation energy (E_d) of free and immobilized cells during cadaverine production. Their results showed that the E_d of immobilized cells was almost 10% higher than that of free cells: 66.7 kJ mol^{-1} and 56.9 kJ mol^{-1}, respectively. The immobilized cells retained 56% activity after 18 reaction cycles, whereas the free cells lost catalytic activity completely after 10 reaction cycles [94]. These results indicate that the application of immobilization techniques was effective for maintaining biocatalyst activity.

In addition to inducible lysine decarboxylase CadA from *E. coli*, the use of constitutive lysine decarboxylase LdcC in the construction of whole – cell biocatalysts was also reported. The resulting strain produced 133.7 g L^{-1} cadaverine in 120 h from a fermentation broth containing 192.6 g L^{-1} L – lysine, with a yield of 99.90%; however, the productivity was only 4.1 g L^{-1} h^{-1} [12]. Li et al. investigated the biotransformation of cadaverine using a codon – optimized lysine decarboxylase from *Klebsiella oxytoca*, which was introduced into *E. coli* MG1655 and overexpressed under the control of a P_{tac} promoter. The resulting strain, LN24, converted L – lysine to cadaverine at a conversion rate of 0.133%/min/g with a yield of 92% [95]. Moreover, the lysine decarboxylase from *H. alvei* AS1.1009 was mutated by error – prone polymerase chain reaction (PCR) and DNA shuffling, and the mutant LDC^{E583G} (which showed 1.48 – fold improved activity in comparison with the wild type) was introduced into *E. coli* JM109 in order to construct a whole – cell biocatalyst for cadaverine production. The resulting biocatalyst produced 63.9 g L^{-1} cadaverine with a conversion yield of 93.4% during 5 h whole – cell bioconversion [96].

3.3 Separation and purification of cadaverine

Cadaverine separation and purification methods are developed based on experience in other diamines. Solvent extraction is the most popularmethod due to low energy consumption, high selectivity, and separation efficiency, as well as simple equipment and operation.

To identify a suitable solvent for extracting cadaverine from fermentation broth, the distribution coefficients for cadaverine in n – butanol, 2 – butanol, 2 – octanol, and cyclohexanol were compared, with the results showing that *n* – butanol was better for cadaverine extraction [75]. Meanwhile, the cadaverine distribution coefficients in 4 – nonylphenol, di – 2 – ethylhexyl phosphoric acid (D2EHPA), Versatic acid 1019, di – nonyl – naphthalene – sulfonic acid (DNNSA), and 4 – octylbenzaldehyde were also investigated. The results showed that 4 – nonylphenol is a suitable extractant, and over 90% of cadaverine can be extracted from fermentation broth after a one – step extraction and back – extraction [97].

After phase separation, cadaverine can be recovered from the cadaverine – containing phase by chromatography, distillation, or precipitation as a salt with suitable acids[9,98]. Of these measures, distillation is a promising method for cadaverine purification on an industrial scale; however, the specific conditions for distillation are dependent on practical factors such as the concentration of cadaverine in the extract phase and the properties of the extractant.

To avoid cadaverine decomposition during distillation at high heating temperatures, a purification process using high boiling point solvents was proposed in a recent patent application[99]. This method involves adding high boiling point solvents, such as C_{14} alkanes or a C_{11}/C_{14} alkane mixture, to the distillation system in order to improve fluidity during reduced – pressure distillation. As a result, nearly 100% of cadaverine can be recovered at the lower heating temperature of 130 ℃, which is 50 ℃ lower than the boiling point of cadaverine.

Although many advances have been made in cadaverine separation and purification, successful commercial application has not yet been achieved. Purity higher than 99.5% is a basic prerequisite for monomers to polymerize effectively; thus, developing the separation and purification technology of bio – based cadaverine is crucial in order to meet the needs of industrial production.

4 APPLICATION OF CADAVERINE IN BIO - BASED POLYAMIDES

4.1 Research progress of bio - based polyamides

Research on bio - based polyamides (nylon) began in the 1940s and is becoming an important focus area in recent years due to economic and environmental factors. According to statistics from the European Bioplastics association, the global production of bio - based plastics was about 1.7 million tons in 2014 and will reach 7.8 million tons by 2019 [100].

At present, the commercialized bio - based polyamides include completely bio - based polyamides (e. g., PA 11 and PA 1010) and partially bio - based polyamides (e. g., PA 610, PA 1012, PA 410, PA 10T, etc.); the classification and producers of each of these are listed in Table 1 [101 - 105]. The main bio - based monomers used are sebacic acid, decan - 1,10 - diamine, and ω - amino - undecanoic acid, which are all prepared from castor oil as shown in Fig. 5 [106 - 108].

Table 1 Classification and producers of commercialized bio - based polyamides.

Classification	Monomer	Materials	Bio - based content (%)	Producers
PA 11 [101]	ω - amino - undecanoic acid	castor oil	100	Arkema, Suzhou HiproPolymers, SABIC
PA 1010 [102]	sebacic acid	castor oil	100	Arkema, DuPont, Evonik
	decan - 1,10 - diamine			
PA 610 [103]	sebacic acid	castor oil	63.5	SABIC, Toray, DuPont, BASF
	hexamethylenediamine	butadiene		
PA 1012 [104]	dodecanedioic acid	alkane	42.9	Arkema, Evonik
	decan - 1,10 - diamine	castor oil		
PA 410 [105]	sebacic acid	castor oil	69.7	DSM
	putrescine	acrylonitrile		
PA 10T [102]	terephthalic acid	benzene	51.8	Evonik, Kingfa
	decan - 1,10 - diamine	castor oil		

Apart from the commercialized bio - based polyamides discussed above, some new types of fully bio - based polyamides are being extensively researched, such as bio - based PA 46, PA 6, and PA 5X. The fully bio - based PA 46 is polymerized from bio - based adipic acid and putrescine. Several synthesis methods have been developed using

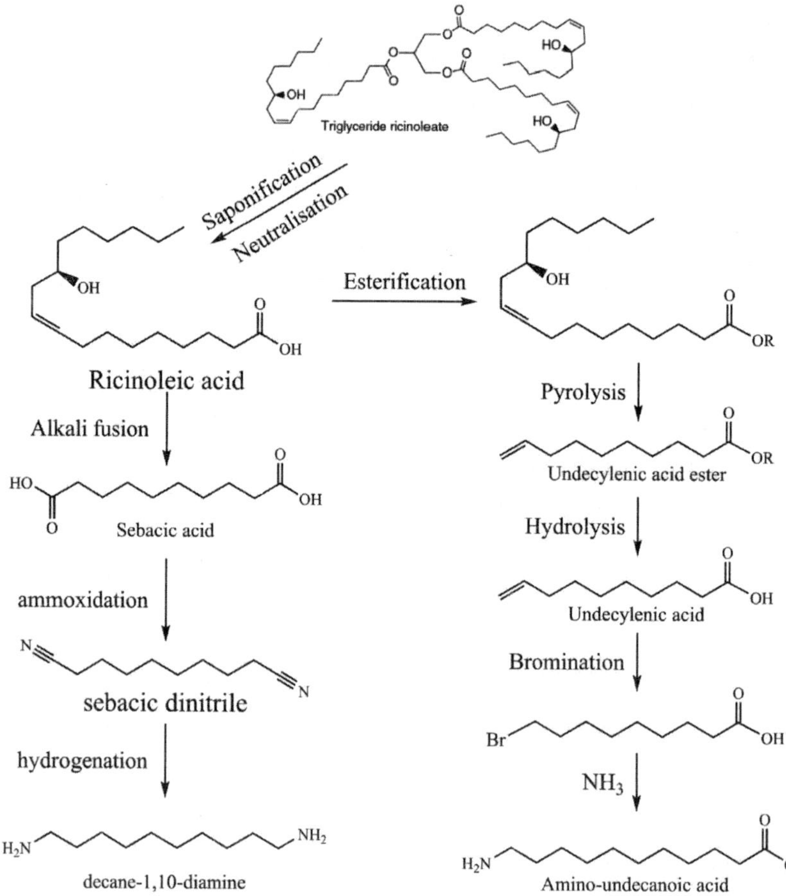

Fig. 5 Transformation of castor oil to sebacic acid, decan – 1, 10 – diamine, and ω – amino – undecanoic acid.

different renewable materials to produce bio – based adipic acid. For example, Draths et al. [109] and van Duuren et al. [110] developed a fermentation approach and an enzymatic catalytic approach, respectively, to synthesize muconic acid from D – glucose, followed by the hydrogenation of muconic acid to adipic acid. Lange et al. [111] reported a method to synthesize levulinic acid from cellulose by acid hydrolyzation; the levulinic acid is then hydrogenated and dehydrated to generate γ – valerolactone, followed by ester exchange, an addition reaction, and a hydrolysis reaction to obtain adipic acid. Boussie et al. [112] invented a method to produce adipic acid via the hydrogenation of glucaric acid, which can be obtained by the catalytic oxidation of glucose. Research on putrescine biosynthesis is also in progress. Schneider et al. have studied putrescine production by the fermentation of glucose using genetically engineered *C. glutamicum*, which achieved a putrescine titer of 19 g L^{-1} at a volumetric productivity of 0.55 g $L^{-1}h^{-1}$ and a yield of 0.16 g g^{-1}[113]. Qian et al. reported a titer of 24.2 g L^{-1} putrescine with a volumetric productivity of 0.75 g L^{-1} h^{-1}, which was obtained by fermentation using engineered *E. coli*[114].

Research on production of the PA 6 monomer caprolactam has also made some progress. Caprolactam can be obtained from L – lysineby catalytic conversion [115] or from γ – valerolactone, which is a component of cellulosic hydrolysate, by acid catalysis [116]. In addition, the production of caprolactam from adipic acid, adiponitrile, 1,3 – butadiene, 6 – aminocaproic acid, adipamide, and muconic acid has been reported [117].

4.2 Research status of bio – based polyamide 5X

In recent years, many studies have been carried out to investigate polyamide synthesis from bio – based cadaverine, instead of from petroleum – based hexanediamine. Usingcadaverine polymerization with appropriate bio – blocks, such as succinate, adipic acid, or sebacic acid, results in the completely bio – based polyamide family PA 5X, such as PA 54, PA 56, and PA 510, respectively. Compared with PA 66, which is obtained from non – renewable fossil fuels, cadaverine – based polyamides also exhibit excellent tensile strength, high melting points, and resistance to organic solvents; thus, they can be used to replace PA 66 in many applications. Table 2 compares several material properties of PA 6, PA 66, PA 56, and PA 510 [75,105,118].

Table 2 Material properties of PA 6, PA 66, PA 56, and PA 510.

	PA6	PA 66	PA 56	PA 510
Melting point(℃)	220	260	253	215
Glass transition temperature(℃)	54	60	55	50
Density($g\ cm^{-3}$)	1.14	1.14	1.14	1.07
Water absorption(%)	3.0	2.8	3.3	1.8

PA 56 has a competitive advantage in the preparation of textile fibers due to its high water absorption and a low glass transition temperature. The high water absorption of PA 56 endows fibers with excellent moisture – wicking performance, which keeps the people wearing the textiles comfortable and prevents static electricity. A low glass transition temperature is conducive to material performance of low temperature resistance, so clothes made from these textiles will not be brittle in high - altitude areas. Moreover, PA 56 has outstanding strength and toughness along with excellent resistance to abrasion, which makes clothes made from such textiles more durable [118].

PA 510, which is synthesized from cadaverine and sebacic acid, possesses the properties of low water absorption, outstanding impact strength at low temperatures, and dimensional stability [75]; thus, it can be applied in fields where conventional PAs are not suitable. PA 510 has wide application prospects in machinery, electronic appliances, radio technology, and other industrial areas and daily necessities, and can be used to manufacture industrial parts, friction - reducing materials, and electrical insulation. In addition, PA 510 is highly suitable for automobiles, aircrafts, and other vehicles because of its low density, which can reduce vehicle weight, resulting in energy saving and environmental protection.

However, no high - purity cadaverine is produced on an industrial scale as yet, because high - performance PA 5X has not yet been commercialized. Studies of other completely bio - based nylons based on cadaverine are still in the early stages [119], but cadaverine - based PA 5X has broad application prospects due to its exceptional performance and environment - friendly characteristics.

5 OUTLOOK AND SUGGESTIONS FOR FUTURE DEVELOPMENT

Studies on the biosynthesis of cadaverine are still in the initial stages, and many factors restrict its large - scale industrial production. The toxic effect of cadaverine on bacterial strains is one of the factors restricting industrialization. Further studies on the mechanisms of cellular damage are needed to provide a basis for the specific engineering of production strains and the optimization of fermentation conditions. In whole - cell catalysis, substance transport in cells and the mechanisms of the catalytic reaction require further study. In addition, from the perspective of fermentation economics, the release of a molecule of carbon dioxide (CO_2) during the process of L - lysine decarboxylation imbalances the carbon cycle and consequently leads to an increase in production costs. Therefore, CO_2 recycling is a key issue in reducing the cost of cadaverine production by whole - cell bioconversion.

The production cost of cadaverine can also be reduced by screening and engineering lysine decarboxylase. At present, the commonly used lysine decarboxylase is the protein CadA from *E. coli*. Although this enzyme has higher decarboxylase activity than other lysine decarboxylases, it suffers from serious substrate and product inhibition. In addition, CadA can only maintain catalytic activity under mild acid conditions (pH 5 – 6.8), which restricts the production of high concentrations of alkaline cadaverine. To meet the requirement of cadaverine separation by solvent extraction, alkalinity (pH > 12) of the fermentation broth (or transformation solution) is needed, while a mild acid condition is needed during production. This apparent contradiction results in excessive consumption of acids and bases during the whole process, which not only increases production cost, but also restricts cleaner production. Therefore, screening novel lysine decarboxylases with alkali resistance or engineering lysine decarboxylases using molecular and synthetic biology approaches, such as site – directed mutagenesis and directed evolution, can be done to reduce substrate and product inhibition and enhance acid/base stability. These steps could greatly enhance cadaverine yield and productivity, and reduce acid – base consumption and production costs.

Acknowledgements

This work was supported by the National Key Research and Development program (2016YFA0204300); the National Nature Science Foundation of China (grant no. 21390200, 31440024); the Project of Science and Technology Department of Gansu Province, China (1304FKCE106); and the Project Sponsored by the Scientific Research Foundation for the Returned Overseas Chinese Scholars (2014), the Ministry of Human Resources and Social Security of the People's Republic of China.

Compliance with ethics guidelines

Weichao Ma, Kequan Chen, Yan Li, Ning Hao, Xin Wang, and Pingkai Ouyang declare that they have no conflict of interest or financial conflicts to disclose.

REFERENCES

[1] Moreau PL. The lysine decarboxylase CadA protects *Escherichia coli* starved of phosphate against fermentation acids. J Bacteriol 2007;189(6):2249 – 61. PMID:17209032doi:10.1128/JB.01306 – 06.

[2] Samartzidou H, Mehrazin M, Xu Z, Benedik MJ, Delcour AH. Cadaverine inhibition of porin plays a role in cell survival at acidic pH. J Bacteriol 2003;185(1):13 – 9. PMID:12486035doi:10.1128/JB.185.1.13 – 19.2003.

[3]Jancewicz AL,Gibbs NM,Masson PH. Cadaverine's functional role in plant development and environmental response. Front Plant Sci 2016;7:870. PMID:27446107doi:10. 3389/fpls. 2016. 00870.

[4]Andersson AC,Henningsson S,Rosengren E. Formation of cadaverine in the pregnant rat. Acta Physiol Scand 1979;105(4):508 – 12. PMID:110032doi:10. 1111/j. 1748 – 1716. 1979. tb00115. x.

[5]Wang Q,Wang Y,Liu R,Yan X,Li Y,Fu H,et al. Comparison of the effects of Mylabris and Acanthopanax senticosus on promising cancer marker polyamines in plasma of a Hepatoma – 22 mouse model using HPLC – ESI – MS. Biomed Chromatogr 2013;27(2):208 – 15. PMID: 22763853doi:10. 1002/bmc. 2777.

[6]Miller – Fleming L,Olin – Sandoval V,Campbell K,Ralser M. Remaining mysteries of molecular biology:The role of polyamines in the cell. J Mol Biol 2015;427(21):3389 – 406. PMID: 26156863doi:10. 1016/j. jmb. 2015. 06. 020.

[7]Becker J,Wittmann C. Bio – based production of chemicals,materials and fuels—Corynebacterium glutamicum as versatile cell factory. Curr Opin Biotechnol 2012;23(4):631 – 40. PMID: 22138494doi:10. 1016/j. copbio. 2011. 11. 012.

[8]Schneider J,Wendisch VF. Biotechnological production of polyamines by bacteria:Recent achievements and future perspectives. Appl Microbiol Biotechnol 2011;91(1):17 – 30. PMID: 21552989doi:10. 1007/s00253 – 011 – 3252 – 0.

[9]Kind S,Wittmann C. Bio – based production of the platform chemical 1,5 – diaminopentane. Appl Microbiol Biotechnol 2011;91(5):1287 – 96. PMID:21761208doi:10. 1007/s00253 – 011 – 3457 – 2.

[10]Qian ZG,Xia XX,Lee SY. Metabolic engineering of *Escherichia coli* for the production of cadaverine: A five carbon diamine. Biotechnol Bioeng 2011; 108 (1): 93 – 103. PMID: 20812259doi:10. 1002/bit. 22918.

[11]Ma W,Cao W,Zhang H,Chen K,Li Y,Ouyang P. Enhanced cadaverine production from L – lysine using recombinant *Escherichia coli* co – overexpressing CadA and CadB. Biotechnol Lett 2015;37(4):799 – 806. PMID:25515797doi:10. 1007/s10529 – 014 – 1753 – 5.

[12]Oh YH,Kang KH,Kwon MJ,Choi JW,Joo JC,Lee SH,et al. Development of engineered *Escherichia coli* whole – cell biocatalysts for high – level conversion of L – lysine into cadaverine. J Ind Microbiol Biotechnol 2015;42(11):1481 – 91. PMID:26364199doi:10. 1007/s10295 – 015 – 1678 – 6.

[13]Velasco AM,Leguina JI,Lazcano A. Molecular evolution of the lysine biosynthetic pathways. J Mol Evol 2002;55(4):445 – 59. PMID:12355264doi:10. 1007/s00239 – 002 – 2340 – 2.

[14]Bakhiet N, Forney FW, Stahly DP, Daniels L. Lysine biosynthesis in Methanobacterium thermoautotrophicum is by the diaminopimelic acid pathway. Curr Microbio 1984;10(4):195 – 8.

doi:10. 1007/BF01627254.

[15] Weinberger S, Gilvarg C. Bacterial distribution of the use of succinyl and acetyl blocking groups in diaminopimelic acid biosynthesis. J Bacteriol 1970;101(1):323 – 4. PMID:5411754.

[16] Misono H, Togawa H, Yamamoto T, Soda K. Meso – α, ε – diaminopimelate D – dehydrogenase: Distribution and the reaction product. J Bacteriol 1979;137(1):22 – 7. PMID:762012.

[17] White PJ. The essential role of diaminopimelate dehydrogenase in the biosynthesis of lysine by Bacillus sphaericus. Microbiology 1983;129:739 – 49. doi:10. 1099/00221287 – 129 – 3 – 739.

[18] Scapin G, Blanchard JS. Enzymology of bacterial lysine biosynthesis. In: Purich DL, editor Advances in enzymology and related areas of molecular biology. New Jersey: John Wiley & Sons, Inc.;2006. p. 279 – 324.

[19] LéJohn HB. Enzyme regulation, lysine pathways and cell wall structures as indicators of major lines of evolution in fungi. Nature 1971; 231 (5299): 164 – 8. PMID: 4397004doi: 10. 1038/231164a0.

[20] Azevedo RA, Lea PJ. Lysine metabolism in higher plants. Amino Acid 2001;20(3):261 – 79. PMID:11354603doi:10. 1007/s007260170043.

[21] Bult CJ, White O, Olsen GJ, Zhou L, Fleischmann RD, Sutton GG, et al. Complete genome sequence of the methanogenic archaeon, Methanococcus jannaschii. Science 1996;273(5278):1058 – 73. PMID:8688087doi:10. 1126/science. 273. 5278. 1058.

[22] Sundharadas G, Gilvarg C. Biosynthesis of α, ε – diaminopimelic acid in Bacillus megaterium. J Biol Chem 1967;242(17):3983 – 4. PMID:4962540.

[23] Kikuchi Y, Kojima H, Tanaka T, Takatsuka Y, Kamio Y. Characterization of a second lysine decarboxylase isolated from *Escherichia coli*. J Bacteriol 1997;179(14):4486 – 92. PMID: 9226257doi:10. 1128/jb. 179. 14. 4486 – 4492. 1997.

[24] Sabo DL, Fischer EH. Chemical properties of *Escherichia coli* lysine decarboxylase including a segment of its pyridoxal 5′ – phosphate binding site. Biochemistry 1974;13(4):670 – 6. PMID:4204273doi:10. 1021/bi00701a006.

[25] Lemonnier M, Lane D. Expression of the second lysine decarboxylase gene of *Escherichia coli*. Microbiology 1998;144(Pt 3):751 – 60. PMID:9534244doi:10. 1099/00221287 – 144 – 3 – 751.

[26] Yamamoto Y, Miwa Y, Miyoshi K, Furuyama J, Ohmori H. The *Escherichia coli* ldcC gene encodes another lysine decarboxylase, probably a constitutive enzyme. Genes Genet Syst 1997;72(3):167 – 72. PMID:9339543doi:10. 1266/ggs. 72. 167.

[27] Kanjee U, Houry WA. Mechanisms of acid resistance in *Escherichia coli*. Annu Rev Microbiol 2013;67:65 – 81. PMID:23701194doi:10. 1146/annurev – micro – 092412 – 155708.

[28] Neely MN, Olson ER. Kinetics of expression of the *Escherichia coli* cad operon as a func-

tion of pH and lysine. J Bacteriol 1996;178(18):5522 –8. PMID:8808945doi:10. 1128/jb. 178. 18.5522 –5528. 1996.

[29]Meng SY, Bennett GN. Nucleotide sequence of the *Escherichia coli* cad operon: A system for neutralization of low extracellular pH. J Bacteriol 1992;174(8):2659 –69. PMID:1556085doi: 10. 1128/jb. 174. 8. 2659 –2669. 1992.

[30]Rauschmeier M, Schüppel V, Tetsch L, Jung K. New insights into the interplay between the lysine transporter LysP and the pH sensor CadC in *Escherichia coli*. J Mol Biol 2014;426(1): 215 –29. PMID:24056175doi:10. 1016/j. jmb. 2013. 09. 017.

[31]Kuper C, Jung K. CadC – mediated activation of the cadBA promoter in *Escherichia coli*. J Mol Microbiol Biotechnol 2005;10(1):26 –39. PMID:16491024doi:10. 1159/000090346.

[32]Haneburger I, Eichinger A, Skerra A, Jung K. New insights into the signaling mechanism of the pH – responsive, membrane – integrated transcriptional activator CadC of *Escherichia coli*. J Biol Chem 2011;286(12):10681 –9. PMID:21216950doi:10. 1074/jbc. M110. 196923.

[33]Popkin PS, Maas WK. *Escherichia coli* regulatory mutation affecting lysine transport and lysine decarboxylase. J Bacteriol 1980;141(2):485 –92. PMID:6767681.

[34]Steffes C, Ellis J, Wu J, Rosen BP. The lysP gene encodes the lysine – specific permease. J Bacteriol 1992;174(10):3242 –9. PMID:1315732doi:10. 1128/jb. 174. 10. 3242 –3249. 1992.

[35]Shi X, Waasdorp BC, Bennett GN. Modulation of acid – induced amino acid decarboxylase gene expression by hns in *Escherichia coli*. J Bacteriol 1993; 175(4): 1182 – 6. PMID: 8381784doi:10. 1128/jb. 175. 4. 1182 –1186. 1993.

[36]Krin E, Danchin A, Soutourina O. Decrypting the H – NS – dependent regulatory cascade of acid stress resistance in *Escherichia coli*. BMC Microbiol 2010;10:273. PMID:21034467doi:10. 1186/1471 –2180 –10 –273.

[37]Fritz G, Koller C, Burdack K, Tetsch L, Haneburger I, Jung K, et al. Induction kinetics of a conditional pH stress response system in *Escherichia coli*. J Mol Biol 2009;393(2):272 – 86. PMID:19703467doi:10. 1016/j. jmb. 2009. 08. 037.

[38]Neely MN, Dell CL, Olson ER. Roles of LysP and CadC in mediating the lysine requirement for acid induction of the *Escherichia coli cad* operon. J Mol Biol 1994;176(11):3278 –85. PMID:8195083doi:10. 1128/jb. 176. 11. 3278 –3285. 1994.

[39]Haneburger I, Fritz G, Jurkschat N, Tetsch L, Eichinger A, Skerra A, et al. Deactivation of the *E. coli* pH stress sensor CadC by cadaverine. J Mol Biol 2012;424(1 –2):15 –27. PMID: 22999955doi:10. 1016/j. jmb. 2012. 08. 023.

[40]Kanjee U, Gutsche I, Alexopoulos E, Zhao B, El Bakkouri M, Thibault G, et al. Linkage between the bacterial acid stress and stringent responses: The structure of the inducible lysine decarboxylase. EMBO J 2011;30(5):931 –44. PMID:21278708doi:10. 1038/emboj. 2011. 5.

[41]Romano A, Trip H, Lolkema JS, Lucas PM. Three – component lysine/ornithine decarbox-

ylation system in *Lactobacillus saerimneri* 30a. J Bacteriol 2013; 195 (6): 1249 – 54. PMID: 23316036doi:10.1128/JB.02070 – 12.

[42] Kind S, Jeong WK, Schröder H, Zelder O, Wittmann C. Identification and elimination of the competing N – acetyldiaminopentane pathway for improved production of diaminopentane by *Corynebacterium glutamicum*. Appl Environ Microbiol 2010; 76 (15): 5175 – 80. PMID: 20562290doi:10.1128/AEM.00834 – 10.

[43] Cacciapuoti G, Porcelli M, Moretti MA, Sorrentino F, Concilio L, Zappia V, et al. The first agmatine/cadaverine aminopropyl transferase: Biochemical and structural characterization of an enzyme involved in polyamine biosynthesis in the hyperthermophilic archaeon *Pyrococcus furiosus*. J Bacteriol 2007;189(16):6057 – 67. PMID:17545282doi:10.1128/JB.00151 – 07.

[44] Revelles O, Espinosa – Urgel M, Fuhrer T, Sauer U, Ramos JL. Multiple and interconnected pathways for L – lysine catabolism in *Pseudomonas putida* KT2440. J Bacteriol 2005;187(21): 7500 – 10. PMID:16237033doi:10.1128/JB.187.21.7500 – 7510.2005.

[45] Kojima S, Kamio Y. Molecular basis for the maintenance of envelope integrity in *Selenomonas ruminantium*: Cadaverine biosynthesis and covalent modification into the peptidoglycan play a major role. J Nutr Sci Vitaminol (Tokyo) 2012;58(3):153 – 60. PMID:22878384doi:10.3177/jnsv.58.153.

[46] Kamio Y, Itoh Y, Terawaki Y, Kusano T. Cadaverine is covalently linked to peptidoglycan in *Selenomonas ruminantium*. J Bacteriol 1981;145(1):122 – 8. PMID:7462141.

[47] Kamio Y, Itoh Y, Terawaki Y. Chemical structure of peptidoglycan in *Selenomonas ruminantium*: Cadaverine links covalently to the D – glutamic acid residue of peptidoglycan. J Bacteriol 1981;146(1):49 – 53. PMID:6783621.

[48] Kamio Y. Structural specificity of diamines covalently linked to peptidoglycan for cell growth of Veillonella alcalescens and *Selenomonas ruminantium*. J Bacteriol 1987;169(10):4837 – 40. PMID:3654585doi:10.1128/jb.169.10.4837 – 4840.1987.

[49] Kamio Y, Nakamura K. Putrescine and cadaverine are constituents of peptidoglycan in *Veillonella alcalescens* and *Veillonella parvula*. J Bacteriol. 1987; 169 (6): 2881 – 4. PMID: 3584075doi:10.1128/jb.169.6.2881 – 2884.1987.

[50] Hirao T, Sato M, Shirahata A, Kamio Y. Covalent linkage of polyamines to peptidoglycan in *Anaerovibrio lipolytica*. J Bacteriol 2000;182(4):1154 – 7. PMID:10648544doi:10.1128/JB.182.4.1154 – 1157.2000.

[51] Burrell M, Hanfrey CC, Kinch LN, Elliott KA, Michael AJ. Evolution of a novel lysine decarboxylase in siderophore biosynthesis. Mol Microbiol 2012; 86 (2): 485 – 99. PMID: 22906379doi:10.1111/j.1365 – 2958.2012.08208.x.

[52] Soe CZ, Telfer TJ, Levina A, Lay PA, Codd R. Simultaneous biosynthesis of putrebactin, avaroferrin and bisucaberin by *Shewanella putrefaciens* and characterisation of complexes with iron

(Ⅲ),molybdenum(Ⅵ)or chromium(Ⅴ). J Inorg Biochem 2016;162:207 – 15.

[53]Schafft M,Diekmann H. [Cadaverine is an intermediate in the biosynthesis of arthrobactin and ferrioxamine E]. Arch Microbiol 1978;117(2):203 – 7. German. PMID:354550doi:10.1007/BF00402309.

[54]Fujita MJ, Nakano K, Sakai R. Bisucaberin B, a linear hydroxamate class siderophore from the marine bacterium *Tenacibaculum mesophilum*. Molecules 2013;18(4):3917 – 26. PMID:23549298doi:10.3390/molecules18043917.

[55]Kadi N,Song L,Challis GL. Bisucaberin biosynthesis:An adenylating domain of the BibC multi – enzyme catalyzes cyclodimerization of N – hydroxy – N – succinylcadaverine. Chem Commun (Camb)2008;(41):5119 – 21.

[56]Barona – Gómez F,Lautru S,Francou FX,Leblond P,Pernodet JL,Challis GL. Multiple biosynthetic and uptake systems mediate siderophore – dependent iron acquisition in *Streptomyces coelicolor* A3(2)and Streptomyces ambofaciens ATCC 23877. Microbiology 2006;152(Pt 11):3355 – 66. PMID:17074905doi:10.1099/mic.0.29161 – 0.

[57] Sidebottom AM, Karty JA, Carlson EE. Accurate mass MS/MS/MS analysis of siderophores ferrioxamine B and E1 by collision – induced dissociation electrospray mass spectrometry. J Am Soc Mass Spectrom 2015;26(11):1899 – 902. PMID:26323615doi:10.1007/s13361 – 015 – 1242 – 7.

[58]Dhungana S,White PS,Crumbliss AL. Crystal structure of ferrioxamine B:A comparative analysis and implications for molecular recognition. J Biol Inorg Chem 2001;6(8):810 – 8. PMID:11713688doi:10.1007/s007750100259.

[59]Meiwes J,Fiedler HP,Zähner H,Konetschny – Rapp S,Jung G. Production of desferrioxamine E and new analogues by directed fermentation and feeding fermentation. Appl Microbiol Biotechnol 1990;32(5):505 – 10. PMID:1367428doi:10.1007/BF00173718.

[60]Imbert M,Béchet M,Blondeau R. Comparison of the main siderophores produced by some species of *Streptomyces*. Curr Microbiol 1995;31(2):129 – 33. doi:10.1007/BF00294289.

[61]Kind S,Kreye S,Wittmann C. Metabolic engineering of cellular transport for overproduction of the platform chemical 1,5 – diaminopentane in *Corynebacterium glutamicum*. Metab Eng 2011;13(5):617 – 27. PMID:21821142doi:10.1016/j.ymben.2011.07.006.

[62]Park SH,Soetyono F,Kim HK. Cadaverine production by using cross – linked enzyme aggregate of *Escherichia coli* lysine decarboxylase. J Microbiol Biotechnol 2017;27(2):289 – 96. PMID:27780956doi:10.4014/jmb.1608.08033.

[63]Kinoshita S,Udaka S,Shimono M. Studies on the amino acid fermentation. Part I. Production of L – glutamic acid by various microorganisms. J Gen Appl Microbiol 2004;50(6):331 – 43. PMID:15965888doi:10.2323/jgam.3.193.

[64] de Graaf AA,Eggeling L,Sahm H. Metabolic engineering for L – lysine production by

Corynebacterium glutamicum. Adv Biochem Eng Biotechnol 2001;73:9 – 29. PMID:11816814doi: 10.1007/3 – 540 – 45300 – 8_2.

[65] Becker J, Klopprogge C, Herold A, Zelder O, Bolten CJ, Wittmann C. Metabolic flux engineering of L – lysine production in *Corynebacterium glutamicum*—Over expression and modification of G6P dehydrogenase. J Biotechnol 2007;132(2):99 – 109. PMID:17624457doi:10.1016/j.jbiotec.2007.05.026.

[66] Blombach B, Schreiner ME, Moch M, Oldiges M, Eikmanns BJ. Effect of pyruvate dehydrogenase complex deficiency on L – lysine production with *Corynebacterium glutamicum*. Appl Microbiol Biotechnol 2007;76(3):615 – 23. PMID:17333167doi:10.1007/s00253 – 007 – 0904 – 1.

[67] Eggeling L, Oberle S, Sahm H. Improved L – lysine yield with *Corynebacterium glutamicum*: Use of dapA resulting in increased flux combined with growth limitation. Appl Microbiol Biotechnol 1998;49(1):24 – 30. PMID:9487706doi:10.1007/s002530051132.

[68] Mitsuhashi S, Hayashi M, Ohnishi J, Ikeda M. Disruption of malate: Quinone oxidoreductase increases L – lysine production by *Corynebacterium glutamicum*. Biosci Biotechnol Biochem 2006;70(11):2803 – 6. PMID:17090916doi:10.1271/bbb.60298.

[69] Takeno S, Murata R, Kobayashi R, Mitsuhashi S, Ikeda M. Engineering of *Corynebacterium glutamicum* with an NADPH – generating glycolytic pathway for L – lysine production. Appl Environ Microbiol 2010;76(21):7154 – 60. PMID:20851994doi:10.1128/AEM.01464 – 10.

[70] Becker J, Zelder O, Häfner S, Schröder H, Wittmann C. From zero to hero—Design – based systems metabolic engineering of *Corynebacterium glutamicum* for L – lysine production. Metab Eng 2011;13(2):159 – 68. PMID:21241816doi:10.1016/j.ymben.2011.01.003.

[71] Pérez – García F, Peters – Wendisch P, Wendisch VF. Engineering *Corynebacterium glutamicum* for fast production of L – lysine and L – pipecolic acid. Appl Microbiol Biotechnol 2016; 100(18):8075 – 90. PMID:27345060doi:10.1007/s00253 – 016 – 7682 – 6.

[72] Mimitsuka T, Sawai H, Hatsu M, Yamada K. Metabolic engineering of *Corynebacterium glutamicum* for cadaverine fermentation. Biosci Biotechnol Biochem 2007;71(9):2130 – 5. PMID: 17895539doi:10.1271/bbb.60699.

[73] Li M, Li D, Huang Y, Liu M, Wang H, Tang Q, et al. Improving the secretion of cadaverine in *Corynebacterium glutamicum* by cadaverine – lysine antiporter. J Ind Microbiol Biotechnol 2014;41(4):701 – 9. PMID:24510022doi:10.1007/s10295 – 014 – 1409 – 4.

[74] Matsushima Y, Hirasawa T, Shimizu H. Enhancement of 1,5 – diaminopentane production in a recombinant strain of *Corynebacterium glutamicum* by Tween 40 addition. J Gen Appl Microbiol 2016;62(1):42 – 5. PMID:26923131doi:10.2323/jgam.62.42.

[75] Kind S, Neubauer S, Becker J, Yamamoto M, Völkert M, Abendroth Gv, et al. From zero to hero—Production of bio – based nylon from renewable resources using engineered *Corynebacterium glutamicum*. Metab Eng 2014;25:113 – 23. PMID:24831706doi:10.1016/j.ymben.2014.

05.007.

[76]Buschke N, Becker J, Schäfer R, Kiefer P, Biedendieck R, Wittmann C. Systems metabolic engineering of xylose – utilizing *Corynebacterium glutamicum* for production of 1,5 – diaminopentane. Biotechnol J 2013;8(5):557 – 70. PMID:23447448doi:10.1002/biot.201200367.

[77]Kikuchi Y, Kojima H, Tanaka T. Mutational analysis of the feedback sites of lysine – sensitive aspartokinase of *Escherichia coli*. FEMS Microbiol Lett 1999;173(1):211 – 5. PMID:10220897doi:10.1111/j.1574 – 6968.1999.tb13504.x.

[78]Van Dien SJ, Iwatani S, Usuda Y, Matsui, K. Theoretical analysis of amino acid – producing *Escherichia coli* using a stoichiometric model and multivariate linear regression. J Biosci Bioeng 2006;102(1):34 – 40. PMID:16952834doi:10.1263/jbb.102.34.

[79]Cheraghi S, Akbarzade A, Farhangi A, Chiani M, Saffari Z, Ghassemi S, et al. Improved production of L – lysine by over – expression of *meso* – diaminopimelate decarboxylase enzyme of *Corynebacterium glutamicum* in *Escherichia coli*. Pak J Biol Sci 2010;13(10):504 – 8. PMID:21848075doi:10.3923/pjbs.2010.504.508.

[80]Ying H, He X, Li Y, Chen K, Ouyang P. Optimization of culture conditions for enhanced lysine production using engineered *Escherichia coli*. Appl Biochem Biotechnol 2014;172(8):3835 – 43. PMID:24682878doi:10.1007/s12010 – 014 – 0820 – 7.

[81]Wang Y, Li Q, Zheng P, Guo Y, Wang L, Zhang T, et al. Evolving the L – lysine high – producing strain of *Escherichia coli* using a newly developed high – throughput screening method. J Ind Microbiol Biotechnol 2016;43(9):1227 – 35. PMID:27369765doi:10.1007/s10295 – 016 – 1803 – 1.

[82]delaVega AL, Delcour AH. Cadaverine induces closing of *E. coli* porins. EMBO J 1995;14(23):6058 – 65. PMID:8846798.

[83]Hamner S, McInnerney K, Williamson K, Franklin MJ, Ford TE. Bile salts affect expression of *Escherichia coli* O157:H7 genes for virulence and iron acquisition, and promote growth under iron limiting conditions. PLoS One 2013;8(9):e74647. PMID:24058617doi:10.1371/journal.pone.0074647.

[84]Iyer R, Delco;ur AH. Complex inhibition of OmpF and OmpC bacterial porins by polyamines. J Biol Chem 1997;272(30):18595 – 601. PMID:9228026doi:10.1074/jbc.272.30.18595.

[85]Tateno T, Okada Y, Tsuchidate T, Tanaka T, Fukuda H, Kondo A. Direct production of cadaverine from soluble starch using *Corynebacterium glutamicum* coexpressing α – amylase and lysine decarboxylase. Appl Microbiol Biotechnol 2009;82(1):115 – 21. PMID:18989633doi:10.1007/s00253 – 008 – 1751 – 4.

[86]Naerdal I, Pfeifenschneider J, Brautaset T, Wendisch VF. Methanol – based cadaverine production by genetically engineered *Bacillus methanolicus* strains. Microb Biotechnol 2015;8(2):

342 – 50. PMID:25644214doi:10. 1111/1751 – 7915. 12257.

[87] Nishi K, Endo S, Mori Y, Totsuka K, Hirao Y, inventors; AJINOMOTO KK, assignee. Method for producing cadaverine dicarboxylate and its use for the production of nylon. European Patent EP1482055. 2006 Jan 3.

[88] de Carvalho CC. Enzymatic and whole cell catalysis: Finding new strategies for old processes. Biotechnol Adv 2011; 29 (1): 75 – 83. PMID: 20837129doi: 10. 1016/j. biotechadv. 2010. 09. 001.

[89] Chen RR. Permeability issues in whole – cell bioprocesses and cellular membrane engineering. Appl Microbiol Biotechnol 2007;74(4):730 – 8. PMID:17221194doi:10. 1007/s00253 – 006 – 0811 – x.

[90] León R, Fernandes P, Pinheiro HM, Cabral JMS. Whole – cell biocatalysis in organic media. Enzyme Microb Tech 1998;23(7 – 8):483 – 500. doi:10. 1016/S0141 – 0229(98)00078 – 7.

[91] Soksawatmaekhin W, Uemura T, Fukiwake N, Kashiwagi K, Igarashi K. Identification of the cadaverine recognition site on the cadaverine – lysine antiporter CadB. J Biol Chem 2006;281 (39):29213 – 20. PMID:16877381doi:10. 1074/jbc. M600754200.

[92] Kim HJ, Kim YH, Shin JH, Bhatia SK, Sathiyanarayanan G, Seo HM, et al. Optimization of direct lysine decarboxylase biotransformation for cadaverine production with whole – cell biocatalysts at high lysine concentration. J Microbiol Biotechnol 2015; 25 (7): 1108 – 13. PMID: 25674800doi:10. 4014/jmb. 1412. 12052.

[93] Ma W, Cao W, Zhang B, Chen K, Liu Q, Li Y, et al. Engineering a pyridoxal 5′ – phosphate supply for cadaverine production by using *Escherichia coli* whole – cell biocatalysis. Sci Rep 2015;5:15630. PMID:26490441doi:10. 1038/srep15630.

[94] Bhatia SK, Kim YH, Kim HJ, Seo HM, Kim JH, Song HS, et al. Biotransformation of lysine into cadaverine using barium alginate – immobilized *Escherichia coli* overexpressing CadA. Bioprocess Biosyst Eng 2015;38(12):2315 – 22. PMID:26314400doi:10. 1007/s00449 – 015 – 1465 – 9.

[95] Li N, Chou H, Yu L, Xu Y. Cadaverine production by heterologous expression of *Klebsiella oxytoca* lysine decarboxylase. Biotechnol Bioproc Engi 2014;19(6):965 – 72. doi:10. 1007/ s12257 – 014 – 0352 – 6.

[96] Wang C, Zhang K, Chen Z, Cai H, Wan H, Ouyang P. Directed evolution and mutagenesis of lysine decarboxylase from *Hafnia alvei* AS1. 1009 to improve its activity toward efficient cadaverine production. Biotechnol Bioproc Eng 2015;20(3):439 – 46. doi:10. 1007/s12257 – 014 – 0690 – 4.

[97] Krzy aniak A, Schuur B, de Haan AB. Extractive recovery of aqueous diamines for bio – based plastics production. J Chem Tech Biot 2013;88(10):1937 – 45. doi:10. 1002/jctb. 4058.

[98] Kind S, Wittmann C, inventors; BASF (China) Company Limited, BASF SE, assignee. Processes and recombinant microorganisms for the production of cadaverine. PCT Patent WO2012114256 A1. 2012 Aug 30.

[99] Liu X, Liu C, Dai D, Qin B, Li N, inventors; Cathay R&D Center Co., Ltd., Cathay Industrial Biotech Ltd., assignee. Purification of cadaverine using high boiling point solvent. PCT Patent WO2014/114000 A1. 2014 Jul 31.

[100] Aeschelmann F, Carus M. Bio – based building blocks and polymers in the world—Capacities, production and applications: Status quo and trends toward 2020. 3rd ed. Hürth: NOVA – Institut GmbH; 2015.

[101] Martino L, Basilissi L, Farina H, Ortenzi MA, Zini E, Di Silvestro G, et al. Bio – based polyamide 11: Synthesis, rheology and solid – state properties of star structures. Eur Polym J 2014; 59: 69 – 77. doi: 10. 1016/j. eurpolymj. 2014. 07. 012.

[102] Wang Z, Hu G, Zhang J, Xu J, Shi W. Non – isothermal crystallization kinetics of Nylon 10T and Nylon 10T/1010 copolymers: Effect of sebacic acid as a third comonomer. Chin J Chem Eng 2016. In press.

[103] McKeen LW. Polyamides (Nylons). In: McKeen LW The effect of temperature and other factors on plastics and elastomers. 3rd ed. Oxford: William Andrew Publishing; 2014. p. 233 – 340.

[104] Wu Z, Zhou C, Qi R, Zhang H. Synthesis and characterization of nylon 1012/clay nanocomposite. J Appl Polym Sci 2002; 83(11): 2403 – 10. doi: 10. 1002/app. 10198.

[105] Rulkens R, Koning C. Chemistry and technology of polyamides. In: Matyjaszewski K, Möller M, editors Polymer science: A comprehensive reference. Amsterdam: Elsevier; 2012. p. 431 – 67.

[106] Karak N. Polyamides, polyolefins and other vegetable oil – based polymers. In: Karak N, editor Vegetable oil – based polymers: Properties, processing and applications. Cambridge: Woodhead Publishing Limited; 2012. p. 208 – 25.

[107] Moorthy JN, Singhal N. Facile and highly selective conversion of nitriles to amides via indirect acid – catalyzed hydration using TFA or AcOH – H2SO4. J Org Chem 2005; 70(5): 1926 – 9. PMID: 15730325doi: 10. 1021/jo048240a.

[108] Azcan N, Demirel E. Obtaining 2 – octanol, 2 – octanone, and sebacic acid from castor oil by microwave – induced alkali fusion. Ind Eng Chem Res 2008; 47(6): 1774 – 8. doi: 10. 1021/ie071345u.

[109] Draths KM, Frost JW. Environmentally compatible synthesis of adipic acid from D – glucose. J Am Chem Soc 1994; 116(1): 399 – 400. doi: 10. 1021/ja00080a057.

[110] van Duuren JB, Brehmer B, Mars AE, Eggink G, Dos Santos VA, Sanders JP. A limited LCA of bio – adipic acid: Manufacturing the nylon – 6, 6 precursor adipic acid using the benzoic acid degradation pathway from different feedstocks. Biotechnol Bioeng 2011; 108(6): 1298 – 306.

PMID:21328320doi:10. 1002/bit. 23074.

[111] Lange JP, Vestering JZ, Haan RJ. Towards 'bio - based' nylon: Conversion of γ - valerolactone to methyl pentenoate under catalytic distillation conditions. Chem Commun (Camb) 2007;(33):3488 - 90. PMID:17700891doi:10. 1039/b705782b.

[112] Boussie TR, Dias EL, Fresco ZM, Murphy VJ, Shoemaker J, Archer R, et al., inventors; Rennovia, Inc., assignee. Production of adipic acid and derivatives from carbohydrate - containing materials. US Patent US8669397 B2. 2014 Mar 11.

[113] Schneider J, Eberhardt D, Wendisch VF. Improving putrescine production by *Corynebacterium glutamicum* by fine - tuning ornithine transcarbamoylase activity using a plasmid addiction system. Appl Microbiol Biotechnol 2012;95(1):169 - 78. PMID:22370950doi:10. 1007/s00253 - 012 - 3956 - 9.

[114] Qian ZG, Xia XX, Lee SY. Metabolic engineering of *Escherichia coli* for the production of putrescine: A four carbon diamine. Biotechnol Bioeng 2009;104(4):651 - 62. PMID:19714672.

[115] Frost JW, inventor; Board of Trustees of Michigan State University, assignee. Synthesis of caprolactam from lysine. US Patent US8367819 B2. 2013 Feb 5.

[116] Chan - Thaw CE, Marelli M, Psaro R, Ravasio N, Zaccheria F. New generation biofuels: γ - valerolactone into valeric esters in one pot. RSC Adv 2013; 3: 1302 - 6. doi: 10. 1039/C2RA23043G.

[117] Beerthuis R, Rothenberg G, Shiju NR. Catalytic routes towards acrylic acid, adipic acid and ε - caprolactam starting from biorenewables. Green Chem 2015; 17: 1341 - 61. doi: 10. 1039/C4GC02076F.

[118] Eltahir YA, Saeed HAM, Chen Y, Xia Y, Wang Y. Effect of hot drawing on the structure and properties of novel polyamide 5, 6 fibers. Text Res J 2014; 84(16): 1700 - 7. doi: 10. 1177/0040517514527378.

[119] Kato K, Masunaga A, Matsuoka H, inventors; Toray Industries Inc., assignee. Polyamide resin, polyamide resin composition, and molded article comprising same. US Patent US 20120016077 A1. 2012 Jan 19.

β – ODAP Accumulation Could Be Related to Low Levels of Superoxide Anion and Hydrogen Peroxide in *Lathyrus sativus* L.

ChengJin Jiao　JingLong Jiang　Chun Li
LanMing Ke　Wei Cheng　FengMin Li　ZhiXiao Li
ChongYing Wang *

摘　要：山黧豆毒素β－ODAP的积累随植株生长发育及环境胁迫而波动。大量研究表明，在植物发育及各种胁迫反应中，以超氧阴离子（$O_2^{\cdot -}$）与过氧化氢（H_2O_2）为主的活性氧（Reactive oxygen species，ROS）起重要作用。为了探明β－ODAP积累与ROS水平是否有联系，本研究对山黧豆叶片β－ODAP与ROS含量进行了分析。结果表明，积累β－ODAP的叶片，往往$O_2^{\cdot -}$与H_2O_2的水平都很低，而$O_2^{\cdot -}$与H_2O_2水平高的叶片很少积累β－ODAP；当用吡啶（pyridine）或脱落酸（ABA）抑制这两种活性氧的产生时，在完整植株或离体幼叶中都发现毒素积累；用氨基三唑（AT）抑制离体幼叶的过氧化氢酶活性时，H_2O_2含量升高，而β－ODAP水平显著下降。另外，给山黧豆幼苗接种根瘤菌时，苗中$O_2^{\cdot -}$与H_2O_2水平升高，而β－ODAP的含量则下降。这些结果说明，毒素β－ODAP的积累与组织中低水平的活性氧呈正相关。

关键词：山黧豆；过氧化氢；超氧阴离子；山黧豆毒素；氧化胁迫；根瘤菌。

Level of the neuroexcitatory β – N – oxalyl – L – α，β – diaminopropionic acid（β – ODAP）in grass pea（*Lathyrus sativus* L.）varies with development and environmental stress. Reactive oxygen species（ROS）（mainly $O_2^{\cdot -}$ and H_2O_2）are frequently reported to play important roles in plant development and in response to various stresses. To investigate the possible inter – relationship between contents of β – ODAP and ROS，grass

* 作者简介：焦成瑾（1967—　），男，甘肃秦安人，现为天水师范学院生物工程与技术学院教授，博士，主要从事酶工程研究。

pea leaves have been analyzed for contents of β - ODAP, $O_2^{\cdot -}$ and H_2O_2. The results showed that leaves containing high levels of β - ODAP, exhibited low levels of $O_2^{\cdot -}$ and H_2O_2, while leaves with high contents of $O_2^{\cdot -}$ and H_2O_2 accumulated little β - ODAP. The application of pyridine or ABA which inhibit the production of $O_2^{\cdot -}$ or H_2 led to an increase in β - ODAP contents in intact or detached young leaves, whereas inhibition of catalase activity using AT(3 - amino - 1,2,4 - triazole), leading to an increase in H_2O_2 content, result in significant decrease in β - ODAP levels of detached young leaves. In addition, inoculation of *Rhizobium* to young seedlings enhanced $O_2^{\cdot -}$ and H_2O_2 levels, but reduced β - ODAP contents in shoots. These results suggest that β - ODAP accumulation could be related to low levels of superoxide anion and hydrogen peroxide in grass pea tissues.

1. INTRODUCTION

The toxic component from the seeds of the important legume crop grass pea (*Lathyrus sativus*) that was considered responsible for human lathyrism has been identified as a neuroexcitatory non - protein amino acid, β - N - oxalyl - L - α, β - diaminopropionic acid (β - ODAP), by two independent groups of investigators (Murti*et al.*, 1964; Rao *et al.*, 1964). Nevertheless, because it has a high degree of adaptability under adverse climatic conditions and does not involve complicated managerial practices, grass pea is still the survival food for the very poor in some regions of Africa and Asia (Misra*et al.*, 1981; Tadesse and Bekele, 2003; VazPatto*et al.*, 2006). Fortunately, a range of varieties with low toxin have been produced after many years of breeding and selection (Campbell *et al.*, 1994). However, there was a great variation in β - ODAP content with environment (Dahiya *et al.*, 1975; Ramanujam et al., 1980; Chen et al., 1992; Siddique *et al.*, 1996; Dixit *et al.*, 1997; Tadesse, 2003), indicating that the synthesis and accumulation of β - ODAP are significantly influenced by external environmental factors.

Previous studies have shown that water stress (Ongena*et al.*, 1990; Xing *et al.*, 2001), zinc, phosphate fertilizer, cadmium, aluminium, iron, nitrogen, salinity, exchangeable cations and even *Rhizobium* (Lambein*et al.*, 1994; Hussain and Chowdhury, 1995; Hussain *et al.*, 1995; Jiao, 2005) affected the concentration of β - ODAP in grass pea seedlings or seeds to varying extent. In addition, the studies consistently showed that large accumulation of β - ODAP occurred mainly in the very young seedlings and the ripening seeds, while very little of it was detected during the late vegetative phase, irrespective of the species (Addis *et al.*, 1994; Jiao *et al.*, 2006). Some attempts to account for the roles of β – ODAP have focused on the stress responses, such as functioning as a carrier molecule for zinc ion under zinc deficiency (Lambein*et al.*, 1994), playing a role in drought tolerance (Xing *et al.*, 2001), and acting as an hydroxyl radical scavenger under oxidative stress (Zhou *et al.*, 2001). However, during normal grass pea development, the fact that large accumulation of β – ODAP specifically occurred during seed germination and seed ripening stages was not explained by all these proposed functions. Our previous study (Jiao *et al.*, 2006) suggested that β – ODAP accumulation and nutrient remobilization *in vivo* may occur simultaneously, the latter can generally take place in a number of plant species and is accompanied by mature or old leaf senescence either during normal growth such as seed germination or under a wide range of environmental stresses (Patel and Rao, 2007; Munns, 2002a, b; Lim and Nam, 2007; Maksymiec, 2007). As a response of the plant to several environmental stimuli (Mittler, 2002), senescence of cells or tissues, even of whole plants was often triggered by ROS (Zimmermann and Zentgraf, 2005). As the β – ODAP level is affected by various stresses, there may be a relationship between β – ODAP accumulation and also changes in the stress related ROS levels.

In order to investigate the possible relationship betweenβ – ODAP accumulation and ROS levels, both β – ODAP and ROS (mainly including $O_2^{\cdot -}$ and H_2O_2) were measured in a series of grass pea samples, which were all reported to contain different contents of β – ODAP, including young leaves and mature leaves (Jiao *et al.*, submitted), shoots from seedlings inoculated with *Rhizobium* (Jiao, 2005), and the leaves treated with ABA (Xiong*et al.*, 2006) and other chemicals, etc.

2 MATERIALS AND METHODS

2.1 Chemicals and reagents

Horseradish peroxidase, catalase, ABA, 3, 3 – diaminobenzidine (DAB), 3 – amino – 1, 2, 4 – triazole (AT), pyridine were purchased from Sigma – Aldrich Chemicals Co. (St. Louis, MO, USA). β – ODAP standard was extracted from grass pea seeds and purified on cation exchange resin according to the procedure previously described by Rao et al. (1964). All other chemical reagents were of analytical reagent (AR) grade.

2.2 Treatment and sampling

2.2.1 Greenhouse experiment and leaves sampling

In greenhouse experiments, grass pea seeds of LZ(1) and LZ(2) were planted in 250 - mm - diameter pots containing the field soil (Jiao *et al.*, 2006). Pots were watered with half - strength Hoagland's solution and thinned to ten plants each pot 10 d after sowing (DAS). Plants were grown for ten weeks under greenhouse conditions with natural light and average day/night temperature of 22/14℃. For each harvest, light green young leaves (first leaves, from shoot top) and dark green mature leaves (first or second leaves, from the shoot base) were collected 10, 15 and 60 DAS, snap - frozen in liquid nitrogen and stored at - 80℃ for β - ODAP, $O_2^{\cdot -}$ and H_2O_2 analysis.

2.2.2 Germination of seeds with Rhizobium and shoots sampling

Rhizobium strain LZ - R, which was specific to grass pea cultivar LZ(2), isolated and purified from the roots of 25 - day - old field - grown plants by previously described procedures (Jiao, 2005), were grown on agar slopes of YMB (yeast mannitol broth) medium (Barrientos *et al.*, 2003). The *Rhizobium* cells were suspended in sterile YMB to a density of 10^7 cells ml^{-1} when inoculation was carried out. The LZ(2) seeds were surface sterilized in 70% ethanol for 30 seconds and then in 0.1% $HgCl_2$ for another 8 min, followed by several rinses in sterile water. The sterilized seeds, together with 0.5 ml of LZ - R cell suspension per seed (for inoculation treatment) or 0.5ml of YMB per seed (for control), were planted in plastic beaker containing sterile vermiculite. All plastic beakers were transferred to a growth chamber with 28℃/24℃ day/night cycle and 70% relative humidity allowing seeds to germinate. The young shoots were harvested 5 days after germination. The fresh weights of 10 shoots (rootsand cotyledons removed) were recorded and were snap - frozen for β - ODAP, $O_2^{\cdot -}$ and H_2O_2 analysis.

2.2.3 Leaves treatment with ABA and sampling

LZ(2) seeds were germinated in moist vermiculite in darkness for two days, exposed to light for another three days, then carefully transplanted to plastic culture pots containing half - strength Hoagland's solution. The culture solution was changed once every four days. The leaves were treated by foliar spraying with 200μM ABA solution containing 0.01% (v/v) tween - 20, eight days after transplanting. Two days later, the young leaves were harvested from plants, snap - frozen in liquid nitrogen and stored at - 80℃ for β - ODAP, $O_2^{\cdot -}$ and H_2O_2 analysis.

2.2.4 Experiments with the detached leaves

Young leaves were cut from the 13 – day – old hydroponic seedlings, rinsed with distilled water, and then placed in petri dish containing 100 μM ABA solution (treatment), or H_2O (control) for 20h at 25℃, with continuous light. To investigate the effects of various inhibitors, the detached young leaves were pretreated separately with 10 mM 3 – amino – 1,2,4 – triazole (AT), or 10 mM pyridine for 6 h, and then exposed to 100 μM ABA for another 20 h. After treatments, the leaves were sampled, snap – frozen in liquid nitrogen and stored at – 80℃ for β – ODAP, $O_2^{\cdot -}$ and H_2O_2 analysis.

2.3 HPLC analysis of β – ODAP

β – ODAP was extracted from different types of tissues, including young leaves, mature leaves, shoots and roots from 5 – day – old seedlings inoculated with rhizobia (Jiao, 2005). All samples, collected from various experiments mentioned above, were washed in distilled water, ground to a fine powder in liquid nitrogen and lyophilized to equilibrate the moisture content. To remove pigments and fatty acids, 20 mg of the freeze – dried material were extracted with 2 ml of acetone, precipitated by centrifugation and subsequently dried under vacuum. The dried material was dissolved in 2 ml of diluted ethanol solution (750 ml/L) and shaken overnight. It was then centrifuged at 17900 g for 15 min, and the supernatant was dried under vacuum, then the residue was dissolved in 0.5 M $NaHCO_3$. The concentrations of β – ODAP in the samples were analyzed by reverse – phase HPLC as 1 – fluoro – 2,4 – dinitrobenzene (FDNB) derivatives using Agilent Technologies 1100 Liquid Chromatography System, adapting the method described by Wang *et al.* (2000) with some minor modification (Jiao *et al.*, 2006). The chromatographic conditions were set up as follows: analytical column: Phenomenex Luna C18 (5 mm, 250 × 4.6 mm) (USA), mobile phase: acetonitrile/ KH_2PO_4 (25 mM), 17/83 (v/v); flow rate: 1 ml/ min; temperature: 40℃; detection wavelength: 360nm. Purified β – ODAP was used as external standard for quantitative analysis.

2.4 Histochemical detection of H_2O_2 and superoxide anion

Leaves or whole plants were infiltrated with 0.01% Tween – 20 containing either 0.1% (w/v) 3,3 – diaminobenzidine (DAB) (Thordal – Christensen, 1997), 10 mM MES (pH 6.5) (detection for H_2O_2) or 0.1% (w/v) nitroblue tetrazolium (NBT) (Huckelhoven*et al.*, 2000), 50 mM potassium phosphate (pH 6.4) (detection for $O_2^{\cdot -}$), incubated in the dark for 8 h at room temperature. The experiments were terminated by immersion of the plants or leaves in boiling ethanol (95%, v/v) for 30 min.

This treatment decolorized the leaves except for the deep brown polymerization product produced by the reaction of DAB with H_2O_2 or the blue formazan product produced by the reaction of NBT with $O_2^{\cdot -}$. After cooling, the leaves or whole plants were preserved in ethanol(75% ,v/v) at room temperature and photographed.

2.5 Spectrophotometric analysis of $O_2^{\cdot -}$ production

The production of superoxide anion was determined using the procedure described by Elstner and Heupel(1976) with minor modification(Lu et al. ,2004). Fresh samples (0.5 g), together with 1ml of 10mM hydroxylammonium chloride, 3 ml of 65mM phosphate buffer(pH 7.8) and 1ml of 0.1M EDTA – Na, were homogenized using a prechilled pestle and mortar, and then centrifuged at 14,000g for 15min at 4℃. To remove the pigment, 1ml of chloroform was added to 1ml of the supernatant, sonicated and centrifuged at 14,000g for 5min at 25℃. 1ml of the supernatant(without pigment) was then mixed with 1ml of 17mM 4 – aminobenzenesulfonic acid and 1ml of 7mM α – naphthylamine, incubated for 10 min at 25℃ and its absorbance at 530nm was determined against a phosphate buffer blank. The total production of superoxide anion was expressed as μmol g^{-1}fwt.

2.6 Spectrophotometric measurement of H_2O_2 content

Hydrogen peroxide wasextracted from plant tissues and detected as described by Zhou et al. (2006) with minor modification. Fresh leaves(0.5g) were frozen in liquid nitrogen and ground to powder in a mortar with pestle, together with 4.5 ml of 5% TCA and 0.12 g activated charcoal. The mixture was centrifuged at 14,000g for 15 min at 4℃. The supernatant was adjusted to pH 8.4 with 17 M ammonia solution and then filtered. The filtrate was divided into 1 ml aliquots. One of these, as a blank was incubated with 1 ml of catalase at room temperatures for 10 min. Both the blank and the treated aliquots were then incubated with 1 ml of the colorimetric reagent for 10 min at 37℃ and the absorbance was determined at 505nm. The colorimetric reagent contained 20 mg of 4 – aminoantipyrine, 20 mg of phenol, 5 mg of horseradish peroxidase(150 U mg^{-1}), dissolved in 100 ml of 100mM acetate buffer(pH 5.6).

2.7 Statistical analysis

The data were statistically analyzed by ANOVA, followed bya Tuckey's test. The analysis of variance for individual parameters was performed on the basis of mean values and a level of significance at $p \leqslant 0.05$.

3 RESULTS AND DISCUSSION

3.1 Levels of β－ODAP, $O_2^{\cdot -}$ and H_2O_2 in leaves from different varieties

In order to investigate whether there is a relationship between contents of β－ODAP and ROS, the leaves from both low toxin variety LZ(1) and high toxin variety LZ(2) at the same developmental stages were collected and measured for β－ODAP, $O_2^{\cdot -}$ and H_2O_2 levels. Histochemical staining of $O_2^{\cdot -}$ and H_2O_2 in the whole 10－day－old plants of both varieties showed that all leaves of LZ(1) contained high levels of $O_2^{\cdot -}$, while only young leaves of LZ(2) exhibited high levels of $O_2^{\cdot -}$ (Fig. 1A). There were no obvious differences in H_2O_2 levels in all leaves of both varieties (Fig. 1B). The results of quantitative determination for both $O_2^{\cdot -}$ and H_2O_2 further confirmed the results of this visual staining (Fig. 2B, C). The $O_2^{\cdot -}$ levels in both young and mature leaves from 10－day－old LZ(1) plants were about 2－fold greater than those from 10－day－old LZ(2) plants (Fig. 2B). In terms of H_2O_2, an obvious difference existed between young leaves but not between old leaves of the two varieties (Fig. 2C). In contrast, β－ODAP contents in both young and old leaves of high toxin line LZ(2) were about two times higher than that in low toxin line LZ(1). These results suggested that low levels of both $O_2^{\cdot -}$ and H_2O_2 may be necessary for β－ODAP accumulation, which was more evident in young leaves than in mature leaves.

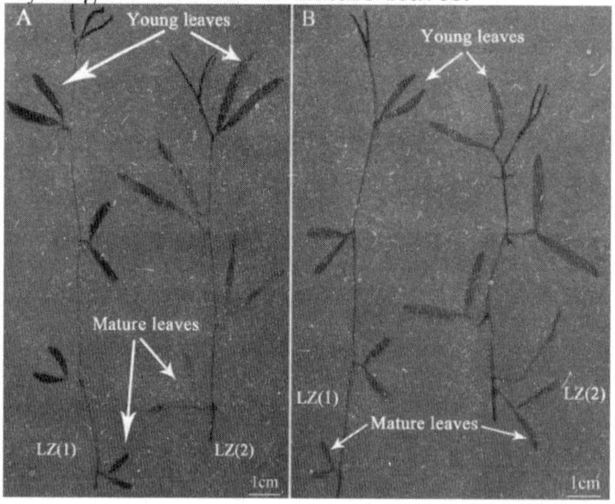

Fig. 1 Histochemical staining of $O_2^{\cdot -}$ by NBT (blue colors) (A) and H_2O_2 by DAB (brown colors) (B) in whole 10－day－old plants grown in greenhouse. LZ(1) and LZ(2) represent the low toxin variety and the high toxin variety, respectively.

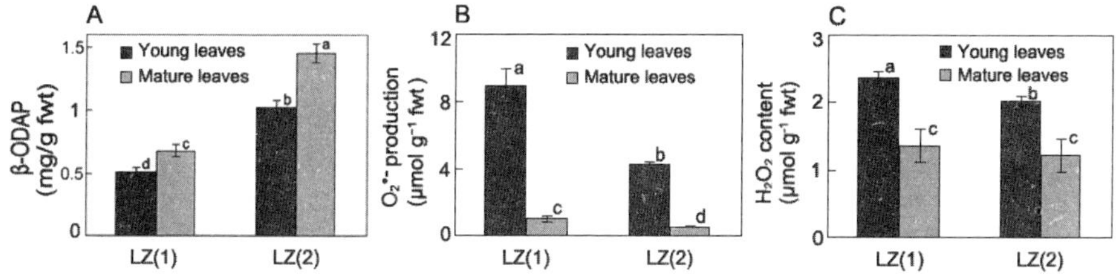

Fig. 2 Contents of β – ODAP(A), $O_2^{\cdot -}$(B), and H_2O_2(C) in the leaves from 10 – day – old plants grown in greenhouse. LZ(1) and LZ(2) represent the low toxin variety and the high toxin variety, respectively. Vertical bars represent S. E. of the means of three replicates. Different letters nearby bars show significant difference at 0.05 levels with ANOVA analysis.

In general, under optimal growth condition, a given level of β – ODAP or ROS in leaves should be a characteristic for a particular metabolic process in grass pea. The reasons of this relationship found between levels of β – ODAP and ROS especially in young leaves are not known at present. Potentially, the high levels of ROS might interfere with β – ODAP synthesis.

3.2 Levels of β – ODAP, $O_2^{\cdot -}$ and H_2O_2 in leaves from the same variety

In the same grass pea variety, the similar relationship between levels of β – ODAP and $O_2^{\cdot -}$ and H_2O_2 was also found in leaves of different developmental stages. Staining results showed that the contents of $O_2^{\cdot -}$ and H_2O_2 in leaves from 60 – day – old plants were significantly higher than those in leaves from 15 – day – old plants whether they were young or mature leaves. In 15 – day – old plants, young leaves were found to contain more $O_2^{\cdot -}$ but not H_2O_2 than mature leaves. In 60 – day – old plants, however, young leaves showed slightly less $O_2^{\cdot -}$ and H_2O_2 than mature leaves (Fig. 3). Quite consistently, the results shown in Fig. 4B and C quantitatively confirmed the staining results. $O_2^{\cdot -}$ accumulation in young leaves of 60 – day – old plants was about 10 – fold greater than those of 15 – day – old plants, while H_2O_2 levels in young leaves of 60 – day – old plants were about three times higher than those in 15 – day – old plants (Fig. 4B and C). These differences in ROS levels were even more pronounced in mature leaves from 15 and 60 – day – old plants (Fig. 4B, C). However, β – ODAP content in both young and mature leaves of 15 – day – old plants were up to 10 – fold greater than those of 60 – day – old plants. Mature leaves of 15 – day – old plants accumulated more β – ODAP than young leaves, while in 60 – day – old plants, the young leaves accumulated slightly more β – ODAP than the mature leaves.

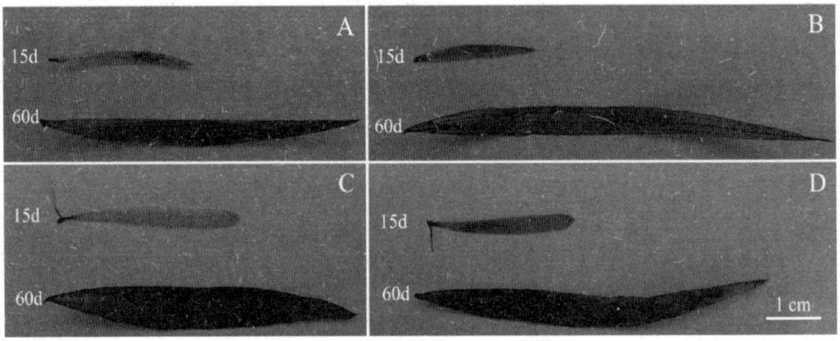

Fig. 3 The histochemical localization of $O_2^{\cdot -}$ by NBT and H_2O_2 by DAB in grass pea leaves. A. young leaves with blue colors showed location of $O_2^{\cdot -}$; B. young leaves with brown colors exhibited the location of H_2O_2; C. mature leaves with blue colors showed the location of $O_2^{\cdot -}$; D. mature leaves with brown colors exhibited the location of H_2O_2.

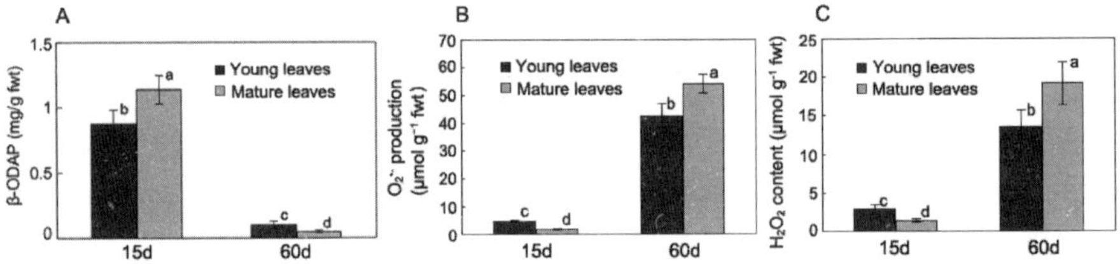

Fig. 4 Levels of β – ODAP (A), $O_2^{\cdot -}$ (B), and H_2O_2 (C) in young and mature leaves from 15 – day – old (15d) LZ (2) or 60 – day – old (60d) LZ (2) plants grown in greenhouse. Vertical bars represent S. E. of the means of three replicates. Different letters nearby bars show significant difference at 0.05 levels with ANOVA analysis.

In plants, ROS are inevitable by – products of metabolic activities of mitochondria, chloroplasts and peroxisomes when they function under either optimal growth conditions or extreme conditions (Taylor et al., 2009). They have emerged as ubiquitous components of signal transduction pathways, controlling a diverse range of physiological functions in a wide spectrum of biological systems (Neill et al., 2002; Foreman et al., 2003; Miller et al., 2009; Garg and Manchanda, 2009). Therefore, a large number of ROS (mainly $O_2^{\cdot -}$ and H_2O_2) generated in both young and mature leaves of 60 – day – old grass pea may be the inevitable result of rapid growth, when the leaves exhibited a strong photosynthesis and assimilation metabolism. While very little accumulation of β – ODAP was observed in these rapidly growing plants, even in the young leaves. On the contrary, in 15 – day – old seedlings, both young and mature leaves showed a relatively weak photosynthesis, of which part of the nutrients was imported from the cotyledons, although the reserves in the cotyledons were nearly exhausted at this stage. This result suggested that β – ODAP accumulation in leaves, especially in young leaves, may depend on both nutritional status and ROS levels in the leaves. High levels of ROS in leaves, at least, may inhibit β – ODAP accumulation.

3.3 Effect of ABA treatment on levels of β – ODAP, $O_2^{\cdot -}$ and H_2O_2 in young leaves

It has been shown that ABA can induce the expression of antioxidant genes encoding superoxide dismutase (SOD) and catalase (CAT) (Hu *et al.*, 2005) and enhance the activities of these antioxidant enzymes in plant tissues (Lu *et al.*, 2009). The seedlings, therefore, were treated with 200μM ABA for 2 days to further investigate whether the difference in β – ODAP content was related to ROS levels. As shown in Fig. 5B and C, the treated seedlings exhibited obvious decreases incontent of $O_2^{\cdot -}$ and H_2O_2 in young leaves, while β – ODAP accumulation in young leaves were clearly increased as expected (Fig. 5A). During the experiment, we noted that the mature leaves began to yellow at 2 days after treatment, indicating the onset of senescence. This result suggested that β – ODAP accumulation in young leaves was related to both the lower levels of $O_2^{\cdot -}$ and H_2O_2 and nutrients imported from pre – senescing mature leaves being induced by ABA, which was also one of the most effective plant hormones in promoting leaf senescence (Yang *et al.*, 2002; Ghanem *et al.*, 2008; Zhang *et al.*, 2009). This procedure was very similar to the way that nutrients were transported from cotyledons to young shoots during seed germination or that degradation metabolites from the aging leaves were imported into filling seeds during seed maturation.

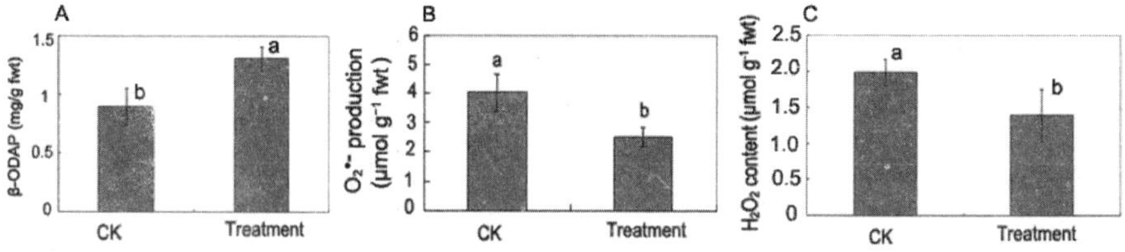

Fig. 5 Changes in contents of β – ODAP(A), $O_2^{\cdot -}$(B), and H_2O_2(C) in young leaves of 13 day – old grass pea LZ(2) plant sprayed with 200 μM ABA. Vertical bars represent S. E. of the means of three replicates. Different letters nearby bars show significant difference at 0.05 levels with ANOVA analysis.

3.4 Reduction of $O_2^{\cdot -}$ or H_2O_2 production can increase β – ODAP level of the detached young leaves

In order to further investigate this possible metabolic relationship between contents of β – ODAP and ROS in leaves, the detached young leaves were also treated with ABA, or combined with AT, or pyridine, which are inhibitors of catalase (Havir, 1992) and NADPH oxidase (Orozco – Cardenas and Ryan, 1999), respectively. It was found that no significant differences in β – ODAP levels of the detached young leaves were detected when treated with ABA (Fig. 6A). However, the treatment with pyridine alone or combined with ABA can result in substantial increase in levels of β – ODAP in young leaves, but only the treatment of AT combined with ABA caused an obvious decrease in β – ODAP as expected (Fig. 6A). The patterns of change in contents of H_2O_2 were just the opposite of β – ODAP contents during treatments with pyridine, ABA + pyridine, or ABA + AT in young leaves (Fig. 6B). Surprisingly, no obvious changes in $O_2^{\cdot -}$ levels were found in all treated leaves, even in leaves treated with pyridine (data not shown). Treatment with AT also did not lead to obvious increase in H_2O_2 levels, for which the reasons are not known. Nevertheless, pyridine or AT together with ABA were observed to be effective in decreasing and increasing H_2O_2, respectively, and the opposite result in β – ODAP contents was also observed in young leaves. Thus, it was also evident that β – ODAP likely tended to accumulate in the leaves with at least low levels of H_2O_2.

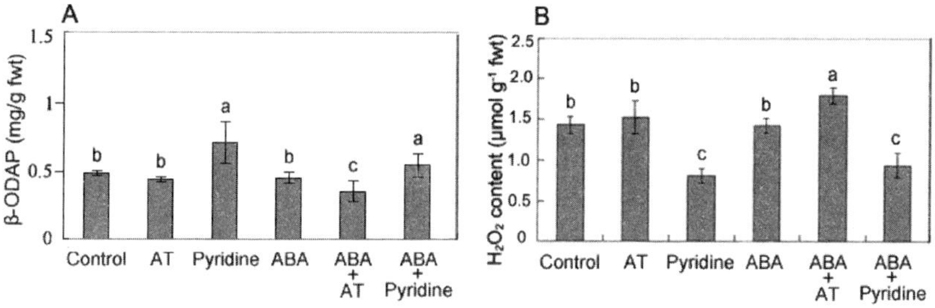

Fig. 6 Effect of different chemicals on contents of β – ODAP(A) and H_2O_2 (B) in detached leaves from 13 – day – old LZ(2) plants. The leaves were treated with H_2O(control) or ABA, and were pretreated with AT or pyridine for 6h, then treated with H_2O or ABA for 20h under continued light in growth chamber. Vertical bars represent S. E. of the means of three replicates. Different letters nearby barsshow significant difference at 0.05 levels with ANOVA analysis.

3.5 Levels of β – ODAP, $O_2^{\cdot -}$ and H_2O_2 in shoots when grown with Rhizobium

The inter – relationship among levels of β – ODAP, $O_2^{\cdot -}$ and H_2O_2 was also observed in young grass pea seedlings inoculated with *Rhizobium*. A previous study (Jiao, 2005) found that β – ODAP contents in 5 – day – old young seedlings grown with *Rhizobium* decreased about 30% based on the dry weight basis compared tc controls (Jiao et al., submitted), although a lower degree of decline was found when based on the fresh weight (Fig. 7A) ($P < 0.05$). It was found that both $O_2^{\cdot -}$ and H_2O_2 contents significantly increased ($P < 0.05$) in shoots of seedlings inoculated with *Rhizobium* compared to controls (Fig. 7B, C). The *Rhizobium* strain LZ – R, used in this experiment, was isolated from the root nodules of a LZ(2) plant and was specific to LZ(2) plant. In the early stages of nodulation, infection of these bacteria to roots first caused defensive reaction of the plant, as other legumes, leading to a systemic increase in ROS levels (Ramu*et al.*, 2002; Kinkema*et al.*, 2006; Chang *et al.*, 2009). We therefore speculated that a decrease in β – ODAP contents in shoots of seedlings inoculated with *Rhizobium* strain LZ – R were related to an increase in levels of $O_2^{\cdot -}$ and H_2O_2.

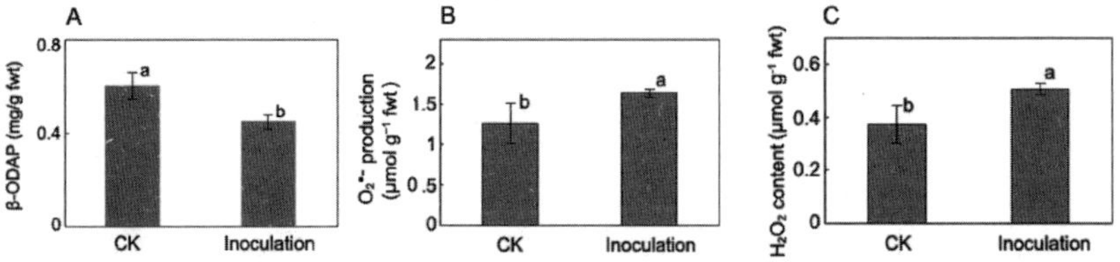

Fig. 7 β – ODAP(A), $O_2^{\cdot -}$(B), and H_2O_2 levels(C) in 5 – day – old LZ(2) seedlings(without roots and cotyledons) inoculated with proper level of rhizobial cells (ca. 1 × 10^7 cells/ml). Vertical bars represent S. E. of the means of three replicates. Different letters nearby bars show significant difference at 0. 05 levels with ANOVA analysis.

In addition, the tissues used in the present experiment all showed the significant changes in the levels of bothβ – ODAP and ROS with different developmental stages. The ripening seeds, however, exhibited no significant changes in levels of both β – ODAP and ROS during maturation, especially in late maturation, but β – ODAP contents were found to increase progressively as the seeds developed, whereas levels of $O_2^{\cdot -}$ and H_2O_2 showed gradual decrease at this developmental stage(data not shown). So, the opposite trends of changes in levels of both β – ODAP and ROS(at least $O_2^{\cdot -}$ and H_2O_2) were also observed in ripening seeds.

Taken together, β – ODAP likely tends to accumulate in those tissues such as young leaves, which generally contain low levels of ROS and apparently more β – ODAP biosynthetic precursor. In terms of β – ODAP synthesis pathway, the biosynthetic precursor of β – ODAP was found to be BIA(β – (isoxazolin – 5 – on – 2 – yl) – alanine) (Lambein*et al.*, 1990; Kuo and Lambein, 1991; Kuo*et al.*, 1994), which is synthesized by cysteine synthase from isoxazlin – 5 – one and O – acetylserine (Ikegami et al., 1993). The latter is also used for cysteine synthesis catalyzed by enzymes of the same family(Wirtz and Hell, 2006). Clearly, BIA synthesis can compete with synthesis of cysteine for the same precursor, O – acetylserine. It should be noted that cysteine and its further metabolite glutathione are key chemicals involved in antioxidant process when ROS levels are very high(López – Martín *et al.*, 2008; Queval*et al.*, 2009). Therefore, the most possible reason for the present relationship between levels of both β – ODAP and ROS can be explained by considering that O – acetylserine is mainly used for synthesis of BIA, and further β – ODAP production, under conditions with lower ROS levels, while it is used as a precursor for cysteine synthesis when ROS contents in tissues (such as mature leaves) are very high. BIA in mature leaves is indeed found under detectable amounts in general (Lambein*et al.*, 1976; Lambein*et al.*, 1990), which contain high levels of ROS, but very low content of β – ODAP such as mature leaves of 60 – day – old plants shown in the present study.

Based on both BIA and cysteine possibly competing for the same synthetic precursor as described above, and cysteine involving in antioxidant process, although the effects of ROS on other amino acids were not examined, nor were cysteine and glutathione detected in our experiment, we suggest that the physiological conditions with low levels of ROS (at least $O_2^{\cdot -}$ and H_2O_2) may be necessary for β – ODAP accumulation in grass pea tissues. Obviously, further study of this accumulation pattern, especially investigation of the metabolic relationship between BIA and cysteine, are needed, because cysteine and its further metabolite methionine are important sulfur – containing amino acids and determinants of grass pea nutritional quality (Lambeinand Ahmed, 2005; Fikre*et al.*, 2008), which is known to be associated with incident of human lathyrism.

4 CONCLUSION

The data presented in this paper for the first time indicated thatβ – ODAP tends to accumulate in the tissues containing low ROS (at least $O_2^{\cdot -}$ and H_2O_2) levels in grass pea, especially in the young leaves either under normal growth or under stresses. This may provide a new insight into the complex mechanisms for β – ODAP accumulation in grass pea tissues. In addition to the nutrient remobilization being possibly involved and BIA possibly competing with cysteine for the same synthetic precursor, lower levels of ROS (at least $O_2^{\cdot -}$ and H_2O_2) may be necessary for β – ODAP accumulation, it would be logical to propose that the present accumulation characteristics of β – ODAP in grass pea tissues may be used to explain why enhanced accumulation of β – ODAP occurred mainly in the very young seedlings and in the ripening seed, as well as why β – ODAP levels varied with a wide range of environmental stresses, many of which were frequently reported to induce a diverse range of physiological events in plants including nutrient remobilization, increase in level of ABA, and ROS burst.

Conflict of interest

The authors declare that there are no conflicts of interest.

Acknowledgements

The authors thank Prof. Lambein and Dr. Kuo for their critical reading and comments on the manuscript. This work was financially supported by National Natural-Science Foundation of China (30625025), Natural Science Foundation of Gansu province, P. R. China (0803RJZA034) and Programs for Changjiang Scholars and Innovative Research Team in University.

REFERENCES

[1] Addis, G., Narayan, R. K. J., 1994. Developmental variation of the neurotoxin, β – N – oxalyl – L – α, β – diaminopropionic acid (ODAP), in *Lathyrus sativus*. Annals of Botany 74, 209 – 215.

[2] Barrientos, L., Badilla, A., Mera M., Montenegro, A., Gaete, G., Espinoza, N., 2003. Performance of Rhizobium strains isolated from *Lathyrus sativus* plants growing in southern Chile. Lathyrus Lathyrism Newsletter 3, 8 – 9.

[3] Campbell, C. G., Mehra, R. B., Agrawal, S. K., Chen, Y. – Z., Abd El Moneim, A. M., Khawaja, H. I. T., Yadov, C. R., Tay, J. U., Araya, W. A., 1994. Current status and future strategy in breeding grasspea (*Lathyrussativus*). Euphytica 73, 167 – 175.

[4] Chang, C., Damiani, I., Puppo, A., Frendo, P., 2009. Redox Changes during the Legume – *Rhizobium* Symbiosis. Molecular Plant 2, 370 – 377.

[5] Chen, Y. – Z., Li, Z. – X., Lu, F. – H., Bao, X. – G., Liu, S. – Z., Liu, X. – C., Zhang, G. – W., Li, Y. – R., 1992. Studies on the screening of low toxic species of*Lathyrus*, analysis of toxins and toxicology. Journal of Lanzhou University (Natural Sciences) 28, 93 – 98.

[6] Dahiya, B. S., Jeswni, L. M., 1975. Genotype and environment interactions for neurotoxic principle (BOAA) in grass pea. Indian Journal of Agricultural Science 45, 437 – 439.

[7] Dixit, G. P., Pandey, R. L., Chandra, S., Pandey, P. S., Asthana, A. N., 1997. Stability of neurotoxin (ODAP) concentration in grass pea. In: R. T. Haimanot& F. Lambein (Eds.), *Lathyrus* and Lathyrism, a decade of progress, University of Ghent, Belgium. pp. 103 – 104.

[8] Elstner, E. F., Heupel, A., 1976. Inhibition of nitrite formation from hydroxylammonium-chloride: a simple assay for superoxide dismutase. Anal. Biochem. 70, 616 – 620.

[9] Fikre, A., Korbu, L., Kuo, Y. – H., Lambein, F., 2008. The contents of the neuro – excitatory amino acidβ – ODAP (β – N – oxalyl – L – α, β – diaminopropionic acid), and other free and protein amino acids in the seeds of different genotypes of grass pea (*Lathyrus sativus*L.). Food Chemistry 110, 422 – 427.

[10] Foreman, J., Demidchik, V., Bothwell, J. H. F., Mylona, P., Mie – dema, H., Torres, M. A., Linstead, P., Costa, S., Brownlee, C., Jones, J. D. G., Davies, J. M., Dolan, L., 2003. Reactive

oxygen species produced by NADPH oxidase regulate plant cell growth. Nature 422:442 – 446.

[11] Garg, N., Manchanda, G., 2009. ROS generation in plants: Boon or bane. Plant Biosystems 143, 81 – 96.

[12] Ghanem, M. E., Albacete, A., Martinez – Andujar, C., Acosta, M., Romero – Aranda, R., Dodd, I. C., Lutts, S., Perez – Alfocea, F., 2008. Hormonal changes during salinity – induced leaf senescence in tomato (*Solanum lycopersicum* L.). J. Exp. Bot. 59, 3039 – 3050.

[13] Havir, E. A., 1992. The *in Vivo* and *in Vitro* Inhibition of Catalase from Leaves of *Nicotiana sylvestris* by 3 – Amino – 1,2,4 – Triazole. Plant Physiology 99, 533 – 537.

[14] Hu, X. – L., Jiang, M. – Y., Zhang, A. – Y., Lu, J., 2005. Abscisic acid – induced apoplastic H_2O_2 accumulation up – regulates the activities of chloroplastic and cytosolic antioxidant enzymes in maize leaves. Planta 223, 57 – 68.

[15] Huckelhoven, R., Fodor, J., Trujillo, M., Kogel, K. H., 2000. Barley Mla and Rar mutants compromised in the hypersensitive cell death response against *Blumeriagraminis* f. sp. hordei are modified in their ability to accumulate reactive oxygen intermediates at sites of fungal invasion. Planta 212, 16 – 24.

[16] Hussain, M., Chowdhury, B., 1995. Agro – ecological factors affecting the concentration of β – N – oxalyl – L – α, β – diaminopropionic acid (ODAP) in*Lathyrus sativus* seeds. In *Lathyrus sativus* and Human Lathyrism: Progress and Prospects (Eds H. Yusuf & F. Lambein), Dhaka, Bangladesh: University of Dhaka. pp. 205 – 209.

[17] Hussain, M., Chowdhury, B., Haque, R., Lambein, F., 1997. Effect of water stress, salinity, interaction ofcations, stage of maturity of seeds and storage devices on the ODAP content of *Lathyrus sativus*. In *Lathyrus sativus* and Human Lathyrism: A Decade of Progress (Eds R. Haimanot& F. Lambein), Addis Ababa, Ethiopia: University of Ghent. pp. 107 – 112.

[18] Ikegami, F., Ongena, G., Sakai, R., Itagaki, S., Kogori, M., Ishikawa, T., Kuo, Y. – H., Lambein, F., Murakoshi, I., 1993. Biosynthesis of β – (isoxazolin – 5 – on – 2 – yl) – alanine, the precursor of the neurotoxin β – N – oxalyl – L – α, β – diaminopropionic acid, by cysteine synthase in *Lathyrus sativus*. Phytochemistry 33, 93 – 98.

[19] Jiao, C. – J., 2005. Studies on accumulation and biological significance of β – ODAP in-*Lathyrus sativus* L. (grass pea). Master Dissertation, Lanzhou University, Lanzhou, China.

[20] Jiao, C. – J., Jiang, J. – L., Ke, L. – M., Cheng, W., Li, F. – M., Li, Z. – X., Wang, C. – Y., 2010. Factors affecting β – ODAP content in *Lathyrus sativus* and their possible physiological mechanisms. (submitted)

[21] Jiao, C. – J., Xu, Q. – L., Wang, C. – Y., Li, F. – M., Li, Z. – X., Wang, Y. – F., 2006. Accumulation pattern of toxin β – ODAP during lifespan and effect of nutrient elements on β – ODAP content in *Lathyrus sativus* seedlings. The Journal of Agricultural Science144(4), 369 – 375.

[22] Kinkema, M., Scott, P. T., Gresshoff, P. M., 2006. Legume nodulation: successful symbi-

osis through short – andlong – distance signaling. Functional Plant Biology 33,707 –721.

[23] Kuo, Y. – H., Khan, J. K., Lambein, F., 1994. Biosynthesis of the neurotoxin β – ODAP in developing pods of*L. sativus*. Phytochemistry 35(4), 911 –913.

[24] Kuo, Y. – H., Lambein, F., 1991. Biosynthesis of the neurotoxin β – N – oxalyl – L – α, β – diamino propionicacid in callus tissue of *Lathyrus sativus*. Phytochemistry 30 (10), 3241 –3244.

[25] Lambein, F., Ahmed, S., 2005. The same goal, a different approach: a new Belgian – Ethiopian project. Lathyrus Lathyrism Newsletter 4, 34 –35.

[26] Lambein, F., Haque, R., Khan, J. K., Kebede, N., Kuo, Y. – H., 1994. From soil to brain: zinc deficiency increases the neurotoxicity of*Lathyrus sativus* and may affect the susceptibility for the motorneurone disease neurolathyrism. Toxicon 32, 461 –466.

[27] Lambein, F., Kuo, Y. – H., Parijs, R. V., 1976. Isoxazolin – 5 – ones chemistry and biology of a new class of Plant products. Heterocycles 4(3), 567 –593.

[28] Lambein, F., Ngudi, D. D., Kuo, Y. – H., 2001. Vapniarca revisited: Lessons from an inhuman human experience. Lathyrus Lathyrism Newsletter 2(1), 5 –7.

[29] Lambein, F., Ongena, G., Kuo, Y. – H., 1990. β – isoxazoline – alanine is involved in the biosynthesis of The neurotoxin β – N – oxalyl – L – 2,3 – diaminopropionic acid. Phytochemistry 29(12), 3793 –3796.

[30] Lim, P. O., Nam, H. G., 2007. Aging and Senescence of the Leaf Organ. Journal of Plant Biology 50(3), 291 –300.

[31] López – Martín, M. C., Romero, L. C., Gotor, C., 2008. Cytosolic cysteine in redox signaling. Plant Signal Behav. 3(10), 880 –881.

[32] Lu, S. – Y., Su, W., Li, H. – H., Guo, Z. – F., 2009. Abscisic acid improves drought tolerance of triploid bermudagrass and involves H_2O_2 – and NO – induced antioxidant enzyme activities. Plant Physiology and Biochemistry 47, 132 –138.

[33] Lu, W., Xu, X. – M., Zhang, R. – X., Dai, X. – B., 2004. Effect ofadding acetic acid on improvement of determination of superoxide anion content in plants. Journal of Nan Jing Normal University(Natural Science) 27, 82 –84.

[34] Maksymiec, W., 2007. Signaling responses in plants to heavy metal stress. Acta Physiol Plant 29, 177 –187.

[35] Misra, B. K., Singh, S. P., Barat, G. K., 1981. Ox – Dapro: the *Lathyrus sativus*neurotoxin. Plant Foods for Human Nutrition(Formerly Qual Plant Plant Foods Hum Nutr) 30, 259 –270.

[36] Mittler, R., 2002. Oxidative stress, antioxidants, and stress tolerance. Trends Plant Sci 7, 405 –410.

[37] Miller, G., Suzuki, N., Ciftci – Yilmaz, S., Mittler, R., 2009. Reactive oxygen species homeostasis and signaling during drought and salinity stresses. Plant, Cell and Environment(Articles

online in advance of print)

[38] Munns, R., 2002a. Salinity, growth and phytohormones. In: Läuchli A, Lättge U, eds. Salinity: environment—plants—molecules. Dordrecht: Kluwer Academic Publishers, 271 – 290.

[39] Munns, R., 2002b. Comparative physiology of salt and water stress. Plant, Cell and Environment 25, 239 – 250.

[40] Murti, W. S., Seshadri, T. R., Venkitasubramanian, T. A., 1964. Neurotoxic compounds of the seeds of*Lathyrus sativus*. Phytochemistry 3, 73 – 78.

[41] Neill, S. J., Desikan, R., Hancock, J. T., 2002. Hydrogen peroxide signaling. CurrOpin Plant Biol 5, 388 – 395.

[42] Ongena, G., Kuo, Y. – H., Lambein, F., 1990. Drought tolerance and neurotoxicity of *L. sativus*seedlings. Archives internationales de Physiologie et de Biochimie 98, 17 – 21.

[43] Orozco – Cardenas, M. L., Ryan, C. A., 1999. Hydrogen peroxide is generated systemati-cally in plant leaves by wounding and systemin via the octadecanoid pathway. Proc Natl Acad Sci (USA) 96, 6553 – 6557.

[44] Patel, K. G., Rao, T. V. R., 2007. Effect of simulated water stress on the physiology of leaf senescence in three genotypes of cowpea (*Vigna unguiculata* (L.) Walp). Indian Journal of Plant Physiology 12, 138 – 45.

[45] Queval, G., Thominet, D., Vanacker, H., Miginiac – Maslow, M., Gakière, B., Noctor, G., 2009. H_2O_2 – activated up – regulation of glutathione in *Arabidopsis* involves induction of genes encodingenzymes involved in cysteine synthesis in the chloroplast. Molecular Plant Advance Access.

[46] Ramanujam, S., Sethi, K. L., Rao, S. L. N., 1980. Stability of neurotoxin content in Khesari. Indian Journal of Plant Breeding and Genetics 40, 300 – 304.

[47] Ramu, S. K., Peng, H. M., Cook, D. R., 2002. Nod factor induction of reactive oxygen species production is correlated with expression of the early nodulin gene rip1 in Medicago truncatula. Mol Plant – Microbe Interact 15, 522 – 528.

[48] Rao, S. L. N., Adiga, P. R., Sarma, P. S., 1964. The isolation and characterization of β – N – oxalyl – L – α, β – diaminopropionic acid: A neurotoxin from the seeds of*Lathyrus sativus* L.. Biochemistry 3, 432 – 436.

[49] Siddique, K. H. M., Loss, S. P., Herwig, S. P., Wilson, J. M., 1996. Growth, yield and neurotoxin (ODAP) concentration of three *Lathyrus* species in Mediterraneantype environments of Western Australia. Australian Journal of Experimental Agriculture 36, 209 – 218.

[50] Tadesse, W., 2003. Stability of grass pea (*Lathyrus sativus* L.) varieties for ODAP content and grain yield in Ethiopia. Lathyrus Lathyrism Newsletter 3, 32 – 34.

[51] Tadesse, W., Bekele, E., 2003. Variation and association of morphological and biochemical characters in grass pea (*Lathyrus sativus* L.). Euphytica 130, 315 – 324.

[52] Taylor, N. L. , Tan, Y. F. , Jacoby, R. P. , Millar, A. H. , 2009. Abiotic environmental stress induced changes in the *Arabidopsis thaliana* chloroplast, mitochondria and peroxisome proteomes. J Proteomics 72,367 – 78.

[53] Thordal – Christensen, H. , Zhang, Z. , Wei, Y. , Collinge, D. B. , 1997. Subcellular localization of H_2O_2 in plants. H_2O_2 accumulation in papillae and hypersensitive response during the barley – powdery mildew interaction. Plant J. 11,1187 – 1194.

[54] Vaz Pattol, M. C. , Skiba, B. , Pang, E. C. K. , Ochatt, S. J. , Lambein, F. , Rubiales, D. , 2006. *Lathyrus* improvement for resistance against biotic and abiotic stresses: From classical breeding to marker assisted selection. Euphytica 147,133 – 147.

[55] Wang, F. , Chen, X. , Chen, Q. , Qin, X. – C. , Li, Z. – X. , 2000. Determination of neurotoxin 3 – N – oxalyl – 2,3 – diaminopropionic acid and non – protein amino acids in *Lathyrus sativus* by precolumn derivatization with 1 – fluoro – 2,4 – dinitrobenzene. Journal of Chromatography A 883,113 – 118.

[56] Wirtz, M. , Hell, R. , 2006. Functional analysis of the cysteine synthase protein complex from plants: structural, biochemical and regulatory properties. J Plant Physiol 163,273 – 286.

[57] Xing, G. – S. , Cui, K. – R. , Li, J. , Wang, Y. , Li, Z. – X. , 2001. Water stress and the accumulation of β – N – oxalyl – L – α, β – diaminopropionic acid in grass pea(*Lathyrus sativus*). Journal of Agricultural and Food Chemistry 49,216 – 220.

[58] Xiong, Y. – C. , Xing, G. – M. , Li, F. – M. , Wang, S. – M. , Fan, X. – W. , Li, Z. – X. , Wang, Y. – F. , 2006. Abscisic acid promotes accumulation of toxin ODAP in relation to free spermine level in grass pea seedlings(*Lathyrus sativus* L.). Plant Physiology and Biochemistry 44, 161 – 169.

[59] Yang, J. , Zhang, J. , Wang, Z. , Zhu, Q. , Liu, L,. 2002. Abscisic acid and cytokinins in the root exudates and leaves and their relationship to senescence and remobilization of carbon reserves in rice subjected to water stress during grain filling. Planta 215,645 – 52.

[60] Zhang, M. , Yuan, B. , Leng, P. , 2009. The role of ABA in triggering ethylenebiosynthesis and ripening of tomato fruit. Journal of Experimental Botany 60,1579 – 1588.

[61] Zhou, B. – Y. , Wang, J. – H. , Guo, Z. – F. , Tan, H. – Q. , Zhu, X. – C. , 2006. A simple colorimetric method for determination of hydrogen peroxide in plant tissues. Plant GrowthRegul. 49,113 – 118.

[62] Zhou, G. – K. , Kong, Y. – Z. , Cui, K. – R. , Li, Z. – X. , Wang, Y. – F. , 2001. Hydroxyl radical scavenging activity of β – N – oxalyl – L – α, β – diaminopropionic acid. Phytochemistry 58,759 – 762.

[63] Zimmerman, P. , Zentgraf, U. , 2005. The Correlation between oxidative stress and leaf senescence during plant development. Mol. Cell. Biol. Lett. 10,515 – 534.

注：本文曾发表在 2011 年 49 卷第 3 期 Food and Chemical Toxicology 期刊上。

后 记

六十年风雨历程，六十年求索奋进。编辑出版《天水师范学院60周年校庆文库》(以下简称《文库》)，是校庆系列活动之“学术华章”的精彩之笔。《文库》的出版，对传承大学之道，弘扬学术精神，展示学校学科建设和科学研究取得的成就，彰显学术传统，砥砺后学奋进等都具有重要意义。

春风化雨育桃李，弦歌不辍谱华章。天水师范学院在60年办学历程中，涌现出了一大批默默无闻、淡泊名利、潜心教学科研的教师，他们奋战在教学科研一线，为社会培养了近10万计的人才，公开发表学术论文10000多篇(其中，SCI、EI、CSSCI源刊论文1000多篇)，出版专著600多部，其中不乏经得起历史检验和学术史考量的成果。为此，搭乘60周年校庆的东风，科研管理处根据学校校庆的总体规划，策划出版了这套校庆《文库》。

最初，我们打算策划出版校庆《文库》，主要是面向校内学术成果丰硕、在甘肃省内外乃至国内外有较大影响的学者，将其代表性学术成果以专著的形式呈现。经讨论，我们也初步拟选了10位教师，请其撰写书稿。后因时间紧迫，入选学者也感到在短时期内很难拿出文稿。因此，我们调整了《文库》的编纂思路，由原来出版知名学者论著，改为征集校内教师具有学科代表性和学术影响力的论文分卷结集出版。《文库》之所以仅选定教授或具有博士学位副教授且已发表在SCI、EI或CSSCI源刊的论文(已退休教授入选论文未作发表期刊级别的限制)，主要是基于出版篇幅的考虑。如果征集全校教师的论文，可能卷帙浩繁，短时间内难以出版。在此,请论文未被《文库》收录的老师谅解。

原定《文库》的分卷书名为“文学卷”“史地卷”“政法卷”“商学卷”“教育卷”“体艺卷”“生物卷”“化学卷”“数理卷”“工程卷”,后出版社建议,总名称用“天水师范学院60周年校庆文库”,各分卷用反映收录论文内容的卷名。经编委会会议协商论证,分卷分别定为《现代性视域下的中国语言文学研究》《“一带一路”视域下的西北史地研究》《“一带一路”视域下的政治经济研究》《“一带一路”视域下的教师教育研究》《“一带一路”视域下的体育艺术研究》《生态文明视域下的生物学研究》《分子科学视域下的化学前沿问题研究》《现代科学思维视域下的数理问题研究》《新工科视域下的工程基础与应用研究》。由于收录论文来自不同学科领域、不同研究方向、不同作者,这些卷名不一定能准确反映所有论文的核心要义。但为出版策略计,还请相关论文作者体谅。

鉴于作者提交的论文质量较高,我们没有对内容做任何改动。但由于每本文集都有既定篇幅限制,我们对没有以学校为第一署名单位的论文和同一作者提交的多篇论文,在收录数量上做了限制。希望这些论文作者理解。

这套《文库》的出版得到了论文作者的积极响应,得到了学校领导的极大关怀,同时也得到了光明日报出版社的大力支持。在此,我们表示深切的感谢。《文库》论文征集、编校过程中,王弋博、王军、焦成瑾、贾来生、丁恒飞、杨红平、袁焜、刘晓斌、贾迎亮、付乔等老师做了大量的审校工作,以及刘勍、汪玉峰、赵玉祥、施海燕、杨婷、包文娟、吕婉灵等老师付出了大量心血,对他们的辛勤劳动和默默无闻的奉献致以崇高的敬意。

《天水师范学院60周年校庆文库》编委会

2019年8月